1

BIOCHEMICAL SOCIETY SYMPOSIA

No. 50

THE MOLECULAR BASIS OF MOVEMENT THROUGH MEMBRANES

BIOCHEMICAL SOCIETY SYMPOSIUM No. 50
held at St. George's Hospital Medical School, London, December 1984

The Molecular Basis of Movement Through Membranes

ORGANIZED AND EDITED BY

P. J. QUINN AND C. A. PASTERNAK

1985
LONDON: THE BIOCHEMICAL SOCIETY

The Biochemical Society,
7 Warwick Court,
London WC1R 5DP, U.K.

Copyright © 1985 by The Biochemical Society: London

ISBN: 0 904498 17 4
ISSN: 0067-8694

Printed in Great Britain by the
University Press, Cambridge

iv

List of Contributors

G. M. Alder (*Department of Biochemistry, St George's Hospital Medical School, London SW17 0RE, U.K.*)

C. L. Bashford (*Department of Biochemistry, St George's Hospital Medical School, London SW17 0RE, U.K.*)

S. Bhakdi (*Institute of Medical Microbiology, University of Giessen, Schubertstrasse 1, D-6300 Giessen, West Germany*)

J. Boonstra (*Department of Molecular Cell Biology, State University Utrecht, Padualaan 8, 3584 CH Utrecht, The Netherlands*)

C. D. Buckley (*Department of Biochemistry, St George's Hospital Medical School, London SW17 0RE, U.K.*)

S. W. Cushman (*Experimental Diabetes, Metabolism and Nutrition Section, Molecular, Cellular and Nutritional Endocrinology Branch, National Institute of Arthritis, Diabetes and Digestive and Kidney Diseases, National Institutes of Health, Bethesda, Maryland 20205, U.S.A.*)

L. H. K. Defize (*Hubrecht Laboratory, International Embryological Institute, Uppsalalaan 8, 3584 CT Utrecht, The Netherlands*)

S. W. de Laat (*Hubrecht Laboratory, International Embryological Institute, Uppsalalaan 8, 3584 CT Utrecht, The Netherlands*)

L. de Meis (*Instituto de Ciências Biomédicas, Departamento de Bioquímica, Centro de Ciências da Saúde, Universidade Federal do Rio de Janeiro, Cidade Universitária, Ilha do Fundão, Rio de Janeiro, CEP 21910, Brazil*)

C. Dingwall (*CRC Molecular Embryology Research Unit, Department of Zoology, University of Cambridge, Downing Street, Cambridge CB2 3EJ, U.K.*)

B. Dobberstein (*European Molecular Biology Laboratory, Postfach 10.2209, 6900 Heidelberg, Federal Republic of Germany*)

D. T. Edmonds (*The Clarendon Laboratory, Parks Road, Oxford, OX1 3PU, U.K.*)

M. Fosset (*Centre de Biochimie du CNRS, Faculté des Sciences, Parc Valrose, 06034 Nice Cedex, France*)

U. Hopfer (*Department of Developmental Genetics and Anatomy, Case Western Reserve University, Cleveland, Ohio 44106, U.S.A.*)

M. Hugues (*Centre de Biochimie du CNRS, Faculté des Sciences, Parc Valrose, 06034 Nice Cedex, France*)

P. L. Jørgensen (*Institute of Physiology, Aarhus University, 8000 Aarhus C, Denmark*)

M. Lazdunski (*Centre de Biochimie du CNRS, Faculté des Sciences, Parc Valrose, 06034 Nice Cedex, France*)

W. R. Loewenstein (*Department of Physiology and Biophysics, University of Miami School of Medicine, P.O. Box 016430, Miami, FL 33101, U.S.A.*)

I. H. Madshus (*Norsk Hydro's Institute for Cancer Research, and The Norwegian Cancer Society, Montebello, Oslo 3, Norway*)

K. J. Micklem (*Department of Biochemistry, University of Oxford, Oxford OX1 3QU, U.K.*)

W. H. Moolenaar (*Hubrecht Laboratory, International Embryological Institute, Uppsalalaan 8, 3584 CT Utrecht, The Netherlands*)

C. Mourre (*Centre de Biochimie du CNRS, Faculté des Sciences, Parc Valrose, 06034 Nice Cedex, France*)

H. J. Müller-Eberhard (*Department of Immunology, Research Institute of Scripps Clinic, La Jolla, California 92037, U.S.A.*)

S. Olsnes (*Norsk Hydro's Institute for Cancer Research, and The Norwegian Cancer Society, Montebello, Oslo 3, Norway*)

C. A. Pasternak (*Department of Biochemistry, St George's Hospital Medical School, London SW17 0RE, U.K.*)

K. Patel (*Department of Biochemistry, St George's Hospital Medical School, London SW17 0RE, U.K.*)

G. Romey (*Centre de Biochimie du CNRS, Faculté des Sciences, Parc Valrose, 06034 Nice Cedex, France*)

D. Sanders (*Department of Biology, University of York, Heslington, York, U.K.*)

K. Sandvig (*Norsk Hydro's Institute for Cancer Research, and The Norwegian Cancer Society, Montebello, Oslo 3, Norway*)

H. Schmid-Antomarchi (*Centre de Biochimie du CNRS, Faculté des Sciences, Parc Valrose, 06034 Nice Cedex, France*)

M. Schramm (*Institute of Pharmacology, Bayer A.G., Postfach 101709, D5600 Wuppertal 1, West Germany*)

I. A. Simpson (*Experimental Diabetes, Metabolism and Nutrition Section, Molecular, Cellular and Nutritional Endocrinology Branch, National Institute of Arthritis, Diabetes and Digestive and Kidney Diseases, National Institutes of Health, Bethesda, Maryland 20205, U.S.A.*)

C. L. Slayman (*Department of Physiology, Yale School of Medicine, New Haven CT, U.S.A.*)

A. Sundan (*Norsk Hydro's Institute for Cancer Research, and The Norwegian Cancer Society, Montebello, Oslo 3, Norway*)

R. Towart (*Miles Laboratories Limited, Stoke Court, Stoke Poges, Bucks., U.K.*)

J. Tranum-Jensen (*Anatomy Institute C, University of Copenhagen, Blegdamsvej 3C, DK-2200 Copenhagen N, Denmark*)

P. T. van der Saag (*Hubrecht Laboratory, International Embryological Institute, Uppsalalaan 8, 3584 CT Utrecht, The Netherlands*)

Preface

The last few years have witnessed rapid advances in our knowledge of the way in which a variety of molecules move across cell membranes. This has been made possible partly by the deployment of techniques for sequencing membrane proteins and processing this information to provide space-filling structural co-ordinates, and partly by refinements of conventional biophysical methods. We are grateful to C. F. Phelps, the retiring organizer of the Biochemical Society Symposia, for recognizing the importance of these developments and for joining us in the planning of a Symposium devoted to that topic. On behalf of other members of the Society, we would like further to record our indebtedness to him for his unstinting service over the past 4 years.

This volume, then, is an account of a Symposium organized under the auspices of the Biochemical Society; it was held at St George's Hospital Medical School, London on 19 and 20 December 1984. The Chairmen of the first day were P. D. Mitchell, F.R.S. and P. F. Baker, F.R.S., who presided over a discussion of the passive, and the pumped, movements of H^+, Na^+, K^+ and Ca^{2+}, and of nutrients such as glucose and amino acids. The emphasis was on plasma membranes rather than on mitochondrial or other intracytoplasmic membranes. The sessions on the second day were chaired by P. N. Campbell and C. A. Pasternak and were devoted to a consideration of the movement of peptides and proteins through membranes, of movement through specialized regions such as nuclear 'pores' and communicating junctions, of modulation of ion movement by growth factors and of movement through pores artificially created by toxins, viruses or complement.

We are particularly honoured as Editors of this Symposium to mark a half century of Biochemical Society Symposia. The list of topics covered is extensive and reflects the development of biochemistry as a leading subject in the life sciences. The list of contributing authors is equally impressive and includes scientists of distinction in every branch of the subject.

Finally, in connection with the organization of the meeting at St George's we would especially like to extend our gratitude to the local organisers, M. J. Clemens and J. M. Graham.

P. J. QUINN
C. A. PASTERNAK

London

Contents

Abbreviations

a.h.p.	After-hyperpolarization potential
BCECF	Bis(carboxyethyl)carboxyfluorescein
DG	1,2-Diacylglycerol
DHP	Dihydropyridine
DMSO	Dimethyl sulphoxide
DOC	Sodium deoxycholate
DSS	Disuccinimidyl suberate
EGF	Epidermal growth factor
EGTA	Ethanedioxybis(ethylamine)tetra-acetic acid
EMC	Encephalomyocarditis (virus)
FCCP	Carbonyl cyanide-p-trifluoromethoxyphenyl hydrazone
HAU	Haemagglutinating unit
HBS	Hepes-buffered saline
IGF-II	Insulin-like growth factor II
IP3	Inositol triphosphate
LDL	Low-density lipoprotein
MAC	Membrane attack complex
PDGF	Platelet-derived growth factor
PEA	*Pseudomonas* exotoxin A
PIP2	Phosphatidylinositol 4,5-bisphosphate
ROC	Receptor-operated channels
SAR	Structure–activity relationship
SDS	Sodium dodecyl sulphate
SITS	4-Acetamido-4′-isothiocyanostilbene-2,2′-disulphonic acid
SV	Simian virus
TAL	Thick ascending limb (nephron)
TEA	Tetraethylammonium
TPA	12-O-Tetradecanoyl-phorbol-13-acetate
VOC	Voltage-operated channels

Biochem. Soc. Symp. **50**, 1–10
Printed in Great Britain

Electrostatic Models for Ion Channels and Pumps

D. T. EDMONDS

The Clarendon Laboratory, Parks Road, Oxford OX1 3PU, U.K.

Introduction

Most of the models used to mimic the behaviour of biological systems at a molecular level are today mechanical. For example, the explanation of the selectivity of an enzyme for a particular substrate is almost invariably that the particular shape of part of the substrate surface is accommodated by a similarly shaped cavity in the enzyme surface, often called the lock and key mechanism.

From the point of view of a physically trained scientist the great disadvantage of mechanical forces is that they are necessarily of very short range. In the example above the substrate need move a distance of only atomic dimensions before all the information contained in the shape of the enzyme cavity is lost. This is a particularly severe handicap when the surfaces of proteins in solution are known to execute thermally excited motion with a root mean square amplitude of 0.1 or 0.2 nm (McCammon & Karplus, 1983).

In contrast to mechanical forces, electrostatic forces have a very long range with the electrostatic potential of a charge falling off with distance R as $1/R$. An electrostatic enzyme–substrate recognition model would have a distribution of electric charges and electric fields at one surface of the substrate that complemented the charges and electric fields at the surface of a particular part of the enzyme so that, positioned together with a particular relative orientation, the electrostatic energy would be a minimum. Two major differences with the mechanical model would be evident, both due to the much longer range of the electrostatic forces. The first is that the sources of the electric field need not be located at the surface but could be buried in the core of the molecule and thus subject to smaller vibration amplitudes and protected from the bombardment of external molecules. The second is that even the surface charges could produce an easily recognizable and distinct field distribution at a distance from the surface, even if the surface is subject to random motion, provided that the range of the motion is small in comparison with the range of the electrostatic fields.

In fact it is likely that in Nature both electrostatic and mechanical recognition are used with electrostatic forces guiding and pre-orientating the substrate at long range (Matthew *et al.*, 1983; Getzoff *et al.*, 1983) whilst the final 'docking' of the substrate is achieved by a combination of both electrostatic and mechanical forces.

Turning now to the particular concern of this paper, namely ion channels and pumps, I am struck by the fact that although the ions, being charged, must be directly influenced by electric fields, the literature is dominated by mechanical gating particles, mechanical selectivity filters and mechanically operated pumps. In an attempt to redress this balance I shall offer a personal view of alternative electrostatic models in the three particular areas of ion channel selectivity, ion channel gating and ion pump operation.

Ion Channel Selectivity

Many of the ion channels found in Nature are highly selective as to the ions they will allow to pass. For example, at the node of Ranvier in frog nerve fibres the sodium channel shows a preference for Na^+ over K^+ of 12:1 whilst the potassium channel exhibits a preference for K^+ over Na^+ greater than 100:1 (Edwards, 1982). Probably the most quoted model for a selectivity filter is that of Hille (1972), who postulates an annular restriction in the path of the ions of dimensions 0.3×0.5 nm lined with negatively charged groups. The selectivity of the filter depends primarily on its mechanical size and partially on the ability of the negatively charged groups to complete hydrogen bonds. The cations are supposed to travel partially hydrated so that for example 3 water molecules with the Na^+ ion fit the annulus and 4 water molecules with the K^+ ion also fit. In view of our present knowledge of the structure of ion channels, namely that the acetylcholine channel is an oligomer of five protein blocks loosely coupled with the pore defined between them (Changeux et al., 1984), or that the sodium channel is most probably defined as a pore between the four largely α-helical sections of a single large protein (Noda et al., 1984), it seems to me unlikely that a mechanical annulus exists with a sufficiently precise shape to distinguish between ions which differ in radius by fractions of an Ångstrom unit. This is particularly so when account is taken of the likely surface motion of amplitude 0.1 or 0.2 nm referred to above and also the relative motions of channel protein subunits.

Alternative models of the selective mechanism in ion channels are charged site models and ordered water models. Both are electrostatic and both stem from consideration of the essentially electrostatic barrier presented to a small ion by a lipid bilayer. An isolated ion of charge Q_I coulombs and radius R_I metres has an electrostatic self energy (or Born charging energy) given in joules by U_s in:

$$U_s = Q_I^2/(8\Pi\epsilon_0 R_I)$$

where $\epsilon_0 = 8.85 \times 10^{-12}$ F/m. For Na^+, U_s is 1.2×10^{-18} J/ion or 722.4 J/mol. The energy may be thought of as residing in the electric field created by the ion which extends to infinity. If, however, the ion is placed in a polarizable fluid such as water the field of the ion tends to align the dipole moments of neighbouring water molecules to point radially away from the ion. The field of the ion now partially terminates on the negative ends of the water dipoles presented to the ion and is thus much more local so that the electrostatic self energy falls drastically and the ion is said to be partially solvated. The electrical polarizability of the interior of a lipid bilayer is very small so that a bare small ion within a bilayer has a high electrostatic self energy and the probability of such an ion spontaneously leaving an aqueous environment to enter the bilayer is quite negligible. Thus the chief function of an ion channel is to provide complementary charges or polarizable dipoles that enable the ion to traverse the lipid bilayer without having to surmount such a large energy barrier (Parsegian, 1969; Edmonds, 1980).

In charged site models much of the solvation of the ion during transit is achieved by oppositely charged sites embedded in the protein wall of the channel and the ion is supposed to traverse the channel by migration between these sites. An advantage of such models is that they are easily made selective by varying the charge $-Q_s$ and effective radius R_s of the charged site. The electrostatic energy of binding to this site for an ion of charge Q_I and radius R_I is given by U_B in:

$$U_B = -Q_s Q_I/(4\Pi\epsilon_0(R_s + R_I))$$

The difference in energy between the water solvated ion and the site bound ion varies with Q_s, Q_I, R_s and R_I, leading to the well known Eisenman (1962) series, giving the relative preferences of a given site for binding ions of different valence and radius.

The disadvantages of charged site models relate to the high electrostatic energy cost of locating the charged sites in a low dielectric constant environment and in the fact that they may not be placed close together without incurring an additional high cost in electrostatic repulsion energy. High selectivity necessitates tight binding at the charged sites and the enforced wide separation of the sites leads to large potential barriers between sites and hence to low transfer rates. In Nature highly selective channels have transfer rates as high as 10^7 ions/s.

These problems are not encountered in ordered water models (Edmonds, 1981a, 1984b) in which the ion solvation is achieved entirely by the electrically neutral water molecules within the channel. However, if the channel water behaved like bulk water there would be no ion selectivity so that it is necessary to assume that the water within the channel is structurally ordered as outlined below. There are many examples known in which water forms ordered layers several molecules thick with the structural order partially dictated by hydrogen bonding to a particular substrate, but the natural material that most closely illustrates some of the properties required by ordered water channel models are the zeolite and clay minerals. In hydrated zeolites an aluminosilicate skeleton defines approximately spherical polyhedra interconnected by tubes. The cavities and tubes contain ordered arrays of water molecules so that, for example, in hydrated zeolite A the large cages contain a distorted dodecahedron of 12 water molecules. Diffusion studies reveal (Barrer & Rees, 1960) that small ions like Na^+ and K^+ move through these structures with high mobility and moreover that only the bare ion moves and is hydrated in transit by the essentially static water molecule arrays.

The ordered water model postulates that the charge structures on the surface of the particular protein rods that define the channel impose a particular order on the water structure and thus provide a multitude of closely spaced selective ion binding sites at the centres of the various water rings in the structure. While retaining its normal tetrahedral bonding geometry, water is capable of forming two small rings, a planar pentagon and a puckered hexagon. An ion at the centre of a water ring has an electrostatic interaction with the electric dipoles of the water molecules of the ring sufficiently strong to re-orient the water dipoles to

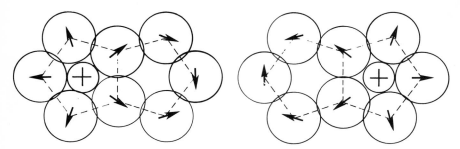

Fig. 1. *Two adjacent planar pentagons composed of water molecules*

Arrows show the orientations of the electric dipole moments of the water molecules as a Na^+ ion moves from the centre of one water ring to the centre of the adjacent water ring. Calculations show that water ring sites such as these with rotated electric dipole moments form low energy selective binding sites for small cations.

point radially away from the ion at the ring centre. Calculations which include ion–water dipole, ion-induced water dipole, ion–water quadruple, water dipole–water dipole and water quadrupole–water dipole interactions, and which include the energy cost of broken water–water hydrogen bonds, reveal that particular water rings provide especially low energy sites for particular ions with energies comparable with the fully hydrated ion (Edmonds, 1980). The planar pentagon rings favour the Na^+ ion while the puckered hexagon rings favour K^+, partially due to the almost exact match in each case of the diameter of the bare ion and the diameter of the free 'hole' in the ring centre. The ions pass through the water channel by passing between the centres of contiguous water rings as illustrated in Fig. 1, which also shows the rotation but not translation of the water molecules that is required to solvate the ion and to provide the selective water ring ion binding sites. An essential difference between the ordered water model and charged site model is that the selective ring sites may be as close together as 0.3 nm without incurring an electrostatic repulsion energy penalty, so that very high transfer rates become possible.

Channel Gating

Besides the spatial order imposed on the water lining the pore by the supporting protein rods, calculations predict that an array of water molecules within a long and thin cylindrical pore in a low dielectric constant sheet will be electrically ordered. The two states with the lowest electrostatic interaction energy retain the same spatial orientation of the water molecules but are such that the electric dipole moments of the molecules have predominantly positive or predominantly negative projections on the channel axis (Edmonds, 1979, 1984*b*). These two electrically polarized states with equal minimum electrostatic energy are sketched in Fig. 2 and below them is sketched the resultant dipolar electric potential distribution that would be measured along the channel axis. A voltage applied across the membrane will create an energy difference between the two polarized states and can thus lead to switching. For example one water structure composed entirely of pentagons is a stacked series of dodecahedrons,

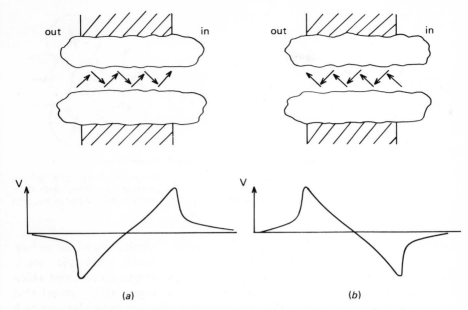

Fig. 2. Diagrammatic section of the proposed ordered water ion channel

The protein rods that define the channel are indicated and the water molecule electric dipoles are suggested by arrows predominantly pointing parallel (*a*) or antiparallel (*b*) to the inward pointing axis of the channel. Below is plotted the axial electrostatic potential distribution resulting from the composite electric dipole moment of the channel.

and such a structure with a diameter about 10 nm and a length of 48 nm has an energy difference between the two oppositely polarized states of 5×10^{-20} J or 12.4 kT when a voltage of 100 mV is applied across it (k is the Boltzmann constant and $T = 290°C$). Effective gating of the ion current through the channel is ensured by the very different current versus voltage characteristics of channels in the two polarizations.

In Fig. 3(*a*) is displayed the current versus voltage characteristic for a sodium channel using a simplified dipolar distribution (Edmonds, 1981*b*) with an external Na^+ concentration of 440 mM and an internal concentration of 50 mM. The switching voltage, at which the electrostatic energy of the water array (under the influence of the membrane voltage V_M and any fields due to neighbouring charges) passes through zero, was chosen as -25 mV as in the squid axon. In Fig. 3(*b*) is displayed as a full line the prediction using exactly the same model parameters but for a potassium channel with outside and inside potassium concentrations of 20 mM and 200 mM respectively and with a switch occurring at a membrane voltage of -50 mV, again for direct comparison with the squid axon. In both these cases the very simple model yields results that resemble those experimentally determined with no need to postulate any gating mechanism other than the switch of the axial dipole moment of the spatially fixed water array.

One particularly interesting difference between this model and the traditional mechanically blocked models is that both configurations of the channel conduct

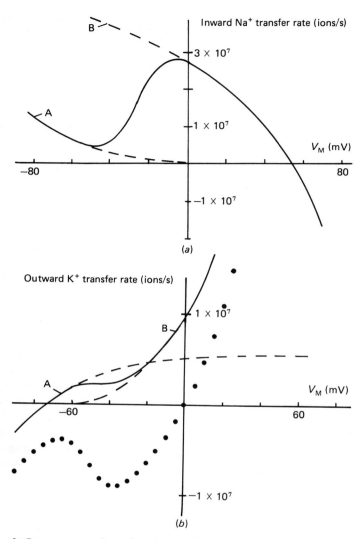

Fig. 3. *Current versus voltage characteristics for model sodium and potassium channels*

(*a*) The current versus voltage characteristic calculated for a model sodium channel with an electric dipolar structure of peak amplitude 150 mV. The external free Na^+ concentration is assumed to be 440 mM and the internal concentration is assumed to be 50 mM to mimic the sodium channel in the squid giant axon. The characteristics marked A and B correspond to the polarizations sketched in Figs. 2(*a*) and 2(*b*) respectively, with the full line giving the total characteristic corresponding to a switch between the two polarizations centred at a membrane voltage $V_M = -25$ mV. (*b*) The predicted characteristic of a potassium channel using the same model parameters but with external and internal free K^+ concentrations of 20 mM and 400 mM and a switching membrane voltage of -50 mV. The dotted line is the prediction if the external concentration is changed to 400 mM but all other parameters remain the same.

ions; it is in a high or a low conductivity state at any membrane voltage rather than being open or shut. In Fig. 3(b) a dotted line shows the prediction of the same simple model with K^+ concentrations both inside and out of 400 mM. Such a characteristic with a negative resistance region is observed experimentally (Segal, 1958; Moore, 1959). To obtain such a characteristic a switch is required between two configurations, both conducting. With an open or shut mechanically blocked model this requires one type of channel to shut when the membrane is polarized and another type of channel to open, with the two types of channel having similar switching voltages. With the electrostatic model the switch is merely between the two polarizations of the same type of channel. Several other phenomena observed in Nature may be explained by this simple electrically gated model (Edmonds, 1981b).

A second interesting feature of the model is that a multiply connected water array is probably unique in its property of being able to solvate, and thus to transmit, heavy ions such as Na^+ and K^+ and also to conduct protons by proton hopping as in ice. The model predicts that an ordered water gated ion channel should also serve as a gated proton channel. Evidence of gated proton conductivity probably through a heavy ion channel has been obtained by Thomas & Meech (1982) for a molluscan neuron.

An Electrostatic Pump

Consider a selective channel with a dipolar electrical structure such as that shown in Fig. 2(a). For this model any potential distribution with positive and negative lobes will suffice and the dipole moment need not switch or be due to the water dipoles. The required fixed dipolar structure could, for example, originate from the known large dipole moment (Hol et al., 1978) of a neighbouring α-helical protein fragment. For definiteness of description I will assume that the channel is selective for K^+ ions. As seen in Fig. 2(a) there is no impediment to entry for K^+ ions from the fluid bathing the outside face but from the inner fluid they must surmount a potential barrier. In fact, calculations using the simplified linear potential show that with a peak potential of 150 mV the probability of entry is some 3000 times higher from the outside than from the inside with equal K^+ concentrations in the fluids bathing the two ends of the channel. Also by assuming equal hemispherical access volumes of radius 0.5 nm at each channel end it is easy to calculate (Edmonds, 1984a) that a K^+ concentration of only 0.85 mM in the outer fluid is sufficient to ensure that the channel is occupied on average half of the time. The corresponding concentration in the inner fluid is very much higher at 6.3 M.

For simplicity in the above calculations a maximum channel occupancy of 1 was assumed. In an attempt to calculate roughly the allowed occupancy of a channel, a model calculation may be performed (Edmonds, 1984a) in which the channel is represented as a cylinder of diameter 1 nm with high dielectric constant ϵ_1 which is embedded in a slab of material of low dielectric constant ϵ_2. Taking $\epsilon_1 = 80$ to represent the water and $\epsilon_2 = 5$ to represent the protein and lipid (Pethig, 1979) I calculate that two univalent cations on the cylinder axis have an electrostatic repulsion energy of 2.2×10^{-20} J or $5.6\ kT$ when separated

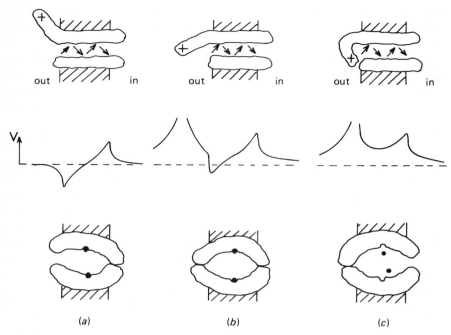

Fig. 4. *Effect of approach of positively charged group to the outer end of the ion channel*

(*a*) The dipolar channel as sketched in Fig. 2(*a*) but with a mobile positively charged group attached to the outer end of the channel support protein. The charged group is assumed sufficiently remote so that an undistorted dipolar axial potential distribution is obtained. Below is sketched an approximately equivalent mechanical model showing high affinity binding sites exposed to the outer fluid only. (*b*) The effect of a closer approach of the charged group to the channel end showing the axial electrostatic potential distribution which is now a composite of that the channel dipoles and the charged group. The potential now shows considerable barriers to ion flow at both ends of the channel, and the mechanical model also shows occluded states. (*c*) Closer proximity of the charged group results in the tipping of the composite potential distribution with the consequent expulsion of cations within the channel to the inside. The mechanical model shows low affinity sites now exposed only to the inside.

by 2 nm. Thus a 5 nm long channel could simultaneously contain 2 or perhaps 3 univalent ions but is highly unlikely to contain more.

Let us now consider the effect produced by a positively charged group approaching the outer end of the channel. In Fig. 4(*a*) is shown the channel in the absence of the charge with its undistorted dipolar potential distribution. In the lower part of Fig. 4(*a*) is the roughly analogous state of the traditional mechanical model of a membrane pump due to Jardetski (1966). In each case the pump exposes high affinity binding sites to K^+ ions from the outer fluid whilst effectively barring entry for K^+ ions from the inner fluid. In Fig. 4(*b*) the positively charged group approaches the outer end of the dipolar channel. The electrostatic potential experienced by ions within the channel is then a simple superposition of the dipolar channel potential and that from the charged group as sketched in the Figure. The ions within the channel are now occluded with little probability of leaving the channel from either end. Even closer approach of the charged group as shown in Fig. 4(*c*) results in a tipping of the whole potential distribution towards the inner end of the channel. Calculation confirms (Edmonds, 1984*a*) that the dwell time of cations within the channel is much reduced and the probability of exit from the inner end tends to unity. Again the roughly equivalent state of the mechanical model is shown below with low affinity sites having access only to the inside. Unlike the mechanical model the changes in access and affinity are readily explained; they require no change either electrical or mechanical in the channel or its support protein but only the passive superposition of two electrostatic potentials.

It is clear from the description above that any mechanism that brought about an alternation between the states shown in Figs. 4(*a*) and 4(*c*) would result in the unidirectional pumping of 2 or 3 K^+ ions each cycle from the outer to the inner fluid even against a large adverse voltage or concentration gradient. An externally applied oscillating membrane voltage would substitute for the mobile charged group and would also result in ion pumping. This is of interest because such an applied oscillating membrane voltage has been shown (Serpersu & Tsong, 1983) to result in an ouabain sensitive Rb^+ ion uptake in human erythrocytes and Rb^+ is known to substitute for K^+ in the Na^+, K^+-ATPase pump of these cells.

An energy input would clearly be required to activate the pump, as is seen from Fig. 4. If the positively charged group is spontaneously attracted to the negatively charged outer end of the channel, even while occupied, then an external agency would be required to remove the charged group (Fig. 4*c* to Fig. 4*a*) after the cations within the channel have been expelled to the inner fluid. In a more complete model this action would be powered by the binding of ATP. A model of Na^+,K^+-ATPase may be constructed by postulating two such dipolar channels in close proximity with a sodium selective channel polarized outward and a potassium channel polarized inward as in Fig. 4(*a*). Many properties of real Na^+/K^+ exchange pumps may be simulated by these means (Edmonds, 1984*a*). I will not develop this further here but will mention one interesting aspect of the simulation. Like the electrostatic gated ion channel model the electrostatic pump channel is never fully open or shut. In the configuration of Fig. 4(*a*) the probability of entry from the inside is small but not zero. This feature becomes important when using the model to simulate well-established but non-physiological modes of action in which the pump is forced to reverse.

I am most grateful to the Royal Society for the award of a Senior Research Fellowship which has bought me time to pursue these interests.

References

Barrer, R. M. & Rees, L. V. C. (1960) *Trans. Faraday Soc.* **56**, 709–721
Changeux, J., Devillers-Thiéry, A. & Chemouilli, P. (1984) *Science* **225**, 1335–1345
Edmonds, D. T. (1979) *Chem. Phys. Lett.* **65**, 429–433
Edmonds, D. T. (1980) *Proc. R. Soc. London Ser. B* **211**, 51–62
Edmonds, D. T. (1981*a*) *Biochem. Soc. Symp.* **46**, 91–101
Edmonds, D. T. (1981*b*) *Proc. R. Soc. London Ser. B* **214**, 125–136
Edmonds, D. T. (1984*a*) *Proc. R. Soc. London Ser. B* **223**, 49–61
Edmonds, D. T. (1984*b*) in *Biological Membranes* (Chapman, D., ed.), vol. 5, pp. 349–387, Academic Press, New York
Edwards, C. (1982) *Neuroscience* **7**, 1335–1366

Eisenman, G. (1962) *Biophys. J.* **2**, 259–323
Getzoff, E. D., Tainer, J. A., Weiner, P. K., Kollman, P. A., Richardson, J. S. & Richardson, D. C. (1983) *Nature (London)* **306**, 287–290
Hille, B. (1972) *J. Gen. Physiol.* **59**, 637–658
Hol, W. G. J., van Duijnen, P. T. & Berendsen, H. J. C. (1978) *Nature (London)* **273**, 443–446
Jardetski, O. (1966) *Nature (London)* **211**, 969–970
Matthew, J. B., Weber, P. C., Salemme, F. R. & Richards, F. M. (1983) *Nature (London)* **301**, 169–171
McCammon, J. A. & Karplus, M. (1983) *Acc. Chem. Res.* **16**, 187–193
Moore, J. W. (1959) *Nature (London)* **183**, 265–266
Noda, M., Shimizu, S., Tanabe, T., Takai, T., Kayano, T., Ikeda, T., Takahashi, H., Nakayama, H., Kanaska, Y., Minamino, N., Kangawa, K., Matsuo, H., Raftery, M. A., Hirose, T., Inayama, S., Hayashida, H., Miyata, T. & Numa, S. (1984) *Nature (London)* **312**, 121–127
Parsegian, A. (1969) *Nature (London)* **221**, 844–846
Pethig, R. (1979) *Dielectric and Electronic Properties of Biological Materials*, chapter 2, Wiley, Chichester
Segal, J. R. (1958) *Nature (London)* **182**, 1370
Serpersu, E. H. & Tsong, T. Y. (1983) *J. Membr. Biol.* **74**, 191–201
Thomas, R. C. & Meech, R. W. (1982) *Nature (London)* **299**, 826–828

Biochem. Soc. Symp. **50**. 11–29
Printed in Great Britain

Steady-State Kinetic Analysis of an Electroenzyme

CLIFFORD L. SLAYMAN and DALE SANDERS*

Department of Physiology, Yale School of Medicine, New Haven, CT, U.S.A., and
*Department of Biology, University of York, Heslington, York, U.K.

Synopsis

Primary active transport of ions through the plasma membranes of plants and fungi is driven by a proton-dependent ATPase, which consists of a membrane-embedded (M_r 104000) polypeptide, forms a β-aspartylphosphate intermediate and is blocked by orthovanadate. It can be extracted from cell membranes and reactivated in native lipid micelles or in exogenous phospholipid vesicles. For the fungus *Neurospora*, vesicle preparations directly display proton-pumping, and can develop membrane potentials ($\Delta\psi$) of 120 mV or pH differences (ΔpH) of 2 units, with a stoichiometry of 1 H^+ transported per ATP molecule split.

In vivo, the proton pump sustains $\Delta\psi$ values of 150–350 mV (cytoplasm negative) and ΔpH values up to 3.5 units (pH$_i$ $\simeq$ 7, with pH$_o$ = 3.5). Since the total proton-motive force thus can exceed 400 mV, compared with a ΔG_{ATP} of 500 mV, the stoichiometry must be 1 H^+/ATP, with little leeway for neutralizing ions. Kinetic analysis of pump-currents measured during forcing of [ATP]$_i$, pH$_o$, pH$_i$, and $\Delta\psi$ yields three main conclusions: (i) again, the stoichiometry is 1 H^+/ATP; (ii) energy conversion occurs during transmembrane charge transfer, which therefore is probably the $E_1 \sim P$—$E_2 \cdot P$ transition (see Na^+,K^+-ATPase); (iii) protons are strongly dissociated at both membrane surfaces, with pK_i $\simeq$ 5.4 versus pH$_i$ = 7.2, and pK_o $\simeq$ 2.9 versus pH$_o$ = 5.8. Considerations of structure and partial-reaction chemistry (by analogy with the Na^+,K^+-ATPase) suggest a kinetically testable model for the transport mechanism: a sequential, double-gated channel, through which the membrane field is transported across the ion, rather than vice versa.

Introduction

There now appear to be three distinct classes of ATP-dependent proton pumps in biological membranes. The first to be defined were the so-called F_o–F_1 enzymes which appear, in electron micrographs of negatively stained material, as ball-and-stalk structures plugged into mitochondrial, chloroplast and bacterial membranes. With 8–10 distinct polypeptide subunits, several of which are present in multiple copies, and aggregate molecular masses of 400–500 kDa, these are by far the most complex of the three classes. The second class reside in the plasma membranes of most plant and fungal cells (i.e., eukaryotic organisms which have evolved cell walls to control osmotic forces), as well as in certain very specialized animal secretory membranes, chiefly the gastric mucosa. These enzymes have turned out to be structurally the simplest, consisting of single polypeptide chains of molecular mass 95–105 kDa. Unlike the other classes, they clearly form high-energy phosphoprotein intermediates during their catalytic cycle, and are diagnostically inhibited by orthovanadate. They have been dubbed E_1–E_2 enzymes. The third class, which is only now emerging as a distinct entity, comprises the organellar ATPases found in plant and fungal vacuolar membranes, in lysosomes, chromaffin granules, platelets, Golgi apparatus, clathrin-coated vesicles, endoplasmic reticulum, and in the acid-secretory vesicles of renal membranes (see review by Maloney & Wilson, 1985). Although their structural affinity is currently in dispute, these enzymes appear unrelated to the E_1–E_2 class, since they are insensitive to orthovanadate. Preliminary molecular mass data indicate they contain multiple subunits with major polypeptides ranging between 55 and 90 kDa and an aggregate molecular weight which could reach several hundred kDa.

The enzyme described below, the plasma membrane proton ATPase from the mycelial fungus *Neurospora*, belongs to the E_1–E_2 class and is closely related not only to the proton pumping enzyme in gastric mucosa, but also to the much better known calcium and sodium/potassium transport systems found in sarcoplasmic reticulum and in the plasma membranes of most animal cells. Two features of the *Neurospora* enzyme, however, afford an almost unique situation for study: first, it is located in the plasma membrane of a large-celled, free-living micro-organism, an organism which—unlike most yeasts, algae or higher plants—lacks large internal vacuoles; and second, it appears to function fully electrogenically, driving not only ion circulation, but also a large current through the *Neurospora* membrane. These features permit the enzyme's behaviour to be studied easily *in vivo* by direct electrophysiological methods, with continuous recording, with time-resolution in the range of milliseconds, and free of electrical complications from intracellular membranes.

The obvious benefits of such circumstances are that certain essential thermodynamic quantities for the transport system can be determined unequivocally: namely, membrane potential and cytoplasmic pH. A not-so-obvious but more important benefit is that the usual chemical-kinetic analyses of the enzyme can be supplemented *in vivo* by electrical-kinetic analysis. That is, net transport of protons through the enzyme can be measured as electric current, and the magnitude of current can be determined as a dynamic function of all three driving potentials: the free energy for phosphate hydrolysis, the transmembrane pH difference and the membrane potential. The last measurement, familiarly known as current–voltage analysis (*I–V* analysis), has been a powerful stock technique in electronics and has been readily adopted by membrane biophysicists interested in channel systems. It has only recently come to be used for studying ion pumps, however, primarily because of the scarcity of suitable preparations. The sodium/potassium pump in most animal cell membranes is weakly or only transiently electrogenic, and the strongly electrogenic proton pump of bacterial and mitochondrial membranes is made almost inacessible to electrophysiological techniques by its presence in small 'cells' (but see Felle *et al.*, 1978, 1980).

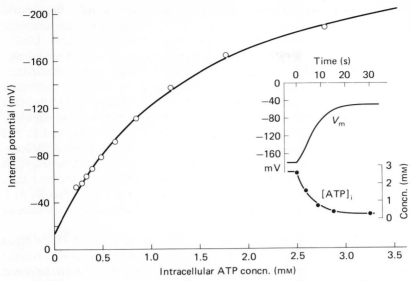

Fig. 1. *Relationship between ATP concentration and membrane potential in Neurospora, consequent to respiratory blockade by cyanide*

Inset: trace of membrane potential (V_m, upper curve) and intracellular ATP concentration (●, lower curve) versus time, with 1 mM-KCN added to the recording medium at 0 s. Results are averaged for seven separate experiments. Main Figure: voltage plotted against ATP values interpolated at corresponding times on the inset curves. The fitted curve drawn is a rectangular hyperbola with the equation V_m (in mV) $= 13 + 274[ATP]_i/(1.52 + [ATP]_i)$. Methods and ATP data were taken from Slayman *et al.* (1973).

It is on the use of current–voltage analysis to delimit the mechanism of proton pumping through the *Neurospora* plasma-membrane ATPase that we shall concentrate most of our attention here.

Physiology of the *Neurospora* Proton Pump

Because *Neurospora* is an obligatory aerobic organism, simple respiratory blockade can be used to deplete metabolic energy. For example, when a blocking concentration of potassium cyanide is rapidly added to the cell bathing medium, pyridine nucleotides become reduced with a halftime of 1–2 s, ATP falls from near 3 mM to 0.3 mM with a halftime of 4 s, and the resting membrane potential collapses from about −200 mV to the range of −10–40 mV lagging slightly behind the fall of ATP. Plots of membrane potential directly versus ATP concentration, for example as shown in Fig. 1, yield saturation curves with $K_{\frac{1}{2}}$ values of 1–2 mM, maximal potentials of about −300 mV, and a voltage offset near −25 mV, at 22–25 °C (Slayman *et al.*, 1973). These values can be taken, in order, to estimate crudely the ATP-K_m of the enzyme, the reversal potential for the current generator, and the (passive) ion-diffusion potential across the pumping membrane. Direct measurements of cytoplasmic pH, by means of H$^+$-selective microelectrodes (Sanders & Slayman, 1982) have yielded values of

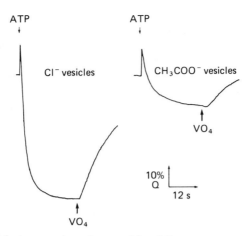

Fig. 2. *Proton uptake by everted membrane vesicles of Neurospora, measured by fluorescence quenching of acridine orange*

Membrane vesicles were prepared by the method of Perlin *et al.* (1984), and suspended in 20 mM-Hepes/KOH buffer at pH 7.2 with 2 μM-acridine orange and 100 mM-KCl (left trace) or potassium acetate (right trace). Fluorescence excitation is at 429 nm, and emission at 600 nm. Downward deflection represents quenching, which occurs as the intravesicular pH falls and the dye is trapped and concentrated inside the vesicles. Initial brief increase in fluorescence upon addition of 1.5 mM-Na$_2$ATP ($+$MgCl$_2$) was an artifact. The observed deflection (left trace) represents acidification by approximately 1.5 units. Note the much reduced extent of quenching in acetate. Orthovanadate inhibits the ATPase and allows accumulated protons to leak out. Data by courtesy of D. S. Perlin & C. W. Slayman.

7.0–7.5, averaging about 7.2 under normal metabolic conditions. Since the extracellular pH in most experiments is 5.8, approximately -80 mV must be added to the figure of -300 in order to estimate the total potential difference against which the pump can work. More complex events take place with sustained cyanide blockade, and recovery of ATP and/or recovery of membrane potential can occur, often accompanied by slow oscillations. We regard only the rapid initial changes as indicating pump behaviour.

Arguments linking these events to proton efflux from intact cells are necessarily indirect, since respiratory blockade affects essentially all energy-requiring metabolic reactions. Net proton efflux does cease in the presence of cyanide and resumes rapidly when the inhibitor is removed (Slayman, 1970), but measured net proton efflux is only a fraction of the membrane current. Direct and unequivocal proof of proton pumping was obtained when plasma membrane vesicles were isolated from *Neurospora* (Scarborough, 1976), and when the purified membrane ATPase could be reconstituted into liposomes. Vesiculated fragments of *Neurospora* membrane reseal 85–90% inside out, so that most ATPase molecules are accessible to ATP from the bathing medium; and the same vesicles readily take up fluorescent indicators, such as the pH probe acridine orange, and the putative voltage probe, oxanol V. Using such preparations, Perlin *et al.* (1984) demonstrated both an ATP-driven net uptake of protons by the vesicles, and an ATP-driven build-up of membrane potential. As is indicated in Fig. 2, proton accumulation (i.e., intravesicular acidification) was maximized in the presence of the permeant anion Cl⁻ and was sharply reduced when chloride was replaced by the impermeant acetate ion. In an entirely complementary fashion, build-up of membrane potential was maximized by acetate ions and minimized by chloride ions; it also occurred much faster than the acid shift. Unfortunately, sustainable steady-state differences reached only 2 pH units or 120 mV, about 30% of the total proton motive force observable in intact cells. This shortfall, plus a rapid dissipation of accumulated protons when vanadate was added to block the ATPase (Fig. 2), indicates leakiness of the membrane vesicles. Completely parallel observations were made using purified ATPase reconstituted into asolectin vesicles.

Chemistry of the Proton Pump

As was mentioned in the Introduction, the plasma membrane ATPase of *Neurospora* belongs to the E_1–E_2 class of transport enzymes, typified by the Na^+, K^+-ATPase of animal cell membranes. An early indication that it could not be an F_0–F_1 enzyme came from the facts (a) that it could not be removed from the membrane by sonication and (b) that it did not show up in negative-staining electron micrographs of *Neurospora* plasma membrane. Eventually, the enzyme was solubilized in active form by extracting the membrane with 0.6% deoxycholate in 45% glycerol (Bowman *et al.*, 1981), or by lysolecithin (Addison & Scarborough, 1981). The plasma membrane ATPases from other fungi have also required detergent extraction: lysolecithin in *Schizosaccharomyces pombe* (Dufour & Goffeau, 1978), and a zwitterionic detergent in *Saccharomyces cerevisiae* (Malpartida & Serrano, 1980), and all have been purified by subsequent density-gradient centrifugation. For *Neurospora*, 45–80% glycerol was used, in the presence of 0.3% deoxycholate. SDS-polyacrylamide gel electrophoresis showed the preparation to contain a single major polypeptide band at 104 kDa, illustrated in Fig. 3. No convincing evidence for a second band (i.e., smaller subunit) has been found, and in this respect the *Neurospora* H^+-ATPase is closer to the Ca^{2+}-ATPases and the gastric mucosal H^+, K^+-ATPase than to the Na^+, K^+-ATPase.

Hydrolytic activity of the enzyme depends absolutely on the presence of phospholipids, but whether the acidic phospholipids (phosphatidyl-serine, -glycerol or -inositol) are crucial, as suggested by Scarborough (1977), is uncertain. Enzyme prepared in micelles (i.e., with sufficient native lipid to activate, but without formation of defined vesicles) reacts with γ-labelled ^{32}P-MgATP to form a stable phosphoprotein intermediate, in which phosphate is bound to an aspartyl residue (Dame & Scarborough, 1981), again firmly pegging the E_1–E_2 class. Under conditions of normal turnover, approximately 20% of the enzyme is in the E ~ P form (J. Kasher & C. W. Slayman, unpublished work).

Although numerous authors have claimed a special role for K^+ ions in the reactivity of H^+-ATPases from plasma membranes of plant cells (see review in Leonard, 1982), and a similar isolated claim has been made for the fungal ATPases (Villalobo, 1982), careful study of the *Neurospora* enzyme has shown only mild stimulation of ATP hydrolysis by various monovalent cations, with no synergy and little specificity (Bowman & Slayman, 1977). [The best cation

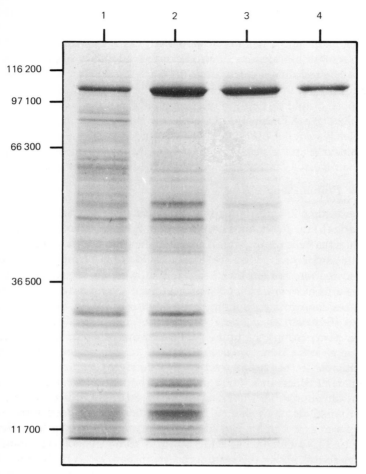

Fig. 3. *SDS-polyacrylamide gel electrophoresis of different fractions from purification of the plasma-membrane ATPase*

Lane 1, plasma membranes; lane 2, membranes washed in 0.1% deoxycholate; lane 3, ATPase solubilized and centrifuged in 0.6% deoxycholate; lane 4, purified ATPase, the peak fraction from the glycerol gradient. Methods and data are from Bowman *et al.* (1981).

is ammonium (10–50 mM), which stimulates by 60–70% of the baseline activity measured in 10 mM-Tris–Pipes buffer.] On this basis alone, the enzyme appears to be purely a proton transport system, clearly differing from the animal cell sodium pump and the gastric mucosal proton pump, both of which seem absolutely to require K^+ ions and also to balance 70–100% of their pumped Na^+ or H^+ with counterflowing potassium. Recent electrical data, indeed, rule out any major obligatory charge-balancing ion movement through the enzyme. Under certain conditions, the pump can actually transport against membrane potentials which exceeded $-400\,mV$ (M. R. Blatt & A. Rodriguez-Navarro, unpublished work), and that accounts for 80% of the free energy available from

ATP hydrolysis (Warncke & Slayman, 1980). The same data, of course, define the H^+:ATP stoichiometry for the enzyme: no integer value greater than unity is possible. Recently, simultaneous measurements of ATP hydrolysis and proton accumulation in *Neurospora* membrane vesicles have confirmed this calculation, with a measured stoichiometry of 0.85–0.95 (D. S. Perlin & B. P. Rosen, unpublished work).

Current–Voltage Kinetics

Any 'electroenzyme' which can drive current through a membrane and through the electric field of that membrane is in principle sensitive to the magnitude of the electric field. If in addition a substrate-dependent change in enzyme velocity affects the electric field, then the reaction diagram for the enzyme is intrinsically multidimensional, with a velocity (= current) axis, substrate axes (1 for each substrate), and a voltage axis. An important implication of this rather formal statement is that we cannot understand the underlying reaction mechanism of the enzyme unless we understand the role of charge displacement within or through the enzyme. And since all primary active transport systems thus far described are ion-transport systems, it is tempting to regard charge displacement as a fundamental feature of active transport. This feature is efficiently investigated, in the membrane environment, by means of current–voltage analysis.

As a starting point for such analysis, we have sketched the hypothetical reaction diagram shown in Fig. 4 for the H^+-ATPase. Since detailed partial reaction chemistry for the fungal enzyme is only beginning to emerge (Addison & Scarborough, 1982; Brooker & Slayman, 1982, 1983), we have extracted the scheme of Fig. 4 from the summary by Karlish *et al.* (1978) for the Na^+, K^+-ATPase in animal cell membranes. Two distinct phosphorylated forms of the enzyme ($E_1 \sim P$ and $E_2 \cdot P$) are included, and one step of the transition between them ($H^+ \cdot E_1 \sim P \rightleftharpoons H^+ \cdot E_2 \cdot P$) is assumed to be the transmembrane movement of charge. We justify this assumption on two grounds: (a) for the Na^+,K^+-ATPase, the E_1–E_2 transition is generally regarded as the actual transport step, and (b) current–voltage analysis of proton pumping in *Neurospora* shows charge-transit to be the energy-transforming step (see below, Fig. 6, and Gradmann *et al.*, 1982), which should therefore be accompanied by the shift from a high-energy to a low-energy phosphoprotein. It is the charge-transit step which must be sensitive to the electric field through the membrane. Two steps, $E_2 \rightleftharpoons E_1$ and $H^+ \cdot E_1 \cdot ATP \rightleftharpoons H^+ \cdot E_1 \sim P \cdot ADP$, are shown as energetically neutral intramolecular rearrangements; and the remaining five steps are included specifically to denote binding and release of the five known ligands (H_o^+, H_i^+, ATP, ADP and P_i).

It is intuitively clear that current–voltage analysis alone can give direct information only about the single reaction step which is envisioned as voltage sensitive, and therefore for experimental treatment the model of Fig. 4 should be reduced to a two-state form with explicit representation only of enzyme states $H^+ \cdot E_1 \sim P$ and $H^+ \cdot E_2 \cdot P$. The appropriate diagram is given in Fig. 5(a) with all other states of the enzyme and all voltage-insensitive reaction steps lumped together in the 'return' (upper) path. If, alternatively, extracellular pH is

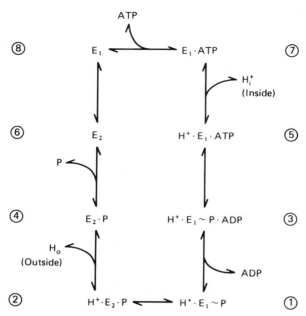

Fig. 4. *Hypothetical reaction sequence for E_1-E_2 type H^+-ATPases, based on sequence described for the Na^+, K^+-ATPase*

Circled numbers are State Indices, used to simplify the subscript designation of reaction parameters.

systematically varied simultaneously with $I-V$ scanning, then binding and release of H_o^+ must also be included, to generate a three-state diagram (Fig. 5b). Finally, if both intracellular and extracellular pH are varied, four- or five-state models are allowed; the five-state model is shown diagrammatically in Fig. 5(c), with two lumped steps: $E_2 \cdot P \rightleftharpoons E_1 \cdot ATP$, and $H^+ \cdot E_1 \cdot ATP \rightleftharpoons H^+ \cdot E_1 \sim P$.

Before proceeding with systematic analysis of such reduced kinetic models, several general points must be made. (i) The major advantage of this approach is to keep the mathematical representations as simple as the nature of experiments will allow. The obvious problem of how hidden reaction states bias interpretation of rate constants for explicit steps (e.g. $H^+ \cdot E_2 \cdot P \rightleftharpoons H^+ \cdot E_1 \sim P$, in all three parts of Fig. 5) is common to all enzyme-kinetic studies, since 'the complete model' is never known. [Physically, this is a question of how to partition reaction rate between the state density and the rate constant at any point in the cycle.] The problem can be handled directly by means of bookkeeping parameters (called reserve factors) which are extracted from the characteristic matrix for each model (Hansen *et al.*, 1981; Gradmann *et al.*, 1982; Sanders *et al.*, 1984). (ii) Inclusion of only a single voltage-dependent step is arbitrary, and has been criticized as overly simple (Läuger, 1984), but it presently has the virtue of accommodating all current–voltage data available on ATP-dependent ion pumps. Possible reasons for including two or more voltage-dependent steps will be discussed later. (iii) Since at present there is no way to specify the actual shape of the electric field through the protein, a function for voltage-dependence must be

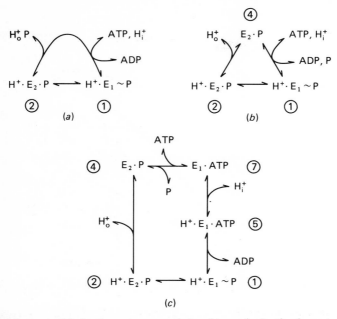

Fig. 5. *Reduced-state models for the proton pump, designed to emphasize the charge-transport step and H⁺ binding/release steps*

State Indices are denoted as in Fig. 4.

assigned arbitrarily. Following Läuger & Stark (1970), we have used a symmetric activation barrier to ions crossing the membrane, which multiplies each zero-voltage rate constant (lower limb of all diagrams in Figs. 4 and 5) by a Botzmann factor: clockwise, $\exp(FV_m/2RT)$; counterclockwise, $\exp(-FV_m/2RT)$. This manoeuvre is equivalent to defining the electrochemical activity of the transported ion, with reference to the centre of the membrane.

Notes on Methods

A variety of strategies have been used to generate current–voltage data in different preparations. Best known, of course, is the repetitive voltage clamp technique applied to electrically triggered, rapidly relaxing phenomena (i.e. action potentials) in excitable membranes. For slow phenomena, ramp, staircase or staggered-pulse programs are more suitable; we use a version of the latter, with voltage-clamp pulses of increasing magnitude, alternating positive and negative, to scan from the resting membrane potential to -300 mV and 0 mV. Pulse durations of 100–200 ms and scan-times of 5–20 s are convenient for steady-state measurements. Because of the mycelial geometry of *Neurospora* at least three microelectrodes are required for a proper measurement: the primary voltage electrode (which provides the clamp signal), a secondary voltage electrode some distance away and the current-passing electrode. Actual clamping current is also measured, between a reference electrode in the cell bath and a

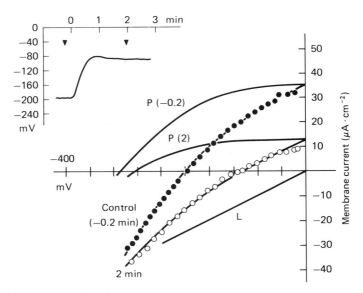

Fig. 6. *Separation of membrane current–voltage curves into pump and leak components*

Inset: single tracing of membrane potential after addition of 1 mM-KCN to the recording medium. Filled triangles designate times (approx 10 s) when *I–V* scans were executed. Main Figure: plotted points are the *I–V* data, i.e., membrane currents calculated, via elementary cable theory, from total input current. Smooth curves, for the pump (P), leak (L) and whole membrane, were drawn by fitting the model of Fig. 5(*a*) to the plotted points. See text for methods. Data are from Gradmann *et al.* (1982). Table of reaction constants (s^{-1}):

	k_{12}°	k_{21}°	K_{21}	K_{12}
Control	1.68×10^{6}	1.08×10^{-1}	2.10×10^{2}	1.44
Cyanide	1.68×10^{6}	1.08×10^{-1}	0.78×10^{2}	1.44

virtual ground point in the recording circuit. Quasilinear cable theory is used to calculate the actual membrane current density from the measured clamping current and the two voltages (Gradmann *et al.*, 1978).

A major problem in analysis of an individual transport system within a whole membrane is to distinguish it from other systems. For current–voltage analysis of the proton pump, two strategies have been adopted. The first is to switch the pump off and on by means of inhibitors, and then examine the difference *I–V* relationship, assuming that only the pump is affected by the inhibitors; for a variety of reasons this approach has proved treacherous (see especially Chapman *et al.*, 1983). A more satisfactory method has been to assign a mathematical form to the ensemble of other transport systems (for simplicity we term the ensemble 'leaks'; in *Neurospora* their forms are quite simple and quite distinct from the curves expected for the proton pump), and then use a statistical optimizing procedure to discriminate between changes of the pump and changes of the leaks under particular experimental conditions.

An example of this procedure is given in Fig. 6. The inset shows a voltage tracing, with the triangles indicating times at which 10 s *I–V* scans were run. At zero time, 1 mм-KCN was admitted to the preparation. Filled circles plot out the control current–voltage curve for the whole membrane, and open circles plot the corresponding curve in KCN. The ensemble leak was assumed to be linear, and the pump was described by the two-state kinetic diagram of Fig. 5(a). The two sets of data were fitted simultaneously, holding all parameters (total of seven) but one in common, and best fits were obtained when the lumped reaction constant for reloading (designated K_{21}, clockwise in the upper path, Fig. 5a) was allowed to vary between the two data sets. All curves drawn in Fig. 6 have been calculated from the fitted parameters. The leak (L), unchanged by cyanide, had a net diffusion potential of zero (intercept at the origin). The single parameter change (K_{21}) in the pump altered the calculated shape of the *I–V* curve for the pump, but most conspicuously diminished the saturating current, which in this case was nearly equal to the short-circuit current.

Interpretations of Current–Voltage Data

Altered energy substrate. Several interesting and important conclusions have emerged from this very simple kinetic analysis of the steady-state electrical behaviour of the *Neurospora* proton pump. The clearest and simplest is that the data cannot be fitted qualitatively with any integer stoichiometric coefficient other than 1, thus independently confirming the ratio H^+ transported:ATP hydrolysed = 1, originally deduced by a thermodynamic argument (see above). A slightly surprising result (to many investigators accustomed to thoughts of EMF values, PMF values and other thermodynamic quantities) is that, at most realizable membrane potentials, kinetic factors rather than thermodynamic factors dominate the behaviour of the proton pump. A single expression of this conclusion is that depolarization caused by ATP withdrawal is 110–120 mV (with no change in resistance) in Fig. 6, but the shift of reversal potential [voltage intercept of curve P(2) minus (P(−0.2)] is less than 30 mV.

The fitted values of the four reaction constants in the two-state model are given in the legend to Fig. 6. Among them, the most important comparison is that the two zero-voltage constants for charge-transport (lower path, Fig. 5a) differ from each other by more than seven orders of magnitude, with k_{21}° (inward) being the smallest reaction constant in the entire set, at 1.1×10^{-1}, and k_{12}° (outward) being the largest, at 1.7×10^{6} s^{-1}. This represents an energetic transition of more than 400 mV, so that, clearly, phosphate bond energy is converted into the proton motive force simultaneously with charge transit of the membrane, as was anticipated in the discussion above. The absolute magnitude of k_{12}° is, as Tanford has pointed out, "uncomfortably large". Although its size might be reduced somewhat by detailed consideration of hidden enzyme states involved in K_{12} and K_{21}, that correction is unlikely to be greater than a factor of 10 (see discussions in Hansen *et al.*, 1981; Gradmann *et al.*, 1982). Furthermore, the value cannot be reduced by exact knowledge of the enzyme density in the membrane, since the figure we have assumed for computation, 10 000 sites·μm^{-2}, is considered maximal for the *Neurospora* membrane. Thus we conclude that once the phosphorylated molecule $H^+ \cdot E_1 \sim P$ is formed and freed of ADP, it undergoes

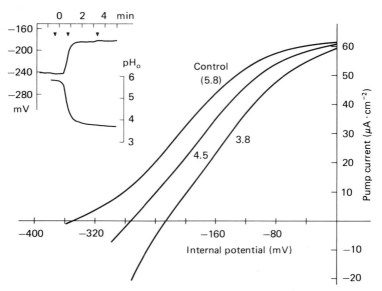

Fig. 7. *Current–voltage curves for the proton pump of Neurospora during changes of extracellular pH*

Inset: curve of membrane potential (upper trace) and pH_o during slow perfusion of the recording chamber with low-pH buffer. Filled triangles designate times when I–V scans were run. Curves drawn in the main Figure were obtained by fitting the model of Fig. 5(a) to I–V data plots. Data are from Slayman & Sanders (1984). Table of pH_o-dependent parameters (s^{-1}):

pH_o	k_{21}^o	K_{12}
5.8	2.1×10^{-2}	3.4×10^3
4.5	23.7×10^{-2}	6.6×10^3
3.8	74.5×10^{-2}	13.4×10^3

rapid energetic decay to displace the proton. Finally, the size of the reaction constant for reloading (K_{21}), 210 s^{-1} in energized cells, estimates the turnover number for the enzyme under voltage-saturating conditions.

Very similar results have been obtained by use of the pump-specific inhibitor orthovanadate, but vanadate action *in vivo* is slow because the ion must first be transported into the cytoplasm and then apparently must compete with a large reservoir of phosphates, before binding with the enzyme. That two kinds of agents, one withdrawing the substrate ATP and one presumed to bind to a form of E_2, should have similar actions on the pump I–V curves is surprising, since the two should act in opposite directions along the reloading pathway.

Altered transport substrate. However, altering concentrations of the pump's other substrate, protons, gives more complex results. Three pump current–voltage curves obtained during a downshift of extracellular pH are shown in Fig. 7. The inset is again a record of membrane potential versus time (upper trace), along with extracellular pH (lower trace), measured by a pH-sensitive microelectrode adjacent to the punctured cell; current–voltage scans are marked by the triangles. Now, as a general rule for analysis of kinetic data, particularly when an experimental manoeuvre may have multiple independent effects (i.e., on the proton pump and on membrane leaks), we have found it useful to do a

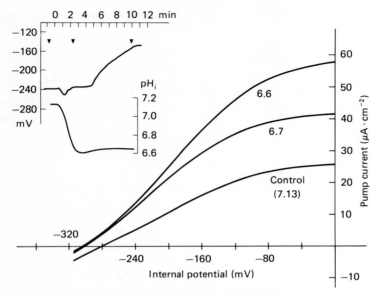

Fig. 8. *Current–voltage curves for the protein pump of Neurospora during changes of intracellular pH*

Details are similar to those given for Fig. 7, except that pH_1 was lowered by adding 5 mM-sodium butyrate to the medium. Data are from Slayman & Sanders (1984). K_{21} is sensitive to pH_i: $= 1.7 \times 10^2$ (7.13), 2.7×10^2 (6.7) and 3.8×10^2 (6.6).

preliminary analysis with the smallest possible model, even though the data and experimental design may allow larger models. Thus, the curves of Fig. 7 were obtained via a two-state analysis, which again (as in the case of cyanide inhibition) extracted a constant membrane leak for all conditions. With lowered pH_o, however, pump behaviour could not be described by changing solely the reaction parameter (K_{12}) which incorporates the extracellular proton binding step, but required also changing the voltage-sensitive reaction constant for proton re-entry (k_{21}^o). This surprising result can be interpreted by systematic consideration of the reserve factors, referred to above (see Hansen *et al.*, 1981), and implies that the extracellular proton binding site must be predominantly dissociated under normal conditions. Two additional properties of the transport enzyme are apparent from the curves drawn. First, the saturating turnover (approximated here by the current intercept at zero potential) is insensitive to extracellular pH; and second, the reversal potential (voltage intercept at zero current) changes approximately as it should for a one-proton pump: $+76$ mV for 1.3 pH unit acidification, and $+47$ mV for 0.7 unit acidification.

A corresponding analysis for altered intracellular pH is shown in Fig. 8. Data were obtained by adding butyric acid (pH 5.8) to the bathing medium, and monitoring with an intracellular pH microelectrode. In this case only a single reaction parameter in the two-state model appeared to change with pH (though the membrane leak also increased greatly with intracellular acidification): again the reloading parameter K_{21}, as can be seen by inspection, since the

Table 1. *Kinetic parameters for a proton pump, deduced from pH effects upon current–voltage relationships*

Common description of pH_o and pH_i effects on pump current–voltage relationships, as shown in Figs. 7 and 8. Parameters were obtained for a four-state kinetic model similar to that in Fig. 5(c), but with the enzyme state $H^+ \cdot E_1 \cdot ATP$ lumped into the whole transition from $E_1 \cdot ATP$ to $H^+ \cdot E_1 \sim P$. k designates assumed single-step rate constants, and K designates pseudo rate constants for several reaction steps lumped together. The superscript $^\circ$ designates constants from which known experimental variables (i.e., membrane potential, protein concentrations) have been extracted.

$K_{71} = K_{71}^\circ [H^+]_i$		
$K_{71}^\circ = 1.24 \cdot 10^{10}$	$K_{17} = 4.54 \cdot 10^4$	$pK_i = 5.44 \ (pH_i = 7.2)$
$k_{12} = k_{12}^\circ \exp(zFV_m/2RT)$	$k_{21} = k_{21}^\circ \exp(-zFV_M/2RT)$	
$k_{12}^\circ = 8.70 \cdot 10^5$	$k_{21}^\circ = 1.31$	
	$k_{42} = k_{42}^\circ [H^+]_o$	
$k_{24} = 2.31 \cdot 10^3$	$k_{42}^\circ = 1.96 \cdot 10^6$	$pK_o = 2.93 \ (pH_o = 5.8)$
$K_{47} = 1.15 \cdot 10^3$	$K_{74} = 2.61 \cdot 10^3$	

curves qualitatively resemble those obtained during ATP withdrawal. Internal pH, unlike external pH, has a strong effect on saturating turnover of the enzyme, and a time dependent effect, as well as the direct effect, can be seen in the 30% increase of short-circuit current which occurs during the 7–8 min required for pH_i to stabilize. Again, the reversal potential changed in accord with thermodynamic expectation.

Comparison of the two pH results raises a very interesting and important problem for the understanding of vectorial enzymes: that is, how the same substrate, $[H^+]_i$ and $[H^+]_o$, acting through a single pathway at different times or in different directions, can have quantitatively very different effects. Raising $[H^+]_o$ does not affect saturating pump velocity, but elevating $[H^+]_i$ increases the pump velocity greatly. It is in putting together such disparate results that the higher-state kinetic models become useful. From the behaviour of the two-state model in Fig. 5(a), it is clear that the three-state model (Fig. 5b), with explicit binding/dissociation of H_o^+, can describe the curves of Fig. 7. It is likewise clear that the symmetric model with explicit binding/dissociation of H_i^+ can describe the curves of Fig. 8. But, can both sets of pH effects be accommodated into a single model with explicit binding/dissociation of H_o^+ *and* H_i^+ and with fixed reaction constants throughout? Although data of this kind are still being analysed, the basic answer to the question is yes.

The largest model allowed by the combined data sets in Figs. 7 and 8 is that in Fig. 5(c), with five explicit kinetic states. But to use that model makes sense only when additional information guarantees the presence of the enzyme form $H^+ \cdot E_1 \cdot ATP$; the mathematical evaluation of pH effects can be done without that form of the enzyme, i.e., by using only a four-state model. The results, then, of fitting the appropriate four-state model to the pump curves of Figs. 7 and 8 are summarized in Table 1. The most important point apparent in the Table is that the pK for each proton binding reaction must be substantially below the pH of the adjacent solution: $pK_i = 5.4$ versus $pH_i = 7.2$; and $pK_o = 2.9$ versus $pH_o \sim 5.8$. Intuitively this result makes sense. When both 'upstream' (internal)

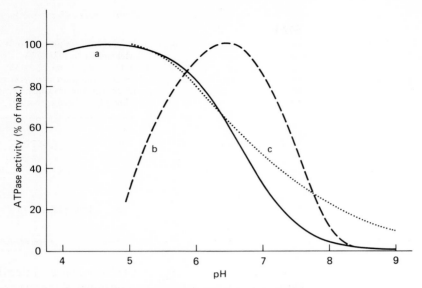

Fig. 9. *Comparison of pH optima for plasma-membrane H^+-ATPases*

a, Theoretical curve for the *Neurospora* enzyme, from reaction constants in Table 1. b, Experimental curve for the purified *Neurospora* enzyme, from Bowman *et al.* (1981). c, Experimental curve for enzyme prepared from *Saccharomyces cerevisiae*, from Peters & Borst-Pauwels (1979).

and 'downstream' (external) binding sites are mostly unloaded, the availability of $H^+ \cdot E_1 \sim P$, but not of $H^+ \cdot E_2 \cdot P$, can be rate limiting to the normal physiological flow of the reaction (clockwise). Therefore, increasing $[H^+]_i$, by increasing H^+ binding, is likely to accelerate the flow, but altering $[H^+]_o$ can have little effect. The arrangement required for $[H^+]_o$ to affect flow in the physiological direction is for $E_2 \cdot P$ to be scarce, so that its level would be shifted proportionate to $[H^+]_o$; that is, the external binding site would need to be largely filled ($pK_o \gg pH_o$).

This model for the plasma-membrane proton pump of *Neurospora*, with the reaction constants given in Table 1, makes a number of quantitative predictions which have been tested and verified in physiological experiments (see Slayman & Sanders, 1984). The model also makes some interesting chemical predictions, of which one has been tested. The pH optimum for the isolated ATPase from the *Neurospora* plasma membrane is near 6.5, with activity falling to about 60% 1 pH unit each side of the optimum (Bowman *et al.*, 1981). Such experiments are done under conditions of zero membrane potential and equal pH on both 'sides' of the enzyme, and those conditions can be combined with the parameter values in Table 1 in order to calculate transport velocity as a function of bulk pH. The result is given by the solid curve (a) in Fig. 9. According to this calculation, transport velocity should increase monotonically with decreasing pH, down to about pH 4.5; and coincidentally it should have a steepest slope near pH 6.5, the measured pH optimum. For comparison, the measured activity of the isolated enzyme is shown as the dashed curve (b) in Fig. 9. We regard

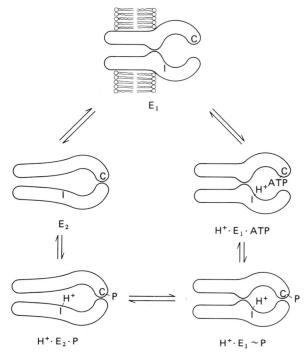

Fig. 10. *Cartoon of a double-gated mechanism to pump protons*

Suggested model is a variation of the peristaltic channel proposed by Läuger (1984), but through changes of molecular conformation, it uses ATP to move the barrier and the electric field rather than the ion.

the calculated curve as the more accurate reflection of normal enzyme chemistry, and the measured curve as a partial reflection of acid-inactivation, which occurs rapidly at low pH. It is perhaps significant in this respect that other investigators (Peters & Borst-Pauwels, 1979) have reported an activity curve for the plasma membrane ATPase of *Saccharomyces* which rises monotonically with falling pH, at least down to pH 5, as shown by the dotted curve (c) of Fig. 9. Since that organism tolerates both lower intracellular pH and lower extracellular pH values than does *Neurospora*, its proton pump must be hardier in acid environments.

Discussion

Up to this point we have said nothing about the physical mechanism of the transport process: that is, about the nature of the conformation change or structural rearrangement which must occur between $H^+ \cdot E_1 \cdot ATP$ and $H^+ \cdot E_2 \cdot P$. Neither the transport kinetic data nor the enzyme kinetic data define that process, which is fundamentally in the domain of structural data.

Direct information about the structure of the fungal H^+-ATPase is just beginning to emerge, but its strong chemical resemblance to the Ca^{2+}-ATPase and the large (α) subunit of the Na^+, K^+ATPase suggests that the fungal enzyme would have strong structural similarity to those enzymes as well. Nicholas (1984) has recently summarized structural data for the α-subunit in a ball-and-stalk cartoon which topographically resembles the F_o–F_1 ATPase, except that it is much smaller, comprised of the single polypeptide embedded in and protruding from the membrane. Approximately 60–70% of the molecule is globular, extends from the membrane on the physiological inner surface, and contains the catalytic site including the aspartylphosphate group; 15–20% comprises α-helices extending through the membrane, and the remaining 10–15% faces the physiological outer surface of the membrane. The array of α-helices, if it is to be likened to the α-helices in bacteriorhodopsin, for example, can be expected to lie approximately perpendicular to the plane of the membrane and to aggregate into one or more presumptive channels.

It is clear that this kind of structure is both too massive and too well anchored to function as a diffusible carrier (i.e., like valinomycin). However, three concepts of transport which are, at least superficially, compatible with the structure have been suggested over the past 5 years. These are (i) the ligand conductor, in which the transported ion hops along an internal surface or array of sites through the protein (Mitchell, 1979); (ii) the peristaltic channel, in which a directed and progressive conformational change or energy wall pushes the ion through a transmembrane pore in the protein (Läuger, 1979, 1984); and (iii) the double-gated channel, in which the transported ion is first gated into the membrane from one side, is trapped or occluded (Glynn & Richards, 1983) with both gates closed, and is then gated out on the opposite side. Thus far the greatest strength of each of these models is geometric and mechanical appeal; all chemical details remain to be fleshed out.

Nevertheless, from a kinetic point of view, all three can be formulated in similar fashion, producing an interesting, rather elegant, and testable modification of the simple kinetic models in Figs. 4 and 5. That modification is most easily presented by reference to the notion of double gating. We might suppose, for example (see Fig. 10), that the uncomplexed enzyme consists of two cavities: one being a channel through the membrane and the other being a space (near the catalytic site) in which ATP and the transport proton bind. In the normal non-transporting condition of the enzyme, the two cavities would be separated by a constriction or gate, which forms the principal dielectric barrier through the protein and sustains essentially the entire electric field.

Once both ATP and H^+ have bound, catalysis (clockwise in Fig. 10) would transfer the γ-phosphate from ATP to the aspartyl group, allowing ADP to be released, and gating shut the catalytic space from the cell cytoplasm. Closing this gate (bottom right in Fig. 10) would necessarily be an energy-requiring step, since the dielectric barrier within the protein is split, with part of the field developed across each gate and therefore part of the field having moved across the occluded ion. The remaining conformational energy in the phosphoprotein would be lost as the gate between the two cavities opens, thus moving the second part of the field across the transport proton and transferring the entire field to the right-hand gate (lower left in Fig. 10). Diffusion would subsequently carry the proton out of the channel, which could then relax back to the non-transporting condition. As a general idea, this model is not particularly novel, but there is one feature of it which has not been emphasized previously and which we think is important. From a physical point of view, the ion is not pumped through the membrane; rather, the membrane is transported past the ion and, specifically, the membrane electric field is driven past the ion. Ligand conduction, to the extent it could be said to occur, would be driven by diffusion.

An elementary kinetic corollary of the double-gated mechanism is apparent: it requires two voltage-dependent steps: one for closing the right-hand gate, and one for opening the left-hand gate. The gates themselves need not bear displaceable charges, but movement of the electric field past the occluded ion makes each step voltage-sensitive. Going back to Fig. 4, in addition to the step $H^+ \cdot E_2 \cdot P \rightleftharpoons H^+ \cdot E_1 \sim P$, which we have supposed to contain reaction constants with Boltzmann terms in V_m, also another step must contain such reaction constants, perhaps $H^+ \cdot E_1 \sim P \cdot ADP \rightleftharpoons H^+ \cdot E_1 \sim P$. It is now our task to explore the testable experimental consequences of a series arrangement of two voltage-dependent steps, on both steady-state and transient current–voltage data. That kind of study, of course, has featured prominently in the developing understanding of passive channels over the past 15 years.

This work has been supported by Research Grant GM-15858 from the National Institute of General Medical Sciences. The authors are indebted to Dr. Dietrich Gradmann of the Institut für Botanik, Universität Göttingen, and Dr. Ulf-Peter Hansen of the Institut für Angewandtek Physik, Universität Kiel for some computations and much helpful discussion, and to Dr Carolyn Slayman of Yale University for permission to use the data in figures 2 and 3.

References

Addison, R. & Scarborough, G. A. (1981) *J. Biol. Chem.* **256**, 13165–13171
Addison, R. & Scarborough, G. A. (1982) *J. Biol. Chem.* **257**, 10421–10426
Bowman, B. J. & Slayman, C. W. (1977) *J. Biol. Chem.* **252**, 3357–3363
Bowman, B. J., Blasco, F. & Slayman, C. W. (1981). *J. Biol. Chem.* **256**, 12343–12349
Brooker, R. J. & Slayman, C. W. (1982) *J. Biol. Chem.* **257**, 12051–12055
Brooker, R. J. & Slayman, C. W. (1983) *J. Biol. Chem.* **258**, 222–226
Chapman, J. B., Johnson, E. A. & Kootsey, J. M. (1983) *J. Membr. Biol.* **74**, 139–153
Dame, J. B. & Scarborough, G. A. (1981) *J. Biol. Chem.* **256**, 10724–10730
Dufour, J. P. & Goffeau, A. (1978) *J. Biol. Chem.* **253**, 7026–7032
Felle, H., Stetson, D. L., Long, W. S. & Slayman, C. L. (1978) in *Frontiers of Biological Energetics* (Dutton, P. L., Leigh, J. S. & Scarpa, A., eds.), pp. 1399–1407, Academic Press, London
Felle, H., Porter, J. S., Slayman, C. L. & Kaback, H. R. (1980) *Biochemistry* **19**, 3585–3590
Glynn, I. M. & Richards, D. E. (1983) *Curr. Top. Membr. Transp.* **19**, 625–638
Gradmann, D., Hansen, U. P., Long, W. S., Slayman, C. L. & Warncke, J. (1978) *J Membr. Biol.* **39**, 333–367
Gradmann, D., Hansen, U.-P. & Slayman, C. L. (1982) *Curr. Top. Membr. Transp.* **16**, 257–276
Hansen, U.-P., Gradmann, D., Sanders, D. & Slayman, C. L. (1981) *J. Membr. Biol.* **63**, 165–190
Karlish, S. J. D., Yates, D. W. & Glynn, I. M. (1978) *Biochim. Biophys. Acta* **525**, 252–264
Läuger, P. (1979) *Biochim. Biophys. Acta* **552**, 143–161
Läuger, P. (1984) *Biochim. Biophys. Acta* **779**, 307–341
Läuger, P. & Stark, G. (1970) *Biochim. Biophys. Acta* **211**, 458–466
Leonard, R. T. (1982) in *Metals and Micronutrients: Uptake and Utilization by Plants* (Robb, D. A. & Pierpont, W. S., eds.), pp. 71–75, Academic Press, London
Maloney, P. C. & Wilson, T. H. (1985) *BioScience* **35**, 43–48
Malpartida, F. & Serrano, R. (1980) *FEBS Lett.* **111**, 69–72
Mitchell, P. (1979) *Eur. J. Biochem.* **95**, 1–20
Nicholas, R. A. (1984) *Biochemistry* **23**, 888–898
Perlin, D. S., Kasamo, K., Brooker, R. J. & Slayman, C. W. (1984) *J. Biol. Chem.* **259**, 7884–7892

Peters, P. H. J. & Borst-Pauwels, G. W. F. H. (1979) *Physiol. Plantarum* **46**, 330–337
Sanders, D. & Slayman, C. L. (1982) *J. Gen. Physiol.* **80**, 377–402
Sanders, D., Hansen, U.-P., Gradmann, D. & Slayman, C. L. (1984). *J. Membr. Biol.* **77**, 123–152
Scarborough, G. A. (1976) *Proc. Natl. Acad. Sci. U.S.A.* **73**, 1485–1488
Scarborough, G. A. (1977) *Arch. Biochem. Biophys.* **180**, 384–393
Slayman, C. L. (1970) *Am. Zool.* **10**, 377–392
Slayman, C. L. (1984) in *Hydrogen Ion Transport in Epithelia* (Forte, J. G., Warnock, D. G. & Rector, F. C., Jr., eds.), pp. 47–56, Wiley-Interscience, New York
Slayman, C. L., Long, W. S. & Lu, C. Y.-H. (1973) *J. Membr. Biol.* **14**, 305–338
Villalobo, A. (1982) *J. Biol. Chem.* **257**, 1824–1828
Warncke, J. & Slayman, C. L. (1980) *Biochim. Biophys. Acta* **591**, 224–233

Biochem. Soc. Symp. **50**, 31–42
Printed in Great Britain

The Apamin-Sensitive Ca²⁺-Dependent K⁺ Channel Molecular Properties, Differentiation and Endogenous Ligands in Mammalian Brain

MICHEL LAZDUNSKI, MICHEL FOSSET, MICHEL HUGUES,
CHRISTIANE MOURRE, GEORGES ROMEY and HEIDY SCHMID-ANTOMARCHI

Centre de Biochimie du CNRS, Faculté des Sciences, Parc Valrose, 06034 NICE CEDEX, France

Synopsis

1. Apamin is a bee venom neurotoxin of 10 amino acids containing two disulphide bridges. 2. Current-clamp and voltage-clamp experiments have shown that apamin externally applied blocks specifically at low concentration (0.1 μM) the Ca²⁺-dependent slow K⁺ conductance which mediates the long-lasting after-hyperpolarization in neuroblastoma cells and rat muscle cells in culture. 3. The apamin-sensitive Ca²⁺-dependent slow K⁺ conductance is voltage-dependent and tetraethylammonium-insensitive. It is distinct from the high conductance Ca²⁺-dependent K⁺ channel revealed by patch-clamp experiments. 4. Biochemical characterization of the apamin receptor in rat striated muscle, neuroblastoma cells, rat synaptosomes, smooth muscles and hepatocytes was carried out with the use of a radiolabelled monoiodo-apamin derivative (¹²⁵I-apamin) of high specific radioactivity (2000 Ci/mmol). 5. The dissociation constant of the apamin–receptor complex is between 15 and 60 pM for all tissue preparations. The density of binding sites is very low: between 1 and 40 fmol/mg of protein. 6. Radiation–inactivation analysis indicates a molecular mass for the apamin receptor of 250000 Da whereas affinity labelling with ¹²⁵I-apamin results in covalent labelling of a single polypeptide chain with a molecular mass of about 30000 Da. 7. Autoradiography of ¹²⁵I-apamin binding sites reveals the presence of Ca²⁺-activated K⁺ channels in many regions of the brain. 8. There is an all-or-none control of the expression of the apamin-sensitive Ca²⁺-dependent K⁺ channel by innervation in mammalian skeletal muscle. 9. There exists an endogenous equivalent of apamin in rat brain.

Apamin, its Structure and its Active Site

Apamin is a neurotoxin extracted from bee venom (Habermann, 1972). It is a polypeptide of 18 amino acids with two disulphide bridges. It is the only polypeptide neurotoxin, as far as we know, that passes the blood-brain barrier. The structure of this toxin is presented in Fig. 1. Analysis of the structure–function relationships of this toxin has shown that two of the 18 amino acids in the sequence have particular importance for the action of the toxin: they are Arg-13 and Arg-14 (Vincent *et al.*, 1975). These two residues seem to be essential

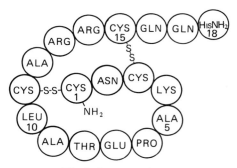

Fig. 1. *Structure of apamin*

elements of the active site of the toxin. Chemical modifications elsewhere in the sequence may decrease the toxicity of the polypeptide but do not suppress its biological activity. Cumulative chemical modifications (of amino group and of the imidazole of His-18, for example) may, however, abolish the activity of the toxin (Vincent *et al.*, 1975). Solid phase synthesis of apamin and analogues has been carried out (Cosand & Merrifield, 1977; Granier *et al.*, 1978). This approach has confirmed that the active site of apamin comprises the two residues Arg-13 and Arg-14.

Apamin has not yet been crystallized successfully. Therefore the exact three-dimensional structure of the toxin remains unknown. However, recent solution analysis of apamin with NMR techniques has suggested that the toxin is highly ordered with an α-helical core and regions of β-type turns (Bystrov *et al.*, 1980; Wemmer & Kallenbach, 1983).

Apamin Blocks Ca^{2+}-Dependent K^+ Channels

Apamin does not seem to interact with receptors of the most classical neurotransmitters (Vincent *et al.*, 1975). The first experiments that lead to the idea that apamin could be a specific inhibitor of a Ca^{2+}-dependent K^+ conductance were carried out on smooth muscles. It was shown that the bee toxin blocked inhibitory responses evoked by ATP, adrenaline or stimulation of non-adrenergic, non-cholinergic nerves in visceral smooth muscles (Vladimirova & Shuba, 1978; Maas & Den Hertog, 1979). It was also shown that apamin caused enhancement of muscle activity and even converted the inhibitory response into an excitatory phase. However, it was only when it was demonstrated by K^+ flux studies that apamin prevented the rise in potassium permeability in guinea-pig hepatocytes and *taenia caeci* Banks *et al.*, 1979; Maas *et al.*, 1980) produced by ATP and noradrenaline that the toxin really became a good candidate as a specific blocker of a Ca^{2+}-dependent K^+ conductance.

A direct demonstration of the specific action of the toxin was obtained using voltage-clamp techniques with both neuroblastoma cells and rat muscle cells in culture (Hugues *et al.*, 1982*a*, *b*). Fig. 2 shows the effect of apamin on the after-hyperpolarization potential (a.h.p.) in neuroblastoma cells. The bee venom

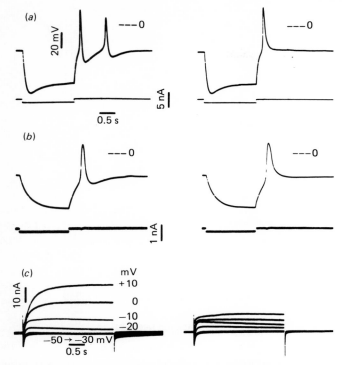

Fig. 2. *Effect of apamin on a.h.p. and ionic currents in neuroblastoma cells*

Selective block by apamin of the a.h.p. after the action potentials of N1E 115 neuroblastoma cells bathed in solutions containing a high concentration of Ca^{2+} (25 mM); voltage-clamp analysis. (*a*) (Left) Multiple spike response evoked by anodal break stimulation in a 25 mM-Ca^{2+}/90 mM-Na$^+$ solution containing 25 mM-TEA. The a.h.p. after the first spike has partially reactivated the Na$^+$ conductance, allowing the initiation of a second spike. (Right) Same cell, 2 min after the application of 0.1 μM-apamin. Note the absence of a.h.p. (*b*) (Left) Ca^{2+} action potential and a.h.p. evoked by anodal break stimulation in a 25 mM-Ca^{2+}, Na$^+$-free solution containing 25 mM-TEA. (Right) Suppression of the a.h.p. after a 2 min application of 0.1 μM-apamin. In (*a*) and (*b*), the zero voltage line is indicated. (*c*) Voltage-clamp analysis of the effect of apamin on neuroblastoma cells bathed in a 25 mM-Ca^{2+}, Na$^+$-free solution containing 25 mM-TEA. Families of membrane currents associated with different step depolarizations from a holding potential (V_h) of -90 mV. (Left) Control currents. (Right) Currents 5 min after the addition of 0.1 μM-apamin. The Ca^{2+} current was not affected; the Ca^{2+}-dependent slow outward current was strongly depressed. The remaining current after 0.1 μM-apamin is due to a non-linear leakage current.

toxin supresses the a.h.p. and it is shown by the voltage-clamp data that this blockade is due to an inhibition of the Ca^{2+}-dependent K$^+$ conductance.

Biochemical Properties of the Apamin Binding Component of the Ca^{2+}-Dependent K$^+$ Conductance

Radiolabelled monoiodoapamin can be prepared at a very high specific radioactivity ($\simeq$ 2000 Ci/mmol) by incorporating iodine on His-18, the C-terminal residue of the toxin (Hugues *et al.*, 1982*c*; Habermann & Fischer, 1979). The mean lethal dose (LD$_{50}$) of the monoiodo derivative of apamin measured

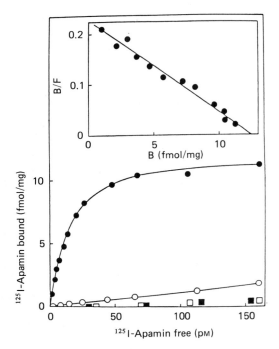

Fig. 3. *Binding of* ^{125}I-*apamin to synaptosomes*

Non-specific binding (○) was determined in the presence of a large excess (1 μM) of unlabelled apamin. Specific binding (●) is the difference between total binding (not shown) and non-specific binding. In control experiments, synaptosomes were omitted in the incubation medium and the binding of ^{125}I-apamin to filters alone was measured in the presence (■) or in the absence (□) of 1 μM-unlabelled apamin. Inset: Scatchard plot of the data; B, bound; F, free; $K_d = 15$ pM; $B_{max.} = 12.6$ fmol/mg of protein.

by intracisternal injection in mice is 4.2 ± 1 μg/kg, as compared with 2 ± 0.5 μg/kg for the native toxin (Hugues *et al.*, 1982c). This decrease of effectiveness by a factor of about 2 after iodination is also observed with a bioassay in which apamin inhibited (and even converted into a contraction) the neurotensin-induced relaxation in segments of guinea-pig proximal colon (Hugues *et al.*, 1982c).

A typical representation of ^{125}I-apamin binding data is present in Fig. 3. One sees in this Figure the main properties of the association of the toxin with the Ca^{2+}-dependent K^+ channel: (i) the specific binding component is much higher than the non-specific binding component, (ii) there is only one family of binding sites (the Scatchard plot is linear), (iii) the affinity of apamin for its receptor is high (it is 15–25 pM for the monoiodo derivative and 10 pM for the native toxin), (iv) the binding of apamin to synaptosomes is characterized by a very low binding capacity [synaptosomes bind only 12–13 fmol of ^{125}I-apamin/mg of protein and therefore contain 150–300 times less apamin-sensitive channels than tetrodotoxin-sensitive Na^+ channels (Hugues *et al.*, 1982c)].

^{125}I-Apamin has now been used to identify apamin-sensitive Ca^{2+}-dependent

K$^+$ channels in a variety of cell types including neuroblastoma cells (Hugues *et al.*, 1982*a*), smooth muscle (Hugues *et al.*, 1982*d*), skeletal muscle cells in culture (Hugues *et al.*, 1982*b*) and hepatocytes (Cook *et al.*, 1983). The high affinity of the toxin for its receptor and the low number of binding sites have been found in all systems. In neuroblastoma or muscle cells in culture the number of apamin-sensitive Ca^{2+}-dependent K$^+$ channel is five to seven times lower than the number of tetrodotoxin-sensitive Na$^+$ channels (Hugues *et al.*, 1982*a*, *b*).

The other properties of the apamin–receptor complex which are of interest are the following: (i) The toxin dissociates very slowly from its receptor. Dissociation rate constants of $1.5 - 4 \times 10^{-4}$ s^{-1} have been found in the different nerve and muscle systems investigated (Hugues *et al.*, 1982*a*, *c*); they correspond to half-lives of dissociations which can be as low as about 60 min. (ii) Non-toxic derivatives of apamin do not bind to the apamin receptor. (iii) The binding of monoiodoapamin to its receptor is sensitive to cations (Hugues *et al.*, 1982*c*; Habermann & Fischer, 1979). K$^+$ and Rb$^+$ at concentrations between 10 μM and 5 mM are able to increase the binding of ^{125}I-apamin to its receptor by a factor of about 2, and each of the many other cations tested (Li$^+$, Na$^+$, guanidinium etc.) completely inhibits ^{125}I-apamin binding in the concentration range 1–100 mM. These results have been interpreted (Hugues *et al.*, 1982*c*) as suggesting the existence of two different binding sites for cations. In this hypothesis, site 1 is specific for K$^+$ and Rb$^+$ and binds these cations with an affinity corresponding to a dissociation constant of about 500 μM. This site is distinct from the binding site of apamin; its occupancy by K$^+$ or Rb$^+$ results in an increased affinity of the receptor for the toxin. Site 2 most probably belongs to the apamin binding site itself; it recognizes every cation tested with a low affinity. This anionic site which serves as a cation binding site is probably the site to which Arg-13 and Arg-14, the two crucial residues of the toxin, bind. It is then not very surprising that molecules which contain guanidinium groups like guanidinium itself, amiloride or neurotensin (which has two contiguous arginine residues like apamin) possess a higher affinity for the apamin receptor than other inorganic cations or positively charged molecules (Hugues *et al.*, 1982*a*, *c*). Among this last class of molecules, quinine and quinidine prevent ^{125}I-apamin binding to the Ca^{2+}-dependent K$^+$ channel with $K_{0.5}$ values of 100–200 μM (Hugues *et al.*, 1982*b*).

Apamin as a Tool to Purify the Apamin-Sensitive Ca^{2+}-Dependent K$^+$ Channel and to Determine its Molecular Weight and its Polypeptide Composition

The apamin receptor is of a polypeptide nature (Hugues *et al.*, 1982*a*, *c*), like the receptor of the many neurotoxins which are known to be specific for the fast Na$^+$ channel (Lazdunski & Renaud, 1982). The problem is to know whether apamin can be used to purify the Ca^{2+}-dependent K$^+$ channel in the way tetrodotoxin, saxitoxin or scorpion toxins have been used for the purification of the Na$^+$ channel (Barchi *et al.*, 1980; Hartshorne & Catterall, 1981; Moore *et al.*, 1982; Barhanin *et al.*, 1983*a*; Norman *et al.*, 1983). Properties of apamin in favour of its successful use to purify the Ca^{2+}-dependent K$^+$ channels

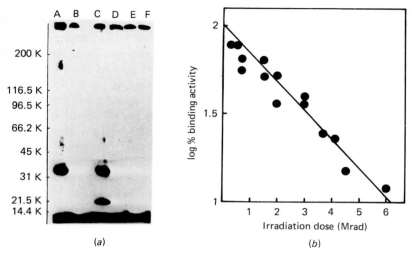

(a)   (b)

Fig. 4. *Structure of the apamin sensitive channel of synaptic membrane*

(a) Sodium dodecyl sulphate gel electrophoresis of synaptic membrane peptide covalently labelled with ^{125}I-apamin after treatment with dissuccinimidyl suberate (DSS). (A) Incubation with ^{125}I-apamin (0.2 nM) in the presence of proteases inhibitors followed by DSS treatment. (B) Same as A but in the presence of unlabelled apamin (1 μM). (C) Incubation with ^{125}I-apamin (0.2 μM). without protease inhibitors, followed by DSS treatment. (D) Same as C but in the presence of unlabelled apamin (1 μM). (E) Control. Same as A but omitting DSS treatment. (F) Same as B but omitting DSS treatment. Control profiles in the presence of protease inhibitors were identical to E and F. (b) Estimation of the molecular size of the apamin-sensitive Ca^{2+}-dependent K^+ channel by the radiation–inactivation technique. The freeze-dried synaptic membranes were irradiated *in vacuo* with high energy electrons at a dose rate of 2 Mrad/min at 10 °C. Semi-logarithmic plot of the data obtained by measuring the decay of the total number of apamin binding sites as function of the irradiation dose shows that the radiation dose after which 37% of the initial specific ^{125}I-apamin binding remains is 2.56 Mrad. By using the empirical equation given by Kepner & Macey (1968), $M_r = 6.1 \times 10^5/D_{37}$, the molecular mass of the apamin receptor was calculated to be $250\,000 \pm 20\,000$ Da.

are its high affinity for its receptor and the slow rate of dissociation of the complex. However, a clear difficulty for visualizing this purification in an optimistic way is due to the very small number of Ca^{2+}-dependent K^+ channels in all preparations which have been assayed up till now. The tetrodotoxin-sensitive Na^+ channel has been purified from brain synaptosomes (Hartshorne & Catterall, 1981; Barhanin *et al.*, 1983*a*) but, as has been seen before, Na^+ channels are 150–300 times more numerous in this preparation than apamin-sensitive Ca^{2+}-dependent K^+ channels, and brain synaptosomes are unfortunately the best source of apamin receptor presently identified.

Available data concerning the structure of the apamin-sensitive channel have been obtained directly on membranes. The relative molecular weight (M_r) of the apamin receptor was determined by using the radiation inactivation technique which was so successful in establishing the M_r of the Na^+ channel (Levinson & Ellory, 1973; Barhanin *et al.*, 1983*b*). This work was carried out in joint collaboration with Dr. Ellory's laboratory in Cambridge. The M_r of the apamin receptor is $250\,000 \pm 20\,000$ (Schmid-Antomarchi *et al.*, 1984) determined by the same technique (Fig. 4*b*).

Affinity labelling of the apamin-sensitive Ca^2-dependent K$^+$ channel was realized successfully by cross-linking the toxin to its receptor on the channel structure by using disuccinimidyl suberate (DSS). To increase chances of success, the experiment was carried out on synaptic membranes, which contain about twice as much apamin-sensitive Ca^{2+} channel as synaptosomes, and at pH 9, where the binding capacity receptor is maximum. The covalent labelling indicates that the apamin receptor in the Ca^{2+}-dependent K$^+$ channel of synaptic membranes is a single polypeptide chain of M_r about 30000 (Fig. 4a) (Hugues et al., 1982e; Schmid-Antomarchi et al., 1984).

It is not known at the present time whether the Ca^{2+}-dependent K$^+$ channel is made of only one type of polypeptide chain ($M_r = 30000$) or whether there are other polypeptide chains which have not been labelled by apamin. If there is only one type of polypeptide chain the channel is then an oligomeric structure containing 8 ± 2 chains of M_r 30000.

Knowing that the M_r of the apamin receptor is near 250000, one can easily calculate that the specific activity of the pure preparation of apamin-sensitive Ca^{2+}-dependent K$^+$ channel will have to be near 4000 pmol/mg of protein. The amount of apamin receptor in synaptosomes being about 12 fmol/mg of protein (Hugues et al., 1982c), the isolation of the pure channel will require a purification by a factor of 300000–400000. We have successfully solubilized the apamin receptor with a complete preservation of the same binding properties it had in the membrane. However, although classical steps of purification have been sufficient to purify the tetrodotoxin-sensitive Na$^+$ channel (Hartshorne & Catterall, 1981; Barhanin et al., 1983a), it will certainly be necessary to use affinity columns containing either apamin or anti-Ca^{2+}-dependent K$^+$ channel antibodies to isolate this channel successfully in the pure form.

The Apamin-Sensitive Ca^{2+}-Dependent K$^+$ Channel is Only One of Several Types of Ca^{2+}-Dependent K$^+$ Channels

Since the development of patch-clamp techniques (Neher et al., 1978) single channel recordings from Ca^{2+}-dependent K$^+$ channels have been reported from various preparations such as bovine chromaffin cells (Marty, 1981), rat myotubes (Barrett et al., 1982; Methfessel & Boheim, 1982), bullfrog ganglion cells (Adams et al., 1982) and rabbit muscle T-tubule membrane fragments reconstituted into planar lipid bilayers (Latorre et al., 1982). In each case, the channel shows a large conductance (100–250 pS) and is highly selective for K$^+$ ions. One of the important problems to solve is whether the a.h.p. which is inhibited by apamin in several cellular systems is due to the same type of Ca^{2+}-dependent K$^+$ channel that has been identified so successfully with patch-clamp techniques. To solve this problem we have therefore carried out detailed work on rat skeletal muscle cells in culture. These cells have the following properties: (i) they have an a.h.p. which is inhibitable by apamin and which is due to a Ca^{2+}-dependent K$^+$ channel (Hugues et al., 1982b), (ii) they have ^{125}I-apamin receptors and (iii) they have Ca^{2+}-dependent K$^+$ channels, which can be identified by patch-clamp techniques (Barrett et al., 1982). Our studies have shown that Ca^{2+}-dependent K$^+$ channels with a large conductance which are identified by patch-clamp techniques are

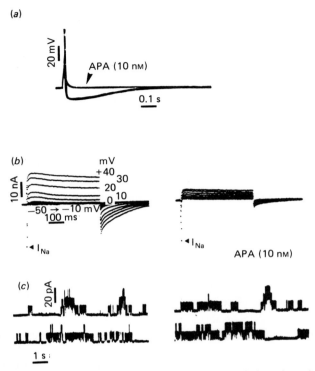

Fig. 5. *Effect of apamin and tetraethylammonium (TEA) on Ca²⁺-dependent K⁺ channels in neuroblastoma cells*

(*a*) Rat myotubes in culture. Selective block of the a.h.p. after a 10 min incubation with 10 nM-apamin (APA). (*b*) Voltage-clamp analysis of the effect of apamin on rat myosacs. Families of membrane currents associated with different step depolarizations from a holding potential of -90 mV: left, control; right, 10 min after the addition of 10 nM-apamin. The fast inward current is not affected (◀); the slow outward current is strongly depressed. This experiment was done in the presence of 20 mM-TEA in order to block TEA-sensitive K⁺ currents. (*c*) Single channel recordings of Ca²⁺-dependent K⁺ channel.

inhibitable by tetraethylammonium (TEA) whereas they are absolutely insensitive to apamin (Romey & Lazdunski, 1984). Conversely, Ca²⁺-dependent K⁺ channels, which generate the a.h.p. and which can be identified by voltage-clamp, are very sensitive to apamin and insensitive to TEA (Romey & Lazdunski, 1984) (Fig. 5). These results, which have also been obtained with neuroblastoma cells, clearly indicate the existence of different types of electrically expressed Ca²⁺-dependent K⁺ channels with a different pharmacology and a different physiological function. The function of the apamin-sensitive Ca²⁺-dependent K⁺ channels is to generate a.h.p.; the function of the Ca²⁺-dependent K⁺ channels identified as large conductance channels by patch-clamp techniques may be to prevent a prolonged depolarization of the cell.

In spite of extensive work, no successful recording of single channel activity from apamin-sensitive Ca²⁺ channels has been obtained up till now. This may be due to the small number of these channels.

Autoradiographic Localization of Apamin-Sensitive Ca^{2+}-Dependent K$^+$ Channels in Rat Brain

Autoradiograms of the apamin binding sites on Ca^{2+}-dependent K$^+$ channels in mammalian brain have been obtained (Mourre *et al.*, 1984). More than 90% of ^{125}I-apamin binding to rat brain section was specific binding.

The apamin-sensitive Ca^{2+}-dependent K$^+$ channel is in many sections of the brain. High grain densities are present in cortex and mainly in the internal pyramidal layer of the cortex, in the olfactory nucleus, in the lateral septal nucleus, in dentate gyrus, Ammon's horn, subiculum, caudate putamen, habenula, geniculate nucleus, medio-ventral thalamus, cochlear nucleus, nucleus of spinal tract of the trigeminal nerve, inferior olive, gracilis and cuneate nuclei and in the granular layer of the cerebellum. Moderate amounts of ^{125}I-apamin binding sites were also found in colliculus and in vestibular and red nuclei. The central grey matter, hypothalamus and cervical nuclei were weakly labelled. The white matter was not labelled.

Developmental Properties of the Ca^{2+}-Dependent K$^+$ Channel in Mammalian Skeletal Muscle and the All-or-None Role of Innervation (Schmid-Antomarchi et al., 1985)

The long-lasting a.h.p. which follows the action potential in rat myotubes differentiated in culture is due to Ca^{2+}-activated K$^+$ channels. These channels have the property to be specifically blocked by the bee venom toxin apamin at low concentrations. Apamin has been used in this work to analyse by electro-physiological and biochemical techniques the role of innervation in the expression of these important channels. The main results are as follows: (i) Long-lasting a.h.p., which follow the action potential in rat myotubes in culture, disappear when myotubes are co-cultured with nerve cells from the spinal cord under the conditions of innervation *in vitro*. (ii) Extensor digitorum longus muscles from adult rats have action potentials which are not followed by a.h.p. but a.h.p. are systematically recorded after muscle denervation and they are blocked by apamin. (iii) Specific ^{125}I-apamin binding is undetectable in innervated muscle fibres but it becomes detectable 2–4 days after muscle denervation, to be maximal 10 days after denervation (Fig. 6a). (iv) Apamin receptors detected with ^{125}I-apamin are present at fetal stages with biochemical characteristics identical to those found in myotubes in culture. The receptor number decreases as maturation proceeds and ^{125}I-apamin receptors completely disappear after the first week of postnatal life, in parallel with the disappearance of multi-innervation (Fig. 6b). All these results taken together strongly suggest an all-or-none effect of innervation on the expression of apamin-sensitive Ca^{2+}-activated K$^+$ channels.

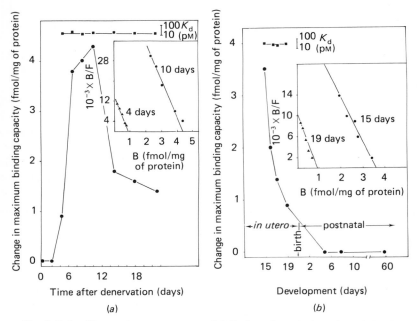

Fig. 6. *Role of innervation in expression of Ca²⁺-dependent K⁺ channels in skeletal muscle*

(*a*) Effect of chronic denervation on lower leg muscle of rat on ¹²⁵I-labelled apamin receptor. Lower legs of adult rats were denervated by section of the sciatic nerve. The contralateral unoperated leg served as control. Time-course of change in maximum binding capacity for ¹²⁵I-apamin after denervation of the lower leg muscle (●). K_d values were obtained at each stage of denervation by Scatchard plot analysis (■). (Inset) Scatchard plot analysis of ¹²⁵I-apamin binding data at two stages after denervation: 4 days (▲) and 10 days (●). K_d and $B_{max.}$ values were derived from three series of experiments at times indicated after denervation. Specific binding at K_d value is between 80 and 90% of total binding. (*b*) Disappearance of ¹²⁵I-apamin receptors in rat skeletal muscle during *in utero* and post-natal life. (●). Time-course of the change in maximum binding capacity for ¹²⁵I-apamin as function of time in rats. K_d values were obtained from Scatchard plot analysis (■). (Inset) Scatchard plots of the ¹²⁵I-apamin binding data at two stages of development; 15 days (●) and 19 days (▲). K_d and $B_{max.}$ values were derived from three series of experiments at indicated times during *in utero* development. The specific binding was about 90% of the total binding under these conditions.

An Endogenous Apamin-Like Factor Modulating Ca²⁺-Dependent K⁺ Channel Activity Exists in Mammalian Brain (Fosset *et al.*, 1984)

An apamin-like factor has been purified from pig brain after acidic extraction of the tissue and several steps of purification on sulphopropyl Sephadex C25 and C18 reverse phase high pressure liquid chromatography (Fosset *et al.*, 1984). The apamin-like activity was followed during the purification procedure by using two biochemical assays: a radio receptor assay (Hugues *et al.*, 1982*c*) and a radioimmunoassay with anti-apamin antibodies (Schweitz & Lazdunski, 1984) and two physiological assays: the measure of the contractile activity of guinea-pig *taenia coli* (Hugues *et al.*, 1982*d*), and the electrophysiological measurement of the ability of fractions to block the a.h.p. in rat skeletal muscle cells (Hugues *et al.*, 1982*b*).

A purification of an apamin-like factor has been achieved. Properties of the factor are the following. (i) It is able to prevent ^{125}I-apamin interaction with its binding site on rat brain synaptosomes. (ii) It antagonizes ^{125}I-apamin recognition by anti-apamin antibodies. (iii) It contracts the intestinal smooth muscle previously released by adrenaline. (iv) It selectively blocks the hyperpolarization that follows its action potential on rat skeletal myotube in culture. (v) It is sensitive to the action of trypsin and insensitive to chymotrypsin digestion.

All these properties are those of apamin itself. These results strongly suggest the presence in pig brain of an endogenous equivalent of apamin. Knowing that a purification procedure gives an activity equivalent to 1.5 ± 0.5 pmol of apamin per pig brain, large scale purification will be necessary to obtain enough apamin-like factor to sequence it. Our results suggest that inactive precursors of the apamin-like factor could also be present in the pig brain; the structure of the native precursor will be known once the cloning of the gene encoding for this precursor is achieved.

References

Adams, P. R., Constanti, A., Brown, D. A. & Clark, R. B. (1982) *Nature (London)* **296**, 746–749

Banks, B. E. C., Brown, C., Burgess, G. M., Burnstock, G., Claret, M., Cocks, T. M. & Jenkinson, D. H. (1979) *Nature (London)* **282**, 415–417

Barchi, R. L., Cohen, S. A. & Murphy, L. E. (1980) *Proc. Natl. Acad. Sci. U.S.A.* **77**, 1306–1310

Barhanin, J., Pauron, D., Lombet, A., Norman, R. I., Vijverberg H. P. M., Giglio, J. R. & Lazdunski, M. (1983a) *EMBO J.* **2**, 915–920

Barhanin, J., Schmid, A., Lombet, A., Wheeler, K. P. & Lazdunski, M. (1983b) *J. Biol. Chem.* **258**, 700–702

Barrett, J. N., Magleby, K. L., & Pallotta, B. S. (1982) *J. Physiol. (London)* **331**, 211–230

Bystrov, v. F., Okhanov, V. V., Miroshnikov, A. I. & Ovchinnikov, Y. A. (1980) *FEBS Lett.* **119**, 113–117

Cook, N. S., Haylett, D. N. & Strong, P. (1983) *FEBS Lett.* **152**, 265–269

Cosand, W. L & Merrifield, R. B. (1977) *Proc. Natl. Acad. Sci. U.S.A.* **74**, 2771–2775

Fosset, M., Schmid-Antomarchi, H., Hugues, M., Romey, G. & Lazdunski, M. (1984) *Proc. Natl. Acad. Sci. U.S.A.* **81**, 7228–7232

Granier, C., Pedroso Muller, E. & Van Rietschoten, J. (1978) *Eur. J. Biochem.* **82**, 293–299

Habermann, E. (1972) *Science* **177**, 314–322

Habermann, E. & Fischer, K. (1979) *Eur. J. Biochem.* **94**, 355–364

Hartshorne, R. P. & Catterall, W. A. (1981) *Proc. Natl. Acad. Sci. U.S.A.* **78**, 4620–4624

Hughes, M., Romey, G., Duval, D., Vincent, J. P. & Lazdunski, M. (1982a) *Proc. Natl. Acad. Sci. U.S.A.* **79**, 1308–1312

Hughes, M., Schmid, H., Romey, G., Duval, D., Frelin, C. & Lazdunski, M. (1982b) *EMBO J.* **9**, 1039–1042

Hughes, M., Duval, D., Kitabgi, P., Lazdunski, M. & Vincent, J. P. (1982c) *J. Biol. Chem.* **257**, 2762–2769

Hughes, M., Duval, D., Schmid, H., Kitabgi, P. & Lazdunski, M. (1982d) *Life Sci.* **31**, 437–443

Hughes, M., Schmid, H. & Lazdunski, M. (1982e) *Biochem. Biophys. Res. Commun.* **107**, 1577–1582

Kepner, G. R. & Macey, R. I. (1968) *Biochim. Biophys. Acta* **163**, 188–203

Latorre, R., Vergara, C. & Hidalgo, C. (1982) *Proc. Natl. Acad. Sci. U.S.A.* **79**, 805–809

Lazdunski, M. & Renaud, J. F. (1982) *Annu. Rev. Physiol.* **44**, 463–473

Levinson, S. R. & Ellory, J. C. (1973) *Nature (London)* **245**, 122–123

Maas, A. D. & Den Hertog, A. (1979) *Eur. J. Pharmacol.* **58**, 151–156

Maas, A. D. J. J., Den Hertog, A., Ras, R. & Van Den Akker, J. (1980) *Eur. J. Pharmacol.* **67**, 265–274

Marty, A. (1981) *Nature (London)* **291**, 497–500

Methfessel, C. & Boheim, G. (1982) *Biophys. Struct. Mech.* **9**, 35–60

Moore, H. P. M., Fritz, L. C., Raftery, M. A. & Brokes, J. P. (1982) *Proc. Natl. Acad. Sci. U.S.A.* **79**, 1673–1677

Mourre, C., Schmid-Antomarchi, H., Hugues, M. & Lazdunski, M. (1984) *Eur. J. Pharmacol.* **100**, 135–136

Neher, E., Sakmann, B. & Steinbach, J. H. (1978) *Pflügers Archiv* **375**, 219–228

Norman, R. I., Schmid, A., Lombet, A., Barhanin, J. & Lazdunski, M. (1983) *Proc. Natl. Acad. Sci. U.S.A.* **80**, 4164–4168

Romey, G. & Lazdunski, M. (1984) *Biochem. Biophys. Res. Commun.* **118**, 669–674

Schmid-Antomarchi, H., Hugues, M., Norman, R. I., Ellory, J. C., Borsotto, M. & Lazdunski, M. (1984) *Eur. J. Biochem.* **142**, 1–6

Schmid-Antomarchi, H., Renaud, J. F., Romey, G., Hugues, M., Schmid, A. & Lazdunski, M. (1985) *Proc. Natl. Acad. Sci. U.S.A.* **82**, 2128–2195

Schweitz, H. & Lazdunski, M. (1984) *Toxicon* **22**, 985–988

Vincent, J. P., Schweitz, H. & Lazdunski, M. (1975) *Biochemistry* **14**, 2521–2525

Vladimirova, I. A. & Shuba, M. F. (1978) *Neurofiziologija* **10**, 295–299

Wemmer, D. & Kallenbach, N. R. (1983) *Biochemistry* **22**, 1901–1906

Biochem. Soc. Symp. **50**, 43–58
Printed in Great Britain

Regulation of Cell-to-Cell Communication by Phosphorylation

WERNER R. LOEWENSTEIN

Department of Physiology and Biophysics, University of Miami School of Medicine, P.O. Box 016430, Miami, FL 33101, U.S.A.

Synopsis

The cyclic AMP-activated protein kinase I, a serine- and threonine-phosphorylating enzyme, regulates cell-to-cell communication. Its deficiency in mutant cells is associated with deficiency of communication. The communication defect is corrected by introduction of the catalytic subunit of the enzyme into the mutant cells. Activation of the enzyme by cyclic AMP in normal cells causes an increase of communication, namely an increase of junctional permeability associated with an increase in the number of membrane particles of gap junction. This upregulation of cell-to-cell membrane channels constitutes a basic mechanism whereby cell communities set their degree of communication. The mechanism is normally put into motion by adenylate cyclase-activating hormones. The mechanism is counteracted by tyrosine-phosphorylating protein kinase (*src* protein), which downregulates junctional permeability, a fast and reversible effect on the channels, independent of the action of the kinase on the cytoskeleton. The two *T* proteins coded by the SV-40 genome cause a similar channel downregulation.

Introduction

I shall deal with a channel of a very different sort from the channels the other speakers at this symposium are dealing with. It is a channel that spans not one, but two cell membranes. This channel was a solution to an ancient evolutionary problem: how to interconnect the cyberneting loops of information of different cells, and to do so without mixing the information sources. The seeds for this problem were sown by evolution herself as she grew, during the first $1\frac{1}{2}$ billion years, a hydrophobic membrane for the cells, a barrier to keep the hard-won informational molecules from drifting apart and to keep nonsense molecules from drifting in. The barrier was instrumental in the selection of cybernetically apt units in an aqueous environment, and was absolutely necessary for Darwinian evolution of cells. But, over the next 2 billion years, the organismic times, it became an obstacle to multicellular organization. The problem was then how to get signal molecules across the hydrophobic barrier to interconnect the units, without abolishing its primitive insulating function.

The problem eventually met with three elegant solutions: three kinds of

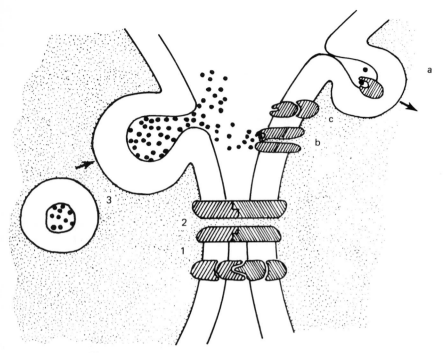

Fig. 1. *Membrane bypasses for intercellular communication*

1, Bridge of interacting molecules spanning the two cell membranes. 2, Intercellular tunnel: a channel spanning the two cell membranes. 3, Bypasses for the less direct humoral forms of communication (via the extracellular liquid): (a) membrane vesicles that fuse with the cell membrane. This bypass serves hydrophilic signals on the way out of cells (exocytosis) and on the way in assisted by a receptor molecule (receptor-mediated endocytosis). In relayed modalities of the humoral communication form (b) a receptor associated with a membrane channel, or (c) a transmembrane molecular bridge, are the bypasses completing the entry route into the target cell.

membrane bypasses that managed to keep the cell membrane leak-proof by hydrophobic lipid–protein or lipid–lipid interactions (Fig. 1). One solution was to imbed the signal molecules into the cell membrane and to let them interact with each other upon cell contact. The signals here complete a molecular bridge between the interiors of two cells. A second solution uses a vehicle with a membrane skin capable of fusing with the cell membrane, to transport the intracellular hydrophilic signals across: a 'wolf in the sheepskin' ruse. The continuity between the cells here is given by the extracellular medium and the ruse may be used twice, on the way out of the cells and on the way in, assisted by a receptor molecule; alternatively, a receptor associated with a membrane channel or a transmembrane molecular bridge completes the way in. A third solution was to tunnel through the two membranes of contiguous cells.

The last was the most hazardous of the three evolutionary strategies. It risked losing the individuality of the information sources. But it was a risk worth taking because it afforded the most direct way of communication. The solution came, as it generally does, through the selection of the right protein: a protein with

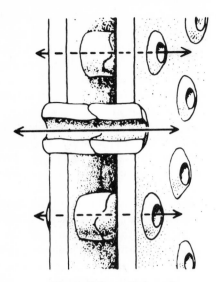

Fig. 2. *Cell-to-cell channel*

The channel is made of two tightly joined transmembrane proteins, one from each apposing cell membrane. The aqueous permeation-limiting bore of the channel is 1.6–2 nm in mammalian cells. (From Loewenstein, 1974.)

a hydrophilic bore of 1.6–2 nm (mammalian cells), wide enough to let through the cytoplasmic inorganic ions, the metabolites, the building blocks of bio-molecular synthesis, and the high energy phosphates, but not the macro-molecules. The channel is made of two protein halves, one contributed by each membrane, tightly joined, forming a rather leakproof aqueous tunnel between cells (Fig. 2).

We find this cell-to-cell channel now in nearly all organized tissues throughout the phylogenetic scale, constituting one of the most basic forms of cellular communication. As a rule, a cell in a tissue is thus connected with several neighbours, and so whole organs or large organ parts are continuous from within.

The channels tend to cluster in the junctional membrane regions where they are recognizable electron-microscopically as clusters of membrane particles, called 'gap junction'. The particles are hexameric: six slanted subunits form a half-channel (Unwin & Zampighi, 1980).

The channels form spontaneously when competent cells are brought into contact. They are then detectable one by one as quantal increments of cell-to-cell conductance (Loewenstein *et al.*, 1978). Typically the total conductance of such a nascent junction increases over a few minutes to a plateau (Ito *et al.*, 1974). This has led to the notion that the communicating junction forms by a progressive accretion of individual cell-to-cell channels (Ito *et al.*, 1974). Electron-microscopically, the number of particles in forming gap junction increases with time (Johnson *et al.*, 1974).

One physiological regulator of the channel is Ca^{2+}. The channel closes within fractions of a second when the concentration of this ion rises above 10 μM in the junctional locale (Rose & Loewenstein, 1976). The six subunits of the gap-junction particle then appear to slide radially towards one another obliterating the bore, a shutter mechanism of the iris diaphragm sort (Unwin & Ennis, 1983). This closure mechanism is activated when a cell springs a leak (Oliveira-Castro & Loewenstein, 1971) or its ion pumps stand still (Rose & Loewenstein, 1976). It is a basic survival mechanism allowing the cell community to disconnect an unhealthy member and to seal itself off.

Another physiologically interesting channel regulator has recently come to light: a cyclic AMP-dependent protein kinase. This phosphorylating enzyme appears to control the formation process of the channel and, thus, to determine the number of (open) channels in the cell junction. This regulation is the subject of my talk.

Jean Flagg-Newton and I accidentally hit upon this mechanism in 1978. We were working with mammalian cell cultures, probing their junctional permeabilities; one week we had forgotten to feed the cultures. We noticed then, to our surprise, that the starved cells had unusually high junctional permeabilities. It goes against the anthropomorphic grain that something should improve by hunger, but this discovery soon set us on the track of cyclic AMP, for it was known that serum-starved cells have increased levels of cyclic AMP. So it came that we joined the fans of this mighty substance.

Upregulation of Cell-to-Cell Channels by Cyclic AMP-dependent Phosphorylation

We performed three kinds of experiments exploring the role of cyclic AMP in cell-to-cell communication. In one kind of experiment we supplied the cells with cyclic AMP. In another we elevated their endogenous cyclic AMP level by treatments with phosphodiesterase inhibitors or adenylate cyclase activators. In a third kind we supplied them with the phosphorylating enzyme subunit. All had the same outcome: an increase in junctional permeability (Flagg-Newton, 1979; Flagg-Newton & Loewenstein, 1981; Flagg-Newton et al., 1981; Radu et al., 1982; Wiener & Loewenstein, 1983).

Cyclic AMP and phosphodiesterase inhibitors

Fig. 3 shows an example in which we supplied dibutyryl cyclic AMP plus the phosphodiesterase inhibitor caffeine to cells of the rat fibroblast B line. We probed the cell junctions with fluorescent-labelled polyamino acids: di- and tri-glutamic acid labelled with lissamine rhodamine B (LRB-Glu-Glu, LRB-Glu-Glu-Glu). These probes were microinjected into one member of a cell pair, and their rates of transit cell-to-cell (junctional transit rate) and cell-to-exterior (loss rate) were measured. The junctional transit rates of both probes increased within 4 h of the treatment, about tripling within 24 h. The loss rates stayed unchanged, clearly showing that the junctional permeability had increased.

Increases of junctional permeability of this sort were typical of cells that make channels with one another readily, as many mammalian cell types do in culture

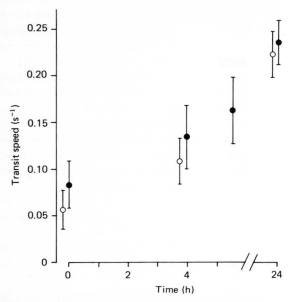

Fig. 3. *Rate of junctional molecular transfer increases by treatment with dibutyryl cyclic AMP (1 mM) and caffeine (1 mM)*

The junctions of isolated cell pairs of the rat B fibroblast line are probed with microinjected LRB-Glu$_2$ (O) and LRB-Glu$_3$ (●), and the junctional transit rate of these fluorescent molecules is measured (SE bars). Treatment starts at time zero. (From Flagg-Newton *et al.*, 1981.)

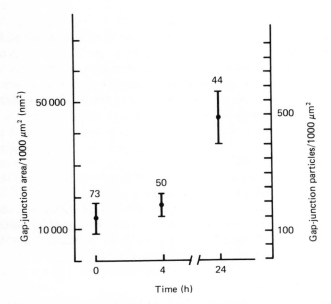

Fig. 4. *Number of gap-junctional membrane particles increases by treatment with dibutyryl cyclic AMP (1 mM) and caffeine (1 mM)*

Mouse 3T3-BalbC cells. Ordinates: right, the number of particles per unit membrane area; left, the area of the particle clusters (gap-junction area) per unit membrane area. Treatment starts at time zero. (From Flagg-Newton *et al.*, 1981.)

(Flagg-Newton *et al.*, 1981; Radu *et al.*, 1982). The most dramatic effects were seen in cells that make channels sparingly: in a cell type that displayed zero junctional permeability in basal conditions, a junctional permeability was 'induced' by the cyclic nucleotide treatment (Azarnia *et al.*, 1981).

The increases in junctional permeability went hand in hand with increases in the number of membrane particles of gap junction. In cells with basal expression of junctional permeability both the number of particles in the gap-junctional clusters and the frequency of the clusters increased upon treatment with cyclic nucleotides (Fig. 4). In the channel-deficient cells, gap-junctional particle clusters emerged where, before the treatment, there were none. Treatment with the protein synthesis inhibitors, cycloheximide or puromycin, but not treatment with the cytokynesis inhibitor, cytochalasin B, blocked that effect (Azarnia *et al.*, 1981; Flagg-Newton *et al.*, 1981).

These results lead us to believe that the increase in junctional permeability reflects an increase in the number of open cell-to-cell channels.

In summary of the results so far we may write the following regulator scheme:

$$cAMP \uparrow \rightarrow P_J \uparrow \tag{1}$$

where the concentration of cyclic AMP in the cells determines the steady-state junctional permeability. A plausible mechanism is an upregulation of the number of open channels, channels newly formed, or the opening of pre-existing ones:

$$cAMP \uparrow \rightarrow \text{number of channels} \uparrow \tag{2}$$

Hormones

As one would expect, the mechanism can be driven by hormones (Radu *et al.*, 1982). Fig. 5 illustrates this for a catecholamine. The cells chosen for this experiment, rat neuroglioma C-6 cells, had β-adrenergic receptors, and a few micromoles of isoprenaline set the channel upregulation into motion. We were probing here with LRB-Glu-Glu, a relatively large molecule with three negative charges. In the basal condition, there were not enough channels to give a detectable transfer to any of the neighbours of the cell injected with that probe (Fig. 5*a*). About 7 h after the hormone treatment (and the consequent rise in the endogenous cyclic AMP concentration) the probe was transferred typically to several (first-order) neighbours (six out of the seven here (Fig. 5*b*), and the number of gap-junctional particles about tripled.

Such enhancement of junctional permeability was shown for other cyclic AMP-elevating hormones, too: for gonadotropin in frog oocyte/follicle junction (Browne & Wiley, 1979) and for prostaglandin E_1 in human lung WI-38 cells (Radu *et al.*, 1982).

Protein kinase

What protein kinase mediates this channel upregulation? The regulatory subunit of the cyclic AMP-dependent protein kinase, the receptor for cyclic AMP, occurs in two forms, I and II. In the holoenzyme, one or the other form

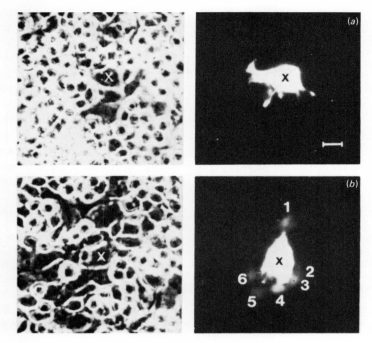

Fig. 5. *Action of a hormone*

Neuroglioma C-6 cells, which have β-adrenergic receptors. LRB-Glu$_2$ is microinjected into a cell (x) in control conditions (a) and 7.1 h after application of 50 μM-isoprenaline (b). The injected cell has seven contiguous neighbours in each case. The fluorescent tracer is transferred to six of these neighbours in the hormone-treated condition, but to none in the control condition. Left, phase contrast micrograph; right, darkfield. Calibration, 15 μM. (From Radu *et al.*, 1982.)

is associated with the catalytic subunit (invariant), which splits off to catalyse the phosphorylation of the protein target:

$$2 \text{ cAMP} + 2 \text{ R-C} \rightarrow 2 \text{ R-cAMP} + 2 \text{ C}$$

$$T + n\text{NTP} \overset{\text{C}}{\rightarrow} T\text{-}P_n + n\text{NDP}$$

where R and C are the regulatory and catalytic subunits, respectively, T the target protein and NTP, a high energy phosphate, commonly ATP (Kuo & Greengard, 1969; Krebs, 1972; Rosen & Krebs, 1981).

To learn which of the two isoenzymes operates in this channel regulation, we resorted to mutant cells deficient in one or the other enzyme form, namely one-step mutants of the Chinese hamster ovary cell line derived by Michael Gottesman and his colleagues (1981). One of the mutants (I$^-$) lacked protein kinase type I and had little of type II, and another mutant (II$^-$) lacked protein kinase II and had type I. Mutant I$^-$ turned out to be clearly deficient in junctional permeability (Fig. 6). Mutant II$^-$, on the other hand, was only slightly less permeable than the wild-type cells (despite a lower cyclic AMP sensitivity of its kinase) (Wiener & Loewenstein, 1983).

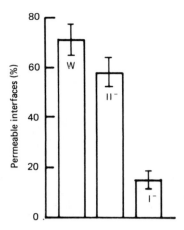

Fig. 6. *Communication deficiency is associated with deficiency of protein kinase type I*

The incidence of LRB-Glu permeable cell interfaces (%, ±s.e.), a sensitive index of junctional permeability, in Chinese hamster ovary cells: wild-type (W), mutants I⁻ and II⁻. The statistical confidence level of the difference between I⁻ and W is $P > 0.0005$ (no significant difference between II⁻ and W). (From Wiener & Loewenstein, 1983.)

Table 1. *Cyclic AMP-dependent protein kinase and communication phenotypes of clones from a reverting I⁻ culture*

1 Activity unit = 1 pmol of P transferred/min per mg of protein. –, Undetectable activity (< 25 units/ml). Kinase II range was 180–260 units/ml in all cases (including those with undetectable kinase I activity), except for W where it was 420. The clones were all derived from the culture of Fig. 7, series 2. At the bottom of the Table are listed the traits of the original wild-type (W) and I⁻ cells. (From Wiener & Loewenstein, 1983.)

Clone	Kinase I (units/ml)	Permeable interfaces* (%)
1	392	71±6
2	360	69±5
3	326	72±7
4	403	78±6
5	381	73±6
6	–	7±2
7	–	13±4
8	–	8±3
9	–	11±4
10	–	6±2
11	–	7±2
W	380	72±6
I⁻	–	12±4

* Incidence of LRB-Glu permeable interfaces (means ± s.e.). The confidence level of the difference between the incidences of I⁺ and I⁻ clones was, in all cases, $P < 0.0001$.

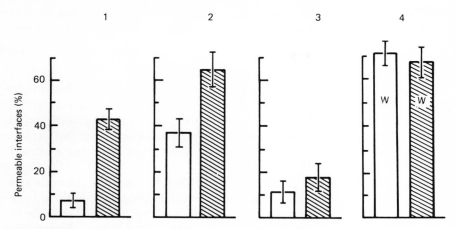

Fig. 7. *Cyclic AMP-dependent protein kinase corrects communication defect in mutant I⁻ cells*

The cells were loaded with the catalytic enzyme subunit, and the incidence of LRB-Glu-permeable cell interfaces was evaluated 4–4.5 h after the loadings (cross-hatched bars). Open bars give the corresponding values of the unloaded controls. In series 1 and 2, I⁻ cells were loaded with active enzyme; in 3, with inactivated enzyme; in series 4 wild-type cells (W) were loaded with active enzyme. Confidence levels of the differences between the values of enzyme-loaded cells and unloaded controls: series 1, $P < 0.002$, series 2, $0.01 > P > 0.005$; series 3, $0.10 > P > 0.05$. (The values of series 2 and 4 can be compared with each other; the cultures had equal densities.) (From Wiener & Loewenstein, 1983.)

Thus, at least, enzyme type I is involved in the channel regulation process and seems sufficient for it; this was the only isoenzyme present in the communication-competent mutant II⁻.

This correlation between enzyme phenotype and communication phenotype also held for backshifts of mutation. After carrying mutant I⁻ in culture for about half a year, clones appeared that had reverted to the normal communication phenotype. This reversion invariably went hand in hand with reversion to the normal enzyme phenotype (Table 1).

Next we asked whether incorporation of the catalytic subunit into the mutant I⁻ would remedy its communication deficiency. Using a chemical permeabilization procedure (Kerrick & Krasner, 1975), we loaded the cells with a purified preparation of the enzyme subunit from bovine skeletal muscle (Wiener & Loewenstein, 1983). Upon loading, the junctional permeability of the mutant cells rose significantly (Fig. 7, 1, 2) and to a level comparable with that of the wild-type cells (compare experiment series 2 with 4). Control loadings of the mutant cells with inactivated enzyme (3) or of wild-type cells with active enzyme (4) gave no change in junctional permeability.

The regulator scheme. General physiological roles

Based on these results, we write the upregulator scheme in the form:

$$\text{cAMP} + \text{R}^\text{I}\text{C} \rightarrow \text{C} \rightarrow \text{T-P} \rightarrow \text{channels} \uparrow \tag{3}$$

where cyclic AMP, by way of the type-I regulatory subunit of the protein kinase, promotes the phosphorylation of a target protein (T) determining the number of (open) cell-to-cell channels. Integrated into cell physiology, the scheme becomes

$$h + r \rightarrow cAMP + R^IC \rightarrow C \rightarrow T\text{-}P \rightarrow channels \uparrow \qquad (4)$$

where h is a hormone and r its receptor.

The scheme now has broad heuristic value. It predicts that, with the right receptor- and protein-kinase endowment, a cell community can regulate its communication by hormones. This regulatory capacity is finely graded and so affords a wide range of cellular controls. We have already seen that certain hormones play such a regulatory role and, no doubt, more hormonal examples will soon become known.

Furthermore, by merely adding a (closed) loop between the channels and hormones, the scheme acquires predictive value regarding the hormonal form of communication: it predicts a self-reinforcement of the hormone response in certain conditions. This follows from the fact that cyclic AMP fits through the channel. Thus, in tissues where this nucleotide is the second messenger of the hormone action, the action may be amplified: the cellular response to the hormone may spread from cell to cell where the phosphodiesterase action is not excessive. Now, the possibility of self-reinforcement comes about because the instruments of such spread, the cell-to-cell channels, are themselves responsive to the nucleotide; they proliferate. Hence, under persistent hormone action, the hormonal communication between cells may be facilitated. Thus,we get a self-regulating cellular amplifier as a product of the interaction between two basic forms of cellular communication. Such amplifiers have interesting physiological potential not only for the dissemination of the hormonal response in organs and tissues, but also as filters of receptor noise and as devices for hierarchical cellular interactions (Loewenstein, 1981).

The primary target of phosphorylation (T in the scheme above) remains to be identified. It is possible that the channel itself is that target. Indeed, the protein of lens gap junction can be phosphorylated *in vitro* by a cyclic AMP-dependent protein kinase (Johnson & Johnson, 1982), and it should soon become known whether this finding *in vitro* is physiologically meaningful. At present it seems equally possible that the critical target of phosphorylation be a step or more removed from the channel; the permeability regulation is relatively slow (hours) and one component, at least, depends on protein synthesis.

Downregulation by Tyrosine Phosphorylation

I turn now to experiments in which we tried to interfere with the cyclic AMP-dependent channel regulation. Will the normal channel regulation be perturbed when phosphorylations of other sort are activated in the cells? We asked this for the case where tyrosine residues, instead of serine or threonine, are phosphorylated. Tyrosine phosphorylation is relatively rare in normal cells, but it can be induced massively by certain transforming viruses, such as the Rous sarcoma virus. The *src* gene of this RNA virus codes for a tyrosine-phosphorylating protein kinase in the host cells, the pp60[src] protein (Collet *et al.*, 1980; Hunter, 1980; Bishop, 1982).

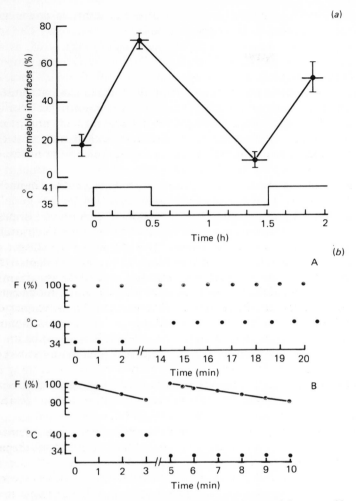

Fig. 8. *The src protein downregulates junctional permeability reversibly*

Quail embryo cells infected with temperature-sensitive Rous sarcoma virus. The temperature (lower ordinates) was shifted between the permissive and the non-permissive levels of the virus, i.e. the levels at which the tyrosyl kinase of the *src* protein is activated and inactivated respectively. (*a*) Incidence of LRB-Glu-permeable cell interfaces ($\pm$s.e.) determined just before each temperature shift (upper ordinates). (*b*) A, LRB-Glu loss rates at the two temperature levels of a typical tight cell; B, LRB-Glu loss rates of a leakier cell, a relatively rare case. The rate constants of the curve are 1.8 and 2.7 min^{-1} at the permissive and the non-permissive temperature levels respectively. F is the cellular fluorescence. (From Azarnia & Loewenstein, 1984*a*.)

src protein action

We used cells containing temperature-sensitive mutant *src* gene. Their pp60^{src} is thermolabile so that its protein kinase activity can be switched from inactive to active form by merely shifting the culture temperature from 41 °C (the non-permissive temperature for the virus) to 33/35 °C, and vice versa. Tyrosine phosphorylation sets in within half hour of a temperature downshift (Radke &

Martin, 1979). We found that the junctional permeability is downregulated within that time (Azarnia & Loewenstein, 1984 a).

Fig. 8 shows the downregulation for the quail embryo cells-$ts68$, as well as the reversibility of the effect: LRB-Glu junctional transfer rose upon temperature upshift, fell upon downshift, and rose again upon upshift (Fig. 8 a). These effects were not due to the temperature changes themselves: the junctional transfer was not significantly changed over that temperature range in normal cells or in cells infected with wild-type virus. Changes in permeability of non-junctional membrane were ruled out as sources of the effects: the loss rates of LRB-Glu were typically unchanged (Fig. 8 b, A) or, in leakier cells, were in the wrong direction for that (Fig. 8 b, B). Thus, as with cyclic AMP-dependent phosphorylation, the change in junctional transfer represents a change in junctional permeability.

The junctional response is fast. The reduction of junctional permeability ensued within 25 min of lowering the temperature to the permissive level—one of the fastest effects of the src gene known. The reversal of the effect was just about as fast and was not affected by inhibitors of protein synthesis. Thus, if we may assume that the permeability loss and reversal reflect symmetric processes, it is likely that the action of pp60^{src} is by way of a modification of the channel open state, rather than by way of a modification of channel protein synthesis. This is not to say that the process of channel assemblage, an assemblage from precursor channel halves or subunits, could not be the target for modification, but it is unlikely that the synthesis (and degradation) of the precursors be the target.

A plausible mechanism is a modification of the channel protein or of an immediate regulator protein by (tyrosine) phosphorylation. A direct mechanism of this sort would be consistent with the swiftness of the junctional response to the pp60^{src} kinase activity. Whether the primary protein target of this viral phosphorylation is the same as that in the normal cyclic AMP-dependent phosphorylation, remains to be seen.

In this regard, we note that the src protein kinase mechanism seems to override the cyclic AMP-dependent protein kinase mechanism. When the virus-infected cells were treated with cyclic AMP, phosphodiesterase inhibitors or the adenylate cyclase activator forskolin, the junctional downregulation by the src protein took place just the same. These treatments then failed to raise the junctional permeability even at the non-permissive temperature (Azarnia & Loewenstein, 1984 a). This dominance is explainable if we may assume that the src-phosphorylated protein responsible for the loss of the cyclic AMP response is more stable than the proteins responsible for the other phenotypic changes in transformation; or, perhaps, more interestingly, if the thermo-sensitive pp60^{src} were slightly leaky, giving a low level of phosphorylation at the non-permissive temperature, enough to prevent the cyclic AMP response but not to produce the other phenotypic changes.

The effect of the src protein seems general. We obtained effects on junctional permeability similar to those of the quail embryo cells-$ts68$, with chick embryo cells-tsLA90, and mouse 3T3 cells-tsLA90 (and vole cells-wt) (Azarnia & Loewenstein, 1984a, b) and Judson Sheridan and Ross Johnson and their colleagues reported such effects on junctional transfer (not junctional permeability) in rat NRK cells-*ts*LA25 before us (Atkinson *et al.*, 1981). Not in all cases was the junctional response as fast as in the quail embryo cells-*ts*68. In the 3T3 cells-*ts*LA90 the response took more than 1 h. This may be due to a difference in 'tightness' of the two viral mutants. LA90 is much tighter than 68; the kinase activity of its pp60^{src} is about one-tenth that of *ts*68 (Joan Brugge, personal communication).

Compared with the cells infected with wild-type viruses, the junctional permeabilities of the cells infected with temperature-sensitive viruses at the permissive temperature were generally lower (Azarnia & Loewenstein, 1984*a*). This difference is not attributable to a higher protein kinase activity of the temperature-sensitive pp60^{src} at the permissive temperature; the activity is actually higher in the wild-type pp60^{src} (Joan Brugge, unpublished results). Rather, a modification in the target protein due to a superphosphorylation might conceivably be the cause of the permeability difference.

Independence of communication response from cytoskeletal response and its relative sensitivity

Among the various cellular alterations caused by pp60^{src} that go under the portmanteau of transformation, are major alterations in the cytoskeleton. These may well have a bearing on the formation and stability of cell junctions. Thus, the question naturally arises whether the change in junctional communication is merely secondary to the cytoskeletal change. Vinculin, the protein that anchors the microfilaments to the cell membrane is tyrosine-phosphorylated (Chen & Singer, 1982), and this phosphorylation is thought to lead to microfilament disorganization and hence to the cells' rounding up (Sefton *et al.*, 1981). These cytoskeletal effects lag hours behind the effect on junctional permeability, but there was always the possibility that earlier, subtler (undetected) cytoskeletal changes might critically interfere with the cell membrane contact needed for channel formation.

This question was set to rest by experiments with a revertant clone of *src*-containing vole cells isolated by Anthony Faras. This clone had reverted to the normal cytoskeletal and morphological phenotype, yet had the high levels of pp60^{src} kinase (and the tumorigenicity) of the fully transformed parent clone (Lau *et al.*, 1979; Nawrocki *et al.*, 1984). Its junctional permeability turned out to be lower than that of normal cells, as low as that of the fully transformed cells (Azarnia & Loewenstein, 1984*b*). The alteration of cell-to-cell channels determined by the *src* gene is thus independent of the alteration of the cytoskeleton.

The results obtained with another revertant clone of vole cells, a revertant that had lost most of the tyrosyl kinase activity, gave an idea of the relative sensitivity of the junctional response. This clone had a remnant of only 2–3 % of the pp60^{src} kinase activity of the fully transformed parent, was cytoskeletally and morphologically normal and lacked tumorigenicity (Nawrocki *et al.*, 1984). It turned out to have a higher junctional LRB-Glu permeability than the fully transformed parent, but its permeability was still lower than that of the normal cells (Azarnia & Loewenstein, 1984*b*). The small kinase remnant apparently sufficed to depress junctional permeability but not to produce the other alterations of transformation. The relatively high sensitivity implied by this result echoes the discussion of the result of the *src* dominance over the cyclic AMP-dependent channel regulation, above. It is interesting in this connexion that pp60[src] is found close to the membrane of cell junction (Willingham *et al.*, 1979; Schriver & Rohrschneider, 1981; Nigg *et al.*, 1982).

Downregulation by SV-40

Azarnia and I explored the action of another transforming virus, the Simian virus 40 (SV-40). Apart from the foregoing considerations concerning phosphorylation, we had another reason for studying the effects of transforming viruses on cell-to-cell communication. We have a long-standing interest in the mechanisms of growth control, an interest prompted by the hypothesis that the cell-to-cell channels are conduits for growth-regulating molecules (Loewenstein, 1966; 1968), and nursed by the finding of a variety of channel-deficient cells exhibiting cancerous growth (Loewenstein, 1979). From the point of view of the hypothesis, it was of immediate interest to see whether the channels were downregulated in the transformed state. The answer we obtained with SV-40 was positive and similar to that obtained with the *src* virus.

Roles of the large and small T proteins

In SV-40 transformation the genetics are more complex than in the *src* case. There are two genes involved in the transformation by the DNA virus: the *A* gene that codes for the large *T* protein, the *T antigen*, and the *F* gene that codes for the little *t* protein, the *t antigen* (Lai & Nathans, 1975; Tooze, 1981). We used viruses with a mutation on the *A* gene that renders *T* thermolabile.

The overall effect of shifting the temperature from the non-permissive to the permissive level was not unlike that obtained with the thermolabile pp60[src]: the junctional permeability fell (reversibly). But the effect was slower. It had lag periods of 4 h or more. Such junctional downregulation in the T^+ state was found in experiments with epithelial cells, primary rat pancreas islet cells, and fibroblasts, rat embryo cells and NREF cells infected with different viruses, *ts*A58, *ts*A239 and *ts*209, all with *A* mutations located between nucleotides 3476 and 3733 on the SV-40 genome (Azarnia & Loewenstein, 1984*c*).

These experiments showed that the *T* antigen causes cell-to-cell channel alteration. Is it a sufficient cause for this? To answer this question, one needs a mutant of the *F* gene. As there are no temperature-sensitive mutants available for that gene, we made do with a deletion. We used a viral DNA constructed by Sompayrac & Danna (1983) by ligating a DNA segment containing a temperature-sensitive mutation on the *A* gene (*ts*A58 or *ts*2009) with a segment containing an *F* gene deletion (*dl* 54-59), and a DNA prepared by William Topp by creating an *F* gene deletion at the Tag I site in *ts*A58 DNA.

Incorporated into rat pancreas islet cells and mouse 10T$\frac{1}{2}$ cells, these DNAs provided us with systems that could be switched $T^- \, t^- \rightleftharpoons T^+ \, t^-$ by temperature shifts. The design was straightforward: a junctional response, namely a reduction of junctional permeability is expected here if, and only if, T is a sufficient cause. The answer was negative. A response ensued only in the T^+t^+ condition, the condition of the pure temperature-sensitive A mutations above (Azarnia & Loewenstein, 1984 c).

In conclusion, both antigens are required for the cell-to-cell channel alteration. This requirement is the same as that for cell transformation by the virus. But it is distinct from the requirement for the cytoskeletal alteration; for that, t is sufficient (Frisque et al., 1980).

There is little to guide us as yet to the general mechanism of action of these SV-40 transforming proteins. For the transforming protein of another papovavirus, the middle T antigen of polyoma virus, it has been shown that this antigen is tyrosine-phosphorylated (Schaffhausen et al., 1982) and binds to pp60$^{c\text{-}src}$ (Courtneidge & Smith, 1983), the normal cellular homologue of viral pp60^{src}, stimulating its tyrosyl kinase activity (Bolen et al., 1984). The T antigen of SV-40 also binds to cellular phosphoprotein (Tevethia et al., 1980) and so may have a role in phosphorylation, but there is no evidence that it stimulates pp60$^{c\text{-}src}$ phosphorylation.

The RSV-infected cells used in the work reported here were gifts of Dr. Roberta Farrell, Dr. Vernon Ingram, Dr. Joan Brugge, Dr. Robert Keane and Dr. Anthony Faras. The SV-40 infected cells were gifts of Dr. Walter Scott, Dr. Lauren Sompayrac and Dr. William Topp. The work in the author's laboratory was supported by a research grant from the National Cancer Institute, U.S. National Institutes of Health.

References

Atkinson, M. M., Menko, A. S., Johnson, R. G., Sheppard, J. R. & Sheridan, J. D. (1981) *J. Cell Biol.* **91**, 573–578

Azarnia, R. & Loewenstein, W. R. (1984*a*) *J. Membrane Biol.* **82**, 191–205

Azarnia, R. & Loewenstein, W. R. (1984*b*) *J. Membrane Biol.* **82**, 207–212

Azarnia, R. & Loewenstein, W. R. (1984*c*) *J. Membrane Biol.* **82**, 213–222

Azarnia, R., Dahl, G. & Loewenstein, W. R. (1981) *J. Membrane Biol.* **63**, 133–146

Bishop, J. M. (1982) *Adv. Cancer Res.* **37**, 1–37

Bolen, J. B., Thiele, C. J., Israel, M. A., Yonemoto, W., Lipsich, L. A. & Brugge, J. S. (1984) *Cell* **38**, 767–777

Browne, C. L. & Wiley, H. S. (1979) *Science* **203**, 182–183

Chen, W.-T. & Singer, S. J. (1982) *J. Cell Biol.* **95**, 205–222

Collett, M. S., Purchio, A. F. & Erikson, R. L. (1980) *Nature (London)* **285**, 167–169

Courtneidge, S. A. & Smith, A. E. (1983) *Nature (London)* **303**, 435–439

Flagg-Newton, J. L. (1979) *Biophys. J.* **25** (2), 297a

Flagg-Newton, J. L. & Loewenstein, W. R. (1981) *J. Membrane Biol.* **63**, 123–131

Flagg-Newton, J. L., Dahl, G. & Loewenstein, W. R. (1981) *J. Membrane Biol.* **63**, 105–121

Frisque, R. J., Rifkin, D. B. & Topp, W. C. (1980) in *Cold Spring Harbor Laboratory Symp. Quant. Biol.*, vol XLIV, pp. 325–331, Viral Oncogenes, Cold Spring Harbor Laboratory

Gottesman, M. M., Singh, T., LeCam, A., Roth, C., Nicolas, J.-C., Cabial, F. & Pastan, I. (1981) in *Protein Phosphorylation. Cold Spring Harbor Conference on Cell Proliferation*, 8A (Rosen, O. M. & Krebs, E. G., eds.) pp. 195–209, CSH, New York

Hunter, T. (1980) *Cell* **22**, 647–648

Ito, S., Sato, E. & Loewenstein, W. R. (1974) *J. Membrane Biol.* **19**, 339–355

Johnson, R., Hammer, M., Sheridan, J. & Revel, J. P. (1974) *Proc. Natl. Acad. Sci. U.S.A.* **71**, 4536–4543

Johnson, K. R. & Johnson, R. (1982) *Fed. Proc. Fed. Amer. Soc. Exptl. Biol.* **41**, 755

Kerrick, W. G. L. & Krasner, B. (1975) *J. Appl. Physiol.* **39**, 1052–1055

Krebs, E. G. (1972) *Current Topics In Cellular Regulation* **5**, 99–133

Kuo, J. F. & Greengard, P. (1969) *Proc. Natl. Acad. Sci. U.S.A.* **64**, 1349–1355
Lai, C.-J. & Nathans, D. (1975) *Virology* **66**, 70–81
Lau, A. F., Krzyzek, R. A., Brugge, J. S., Erikson, R. L., Schollmeyer, J. & Faras, A. J. (1979) *Proc. Natl. Acad. Sci. U.S.A.* **76**, 3904–3908
Loewenstein, W. R. (1966) *Ann. N.Y. Acad. Sci.* **137**, 441–472
Loewenstein, W. R. (1968) *Devel. Biol.* **19** (Suppl. 2), 151–183
Loewenstein, W. R. (1974) in *Cell Membranes: Biochemistry, Cell Biology and Pathology* (Weissmann, G. & Claiborne, R., eds.), pp. 105–114, H.P. Publishing Co. Inc., New York
Loewenstein, W. R. (1979) *Biochim. Biophys. Acta* **560**, 1–65
Loewenstein, W. R. (1981) *Physiol. Rev.* **61**, 829–913
Loewenstein, W. R., Kanno, Y. & Socolar, S. J. (1978) *Nature (London)*, **274**, 133–136
Nawrocki, J. F., Lau, A. F. & Faras, A. J. (1984) *Molec. Cell Biol.* **4**, 212–215
Nigg, E. A., Sefton, B. M., Hunter, T., Walter, G. & Singer, S. J. (1982) *Proc. Natl. Acad. Sci. U.S.A.* **79**, 5322–5326
Oliveira-Castro, G. M. & Loewenstein, W. R. (1971) *J. Membrane Biol.* **5**, 51–77
Radke, K. & Martin, G. S. (1979) *Cold Spring Harbor Symp. Quant. Biol.* **44**, 975–982
Radu, A., Dahl, G. & Loewenstein, W. R. (1982) *J. Membrane Biol.* **70**, 239–251
Rose, B. & Loewenstein, W. R. (1976) *J. Membrane Biol.* **28**, 87–119
Rosen, O. M. & Krebs, E. G. (eds.) (1981) *Protein Phosphorylation*, Cold Spring Harbor Conference on Cell Proliferation, vol. 8 A & B, CSH, New York
Schaffhausen, B. S., Dorai, H., Arakere, G. & Benjamin, T. L. (1982) *Mol. Cell. Biol.* **2**, 1187–1198
Schriver, K. & Rohrschneider, L. (1981) *J. Cell Biol.* **89**, 525–535
Sefton, B. M., Hunter, T., Ball, E. H. & Singer, S. J. (1981) *Cell* **24**, 165–174
Sompayrac, L. & Danna, K. J. (1983) *J. Virol.* **46**, 620–625
Tevethia, S. S., Greenfield, R. S., Flyer, D. C. & Tevethia, M. J. (1980) *Cold Spring Harbor Symp. Quant. Biol.* **44**, 235–242
Tooze, J. (1981) (ed.) *DNA Tumor Viruses Molecular Biology of Tumor Viruses*, part 2, 2nd edn., Cold Spring Harbor Laboratory Publication
Unwin, P. N. T. and Ennis, P. D. (1983) *J. Cell Biol.* **97**, 1459–1466
Unwin, P. N. T. & Zampighi, G. (1980) *Nature (London)*, **283**, 545–549
Wiener, E. C. & Loewenstein, W. R. (1983) *Nature (London)*, **305**, 433–435
Willingham, M. C., Jay, G. & Pastan, I. (1979) *Cell* **18**, 125–134

Biochem. Soc. Symp. **50**, 59–79
Printed in Great Britain

Molecular Basis for Active Na,K-Transport by Na,K-ATPase from Outer Renal Medulla

PETER L. JØRGENSEN

Institute of Physiology, Aarhus University, 8000 Aarhus C, Denmark

Synopsis

Active transcellular transport of NaCl in thick ascending limb of Henle's loop (TAL) consists of primary active transport of Na^+ and secondary active transport of Cl^-. The Na,K-pump maintains a low electrochemical potential for Na^+ in the cytoplasm. The Na,K,Cl-cotransport system couples the entry of Cl^- across the lumen membrane to entry of Na^+ along its gradient. In mammalian kidney the TAL is concentrated in the red outer medulla and this tissue has exceedingly high concentrations of Na,K-pump sites, comparable with those in salt glands of birds and fishes. With the red outer medulla as starting material we developed procedures for purification of the Na,K-pump in membrane-bound form. The pump protein can be immobilized in two-dimensional membrane crystals by stabilization of the E_2 form with vanadate or phosphate. The minimum functional protein unit of both membrane-bound and soluble Na,K-ATPase is an $\alpha\beta$-dimer (α-subunit M_r 84000–106000 and β-subunit M_r 32000–38000 plus carbohydrate). Binding data, proteolytic digestion and fluorescence analysis show that both Na,K-induced and ATP-dependent conformational transitions of the α-subunit can be demonstrated in preparations consisting of soluble and fully active $\alpha\beta$-units. This is consistent with the notion that a single $\alpha\beta$-unit can catalyse the whole series of intermediary reactions and cation transport in the membrane. The Na,K,Cl-cotransport system can be identified by assaying bumethanide-sensitive ion fluxes in membrane vesicles from the outer renal medulla. Using equilibrium binding with [³H]bumethanide as an assay the cotransporter has been partially purified to capacities of 600–1000 pmol/mg of protein. A polypeptide of M_r 34000 is covalently labelled with [³H]bumethanide by direct photolysis. The [³H]bumethanide binding protein is not extracted by nonionic detergent except at high ionic strength, suggesting that the Na,K,Cl-cotransport protein is associated with cytoskeleton components. This association may be important for control of the entry of NaCl into the cytoplasm and for cellular regulation of the rate of active transport of NaCl across the tubule cells in the TAL.

Introduction

This paper considers basic questions concerning the protein structure of the Na,K-pump and conformational changes related to active Na,K-transport. We will first look at the structure and organization of α-subunit and β-subunit of

Na,K-ATPase in the membrane and at experiments designed to examine whether the Na,K-pump consists of a monomer $\alpha\beta$-unit or if association within oligomeric $(\alpha\beta)_2$ structure is required for catalysis of active Na,K-transport.

Another basic problem of importance for understanding the molecular basis for active Na,K-transport is the nature and the location of the motion within the pump protein. A relatively massive structural transition between E_1 and E_2 forms of the α-subunit of Na,K-ATPase was detected as a difference in exposure of bonds to tryptic cleavage in NaCl or KCl medium (Jørgensen, 1975). In addition to the bonds exposed to proteolysis there is evidence that the structural change involves tryptophan residues (Karlish & Yates, 1978), thiol groups, ionizable groups and intramembrane segements (Jørgensen, 1982). Circular dichroism spectroscopy shows that at least 80 amino acid residues are involved in a change in secondary structure from α-helix to β-sheet accompanying E_1 to E_2 transition (Gresalfi & Wallace, 1984).

It is not known if the E_1–E_2 transition involves α–α-subunit interaction or if a single $\alpha\beta$-unit can undergo the conformational change. To examine this, E_1–E_2-transitions were studied in soluble and fully active preparations of Na,K-ATPase consisting predominantly of $\alpha\beta$-units. Selective tryptic and chymotryptic modification have been established as valuable tools for examining structure–function relationships of the α-subunit (Jørgensen, 1974, 1975, 1982). In studies of the localization of the E_1–E_2 transitions within the α-subunit and their relationship to cation transport, selective proteolytic cleavage of the membrane-bound Na,K-ATPase with trypsin and chymotrypsin (Jørgensen, 1974, 1975) have been combined with selective chemical labelling and studies of the function and properties of cleaved fragments of the α-subunit.

Structure of the Pure Renal Na,K-Pump

The purification procedure. Membranes from the thick ascending limb of Henle in the inner stripe of the bright red outer medulla form the starting material. The principle is to keep the Na,K-pump embedded in the membrane while extraneous proteins are extracted by incubation with a low concentration of sodium dodecyl sulphate (NaDodSO$_4$) in the presence of ATP (Jørgensen, 1974). NaDodSO$_4$ selectively removes extraneous proteins and the membrane-bound Na,K-ATPase can be recovered in a single isopycnic centrifugation in a zonal rotor. The preparation is 90–100% pure with respect to the content of the specific proteins, the α-subunit with M_r 93000–106000 and the β-subunit with M_r 36000 plus carbohydrate, and with respect to phosphorylation and binding of the inhibitors ouabain and vanadate (Jørgensen, 1974, 1982).

Soluble and fully active $\alpha\beta$-units are prepared by mixing pure membrane-bound Na,K-ATPase with dodecylocta-ethylene-glycol-monoether $(C_{12}E_8)$ in proper conditions (Brotherus *et al.*, 1983). The soluble $\alpha\beta$-units each bind one molecule of ATP (Jensen & Ottolenghi, 1983) and reconstitute directly into phospholipid vesicles to catalyse active Na,K-transport at high rates (Brotherus *et al.*, 1983).

Ultrastructure of the Na,K-pump. The Na,K-pump was first observed as distinct surface particles by electron microscopy after negative staining of pure membrane-bound Na,K-ATPase (Maunsbach & Jørgensen, 1974; Deguchi *et*

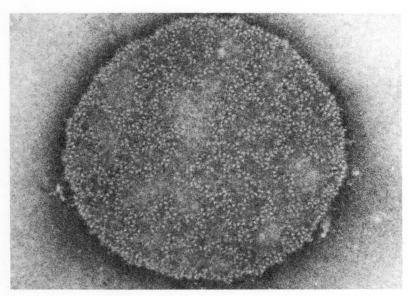

Fig. 1. *Electron microscopy after negative staining with potassium phosphotungstate of unfixed membrane fragments consisting of equal amounts of Na,K-pump protein and lipid*

The membranes appear as rounded discs that are studded with particles occurring with a frequency of $12500/\mu m^2$. Particles are arranged in clusters and strands. Magnification $\times 240000$.

al., 1977). As shown in Fig. 1, the membranes appear as rounded discs that are studded with a uniform population of particles ($12500/\mu m^2$) with diameters up to 5 nm. The protein can also be visualized as intramembrane particles after freeze–fracture replication (Deguchi *et al.*, 1977). The orientation of the protein remains asymmetric in the pure membrane-bound Na,K-ATPase with particle-rich and particle-poor fracture faces. The distribution of particles is similar to that seen after fracture of basolateral cell membranes in the intact tubule cells of the TAL and in right-side-out vesicles of the plasma membrane (Skriver *et al.*, 1983). In contrast, random orientation of Na,K-pump proteins is seen in reconstituted phospholipid vesicles prepared by mixing soluble Na,K-ATPase and excess phospholipid (Skriver *et al.*, 1980). The intramembrane particles with diameter 9 nm correspond to Na,K-pump molecules because the frequency of particles is proportional to the amount of Na,K-ATPase used in reconstitution and the density of particles is linearly related to the capacity for active Na-transport over a range of 0.2–16 particles per vesicle (Skriver *et al.*, 1980).

Molecular shape and subunit structure in membrane crystals. Na,K-pump molecules floating in the lipid bilayer are tightly packed in the disc-shaped membrane fragments of the pure preparation (Fig. 1). The concentration of α-subunit in the lipid phase amounts to 6–8 mM, corresponding to almost 1 g of protein/ml of lipid phase. In these conditions of supersaturation, stabilization of the protein in the E_2-conformation in vanadate or phosphate medium favours formation of crystalline arrays. The protein particles associate in a two-dimensional lattice with bond energies that are sufficient to compensate for the

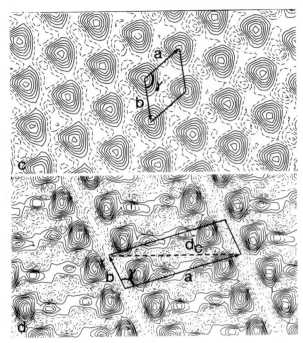

Fig. 2. *Computer-reconstructed images of two dimensional crystals formed in vanadate (c) or in phosphate (d)*

Selection of well-ordered crystalline arrays was made by optical diffraction. Images suitable for further analysis were densitometered at 20 μm intervals. Projection maps were calculated by using the Fourier transform amplitudes and phases collected at the reciprocal lattice points (Herbert *et al.*, 1982). The protein-rich regions (positive regions) are drawn with unbroken contour lines; negative stain regions have dashed lines. In the reconstructed images 1 mm corresponds to 0.28 nm. The unit cell dimensions are: (c), a = 53 nm, b = 5.1 nm, γ = 120°; (d), a = 13.5 nm, b = 4.4 nm, γ = 101°. [From Jørgensen et al. (1982).]

loss of translational and rotational entropy of the freely moving particle. Within hours after start of the incubation, linear polymers (some of them dimeric) are observed in the membranes. Later these polymers associate laterally to form extensive two-dimensional arrays (Skriver *et al.*, 1981; Jørgensen *et al.*, 1982).

Computer based image processing of electron micrographs of the negatively stained crystals shows that the symmetry in the vanadate crystal is p1 and that the unit cell contains one $\alpha\beta$-unit (Fig. 2). (p1 and p21 indicate the lattice symmetry of the two-dimensional crystal, using the nomenclature of two-sided plane groups with one axis chosen to be perpendicular to the plane of the crystal.) The part of the particle protruding above the plane of the bilayer is a compact structure with diameter close to 5 nm. The unit cell in the arrays grown in phosphate medium has two strong positive regions and contains two $\alpha\beta$-units. The symmetry is p21 and two $\alpha\beta$-units occupy one unit cell. These regions are subdivided into one large and one smaller peak possibly corresponding to α-subunits and β-subunits, respectively (Hebert *et al.*, 1982). In support of this interpretation, tryptic cleavage of the α-subunit alters the shape of the body in the crystal while the smaller peak or hook is unaffected (Mohraz & Smith, 1984). The hook representing the β-subunit can also be observed in monomer crystals when the resolution is sufficiently high (Mohraz & Smith, 1984). The $\alpha\beta$-unit is therefore the minimum asymmetric unit of Na,K-ATPase in the membrane, but the protein crystallizes in two principally different forms, a monomer p1 and a dimer p21 form. It is probable that the dimer represents a transitory stage in the formation of the monomeric crystal since linear polymers consisting of paired protein units are seen during the formation of vanadate-induced crystals.

Using the same techniques, the Ca-ATPase from sarcoplasmic reticulum also forms two-dimensional crystals (Dux & Martonosi, 1983). In parallel to the larger molecular weight due to the presence of the β-subunit, the α-lattice dimension of Na,K-ATPase is 40% larger than that of Ca-ATPase as it accommodates a hook-like structure which the Ca-ATPase lacks (Hebert et al., 1982; Mohraz & Smith, 1984).

Length of the Na,K-pump molecule. Formation of paired membranes in thin sections of the crystalline membrane-bound Na,K-ATPase has allowed estimation of two important parameters of the Na,K-pump molecule. The total length of the molecule is 10.5–12.5 nm. The length of extensions of the protein into the cytoplasmic phase is 5 nm but the molecule extends only 1–3 nm beyond the extracytoplasmic surface (Zampighi et al., 1984). These estimates are in agreement with studies using labelling of the protein surface, showing that the mass of the protein is asymmetrically distributed across the bilayer with three to four times more surface area on the cytoplasmic side of the membrane than on the extracellular side (Sharkey, 1983). Fluorescence data also gave distances of 7–8 nm between binding sites for ATP at the cytoplasmic surface and the site for ouabain at the extracellular surface (Cantley, 1981; Moczydlowski & Fortes, 1981).

Analysis of crystals thus provided the first direct observations of molecular shape, diameter and length of the Na,K-pump. It should be remembered that negative staining reveals only hydrophilic features at the surface, while structural detail of lipid embedded portions or the path of polypeptide chains remain obscure. Further information requires elimination of the stain and the use of low dose electron microscopy of more extensive and regular two-dimensional crystals. Another perspective is crystallization in three dimensions. Successful techniques are developed for small membrane proteins (Michel et al., 1980), but so far not for the large ion pump molecules.

Subunit structure. The $\alpha\beta$-unit is the minimum functional unit of pure renal Na,K-ATPase both in the membrane-bound form (Jørgensen, 1982; Hebert et al., 1982) and in the soluble, fully active state in $C_{12}E_8$ (Brotherus et al., 1981); but the possibility that formation of $(\alpha\beta)_2$-complexes is required for active Na,K-transport is not excluded. Each soluble $\alpha\beta$-unit binds one ATP molecule; the stoichiometry of ATP binding in the membrane-bound state is more uncertain. The soluble enzyme reconstitutes to active pumps in phospholipid vesicles (Brotherus et al., 1983). However, direct evidence for competence of the $\alpha\beta$-unit as Na,K-transport system has not been obtained. Transport across phase boundaries cannot be measured in detergent solution and the molecular size cannot be determined in phospholipid vesicles. One way to address this problem

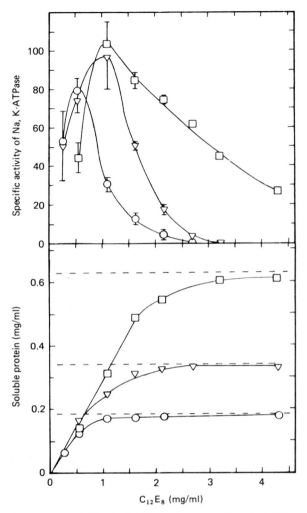

Fig. 3. *Specific activity of soluble Na,K-ATPase (upper graph) and solubilization of the protein of membrane-bound Na,K-ATPase (lower graph) at three different concentrations (○, 0.18 mg/ml; ▽, 0.34 mg/ml; □, 0.62 mg/ml)*

Membrane-bound Na,K-ATPase was sedimented at 100000 rev./min for 10 min in the Beckman Airfuge and mixed at 20 °C with increasing concentrations of $C_{12}E_8$ in 130 mM-NaCl/20 mM-KCl/10 mM-TES/1 mM-Tris-EDTA/1 mM-dithiothreitol, pH 7.5. After incubation for 30 min, 40 μl was taken out for Na,K-ATPase assay and supplemented with 4 μl of 100 mM-Mg-ATP. After 20, 40 or 60 s at 20 °C the reaction was stopped with 1 ml of ice-cold 0.5 M-HCl containing 30 mg of ascorbic acid, 5 mg of ammonium heptamolybdate and 10 mg of NaDodSO$_4$. The tubes were transferred to an ice bath. For colour development, 1.5 ml, containing 30 mg of sodium meta-arsenite/30 mg of sodium citrate/30 μl of acetic acid, was added. The tubes were heated for 10 min at 37 °C and absorbance was read at 850 nm. Protein was measured by the Lowry method after precipitation with trichloroacetic acid. As standard we used membrane-bound Na,K-ATPase in which the protein concentration had been determined by quantitative amino acid analysis (Brotherus *et al.*, 1983).

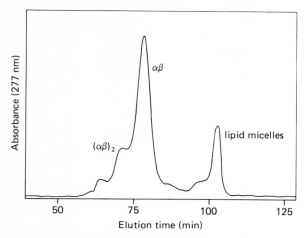

Fig. 4. *Distribution of the proteins of soluble Na,K-ATPase on αβ-units, (αβ)₂-units and higher aggregates $(\alpha\beta)_{2+n}$ after chromatography*

Chromatography was on a 7.5 mm × 600 mm TSK-GEL G 3000 SW column equilibrated with 2 mg/ml $C_{12}E_8$/25 mM-Tris–acetate/150 mM-KCl, pH 7.0, and elution was at 0.18 ml/min. The sample consisted of 100 μl of soluble Na,K-ATPase (0.8 mg of protein/ml). Assuming Gaussian distribution the area of αβ%-peak comprised 82% of total absorbance at 277 nm. As determined by low angle laser light scattering on an identical preparation of soluble Na,K-ATPase the molecular masses were: αβ, 123 + 8 kDa; $(\alpha\beta)_2$, 286 + 30 kDa; $(\alpha\beta)_{2+n}$, 1740 + 230 kDa (Hayashi *et al.*, 1983).

is to examine whether transitions between E_1 forms and E_2 forms of the α-subunit occur in the soluble monomeric αβ-unit or if this requires α–α-subunit interaction.

Soluble and Fully Active αβ-Units

Solubilization of pure membrane-bound Na,K-ATPase in $C_{12}E_8$ yields soluble and fully active αβ-units. Fig. 3 shows that the activity of soluble Na,K-ATPase was close to 100% of that of membrane-bound enzyme at protein concentrations 0.3–0.6 mg/ml and a $C_{12}E_8$/protein ratio of 2–3. In this experiment the Na,K-ATPase activity was assayed by adding Mg-ATP to the detergent solution and comparing activity with that in membrane-bound preparations that were assayed in parallel without detergent. Sedimentation velocity analysis immediately after solubilization shows that this preparation sediments as a major component 85–90% with $s_{20w} = 7$–7.4, corresponding to M_r 140000–170000, a protomeric αβ-unit. The Na,K-ATPase activity is stable in the cold, but after 20–30 h, e.g. after passing over a column or after ultracentrifugation to achieve equilibrium, secondary aggregation to dimers $(\alpha\beta)_2$ or trimers $(\alpha\beta)_3$ is observed (Brotherus *et al.*, 1981, 1983). This secondary aggregation is likely to be due to associations between detergent molecules, a serious complication that may be difficult to distinguish from specific protein–protein interactions. This time dependent secondary aggregation of αβ-units to dimer, trimer or higher complexes can clearly be demonstrated in h.p.l.c. experiments with TSK 3000SW gel filtration columns (Hayashi *et al.*, 1983). Fig. 4 shows that chromatography

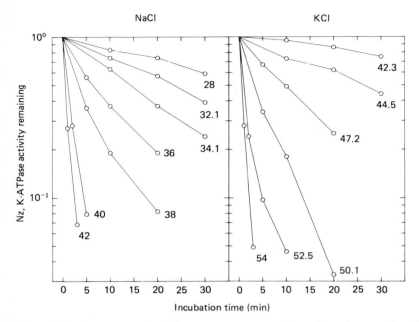

Fig. 5. *Thermoinactivation of soluble Na,K-ATPase in NaCl or KCl as a function of incubation time*

The temperatures used are shown at the ends of the curves. Membrane-bound Na,K-ATPase was suspended at 0.5 mg of protein/ml in TES (30 mM)/EDTA (3 mM)/dithiothreitol (2 mM), pH 7.5, and 20 mM-NaCl (left) or 20 mM-KCl (right). A portion (800 μl) of this mixture was mixed with 800 μl of $C_{12}E_8$ (1.2 mg/ml) and incubated for 30 min at 20 °C. After centrifugation for 10 min at 100000 rev./min in the Beckman Airfuge 200 μl aliquots were incubated in thermostatted water baths at temperatures in the range 28–54 °C. After 2–30 min, 35 μl was taken out for assay at 20 °C and supplemented with 10 μl of 650 mM-NaCl/100 mM-KCl and 5 μl of 100 mM-ATP/100 mM-MgCl$_2$. After 20, 40, or 60 s at 20 °C the reaction was stopped and P$_i$ was determined as in Fig. 3.

on TSK GEL 3000SW columns gave elution patterns similar to those obtained by Hayashi *et al.* (1983), who used our protocol for solubilization of Na,K-ATPase from outer renal medulla. The elution pattern in Fig. 4 shows that 82% of protein was eluted as $\alpha\beta$-units immediately after solubilization, in agreement with the result of sedimentation velocity analysis. After 10–20 h the chromatograms show secondary aggregation to $(\alpha\beta)_2$ and higher oligomers (Hayashi *et al.*, 1983).

The soluble preparation reconstitutes directly into phospholipid vesicles by the freeze–thaw technique to catalyse Na,K-transport at high rates (Brotherus *et al.*, 1983). It binds one molecule of ATP per $\alpha\beta$-unit (Jensen & Ottolenghi, 1983) without evidence for co-operative interaction, while negative co-operativity of nucleotide binding sites is observed in the membrane-bound Na,K-ATPase where $\alpha\beta$-units are tightly packed.

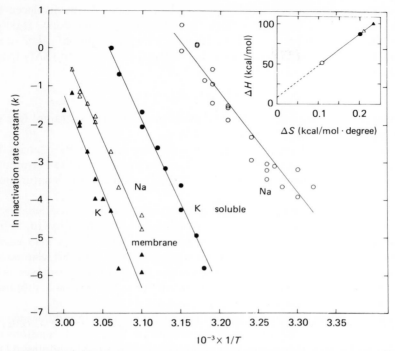

Fig. 6. *Arrhenius plot of inactivation rate constants (ln k) versus temperature (1/T, K)*
The rate constants were calculated for membrane-bound Na,K-ATPase in KCl (▲) or NaCl (△) and for soluble Na,K-ATPase in KCl (●) or NaCl (○). The inset shows a plot of $\Delta H^{\ddagger}$ versus $\Delta S^{\ddagger}$ for the four preparations at 37 °C. Calculations are explained in the text.

E_1–E_2 Transitions in Soluble $\alpha\beta$-Units

During attempts to study proteolysis of soluble Na,K-ATPase in $C_{12}E_8$ at 37 °C we soon realised that heat can be used to demonstrate conformational transitions in Na,K-ATPase, because the E_1 form of the protein is more thermolabile than the E_2 form. As shown in Fig. 5, the soluble Na,K-ATPase was much more sensitive to heat in NaCl medium than in KCl. For example, the rate constant for inactivation in NaCl at 42 °C (-1.31 min^{-1}) is severalfold larger than in KCl (-0.0051 min^{-1}). In a higher temperature range (55–60 °C) the membrane-bound Na,K-ATPase is also more sensitive to heat in NaCl than in KCl medium, as previously demonstrated by Fischer (1983), but this difference is small when compared with the pronounced difference in thermosensitivity of the soluble forms.

The inactivation rate constants were determined at a series of different temperatures for each enzyme species membrane-bound and soluble, in media containing NaCl or KCl. Plotting lnK versus $1/T$ as in Fig. 6 yielded linear Arrhenius plots with a slope equal to $-E/R$ and a vertical intercept equal to lnA, where E is the activation energy (kcal/mol) and A is the frequency factor (s^{-1}). The lines for the membrane-bound Na,K-ATPase in NaCl or KCl were parallel with the line for the rate constants of soluble Na,K-ATPase in KCl. The line for the rate constants of soluble Na,K-ATPase in NaCl was displaced to the right, with smaller values for both the intercept and the slope. Having obtained the activation energy and the frequency factor permitted calculation of the enthalpy of activation ($\Delta H^{\ddagger}$) and the entropy of activation ($\Delta S^{\ddagger}$) from the equations:

$$E = RT + \Delta H^{\ddagger}; \quad \text{and} \quad A = RT/N \, e \exp \Delta S^{\ddagger}/R$$

The data for the four lines in the Arrhenius plots falls on a straight line in the 'compensation plot' of the inset of Fig. 6.

This means that enthalpy changes, $\Delta H^{\ddagger}$, are compensated by changes in entropy, $\Delta S^{\ddagger}$. The linearity of this function suggests that both the Na-bound forms and the K-bound forms passes identical transition states in the denaturation process.

One can assume that transition from the native state to a disordered denatured state is an unfolding process consisting of progressive exposure of hydrophobic groups and peptide bonds to solvent. The increase in heat lability may result from a structural change causing increasing exposure of hydrophobic residues to solvent water. With respect to thermoinactivation the E_1Na form behaves as if it has properties corresponding to a state between that of the native membrane-bound enzyme on one side and the denatured state on the other side.

In a subsequent series of experiments it was shown that this difference in thermolability is due to stabilization of the alternative E_1–E_2 conformations of soluble Na,K-ATPase. It was seen that formation of E_1KADP by adding ADP ($K_{\frac{1}{2}}$ 8–10 μM) increased the rate of thermoinactivation in KCl medium to the level observed in NaCl for the E_1Na form. Ligands stabilizing E_1 are therefore heat-labilizing, whereas ligands known to stabilize E_2 forms are heat-stabilizing. The thermoinactivation experiments thus support the conclusion from labelling with iodonaphthylazide (Karlish et al., 1977) and tryptophan fluorescence experiments (Karlish & Yates, 1978) that transition from E_1 to E_2 is accompanied by movement of residues from relatively hydrophilic to hydrophobic environments.

These observations suggest that the inactivation of soluble Na,K-ATPase during hydrolysis of ATP would depend on the conformational equilibrium. To test this, inactivation rates were examined during turnover at different ATP concentrations. Owing to the potassium–nucleotide antagonism a large fraction of the enzyme will be stabilized in the E_2K form at low ATP concentration. At high ATP one would expect the equilibrium to be poised in direction of the E_1 forms with higher rates of inactivation of soluble Na,K-ATPase. We found that the rate of inactivation was increased when the ATP concentration was raised from 2 μM to 100 μM and conclude from this experiment that the Na,K-ATPase reaction of soluble monomeric Na,K-ATPase involves E_1–E_2 transitions.

Our data show that the two conformations can be distinguished by differences in susceptibility of Na,K-ATPase to irreversible temperature dependent inactivation. A linear relationship between enthalpies ($\Delta H^{\ddagger}$) and entropies ($\Delta S^{\ddagger}$) of activation for the E_1 and E_2 forms suggest that this inactivation proceeds through a common transition state. The E_1 form of soluble Na,K-ATPase behaves as if it is a partially unfolded protein with properties intermediate between those of the native membrane-bound enzyme and the denatured protein. According to this interpretation E_2 forms have fewer hydrophobic residues exposed to solvent than E_1 forms. This is in agreement with the demonstration of preferential labelling of the E_2 form of membranous enzyme from within the lipid bilayer with the hydrophobic [125I]iodonaphthylazide (Karlish et al., 1977). The increase in tryptophan fluorescence accompanying transition from E_1 to E_2 has also been explained by movement of tryptophan residues from a relatively hydrophilic to a relatively hydrophobic microenvironment (Karlish & Yates, 1978).

It is unlikely that interaction of the detergent $C_{12}E_8$ with the nucleotide binding site in the E_1 form could contribute to inactivation, since protection by ADP or ATP is not seen in NaCl medium and addition of ADP to the KCl medium increases the rate of inactivation to the level with NaCl in the medium. A more obvious possibility for explaining the inactivation is that the exposure of hydrophobic residues to solvent water in the E_1 forms may be related to inadequate length of the dodecyl alkyl chain of $C_{12}E_8$ for stabilization of hydrophobic protein domains that normally associate with alkyl chains of 16–20 carbon atoms.

Proteolytic Cleavage and Fluorescence Changes in Soluble $\alpha\beta$-Units

Fig. 7 demonstrates fluorescence analysis of conformational transitions in soluble Na,K-ATPase (to the left fluorescence from fluorescein, to the right tryptophan fluorescence). It is clear that Na,K-dependent transitions and changes accompanying vanadate and ouabain binding could be demonstrated as in the membrane-bound enzyme. A titration of these responses shows that apparent affinities for Na^+ and K^+ are identical for soluble and membrane-bound enzyme as measured by fluorescence from fluorescein.

Working at relatively low temperature (20 °C) to avoid inactivation we were able to demonstrate the distinct inactivation patterns of E_1 and E_2 with trypsin and chymotrypsin, although they were not as clearcut as in the membrane-bound enzyme. To demonstrate this we had to raise the protein concentration to 0.7 mg/ml in the soluble preparation. In NaCl the rate of inactivation by chymotrypsin was much faster than in KCl. In KCl medium, addition of ATP gave a biphasic pattern of tryptic inactivation as in the membrane-bound state. Examination of cleavage shows that it is possible to demonstrate the shift between the two forms, trypsin giving the 58K and 46K fragments in the E_2K form and the 78K fragment in the E_1K ATP form, not shown.

Since we had to raise the protein concentrations to 0.7 mg/ml to demonstrate conformational transitions in soluble Na,K-ATPase with proteolytic enzymes as tools, it was important to re-examine the sedimentation properties. The requirement for a higher protein concentration could mean that aggregation of $\alpha\beta$-units was required to demonstrate the conformational transitions. However, sedimentation velocity analysis shows that this is not the case. At the high protein concentration both the unlabelled and the fluorescein-labelled Na,K-ATPase is 90% homogeneous and sediments with $s_{20,w} = 6.7$–6.9, corresponding to M_r around 160000.

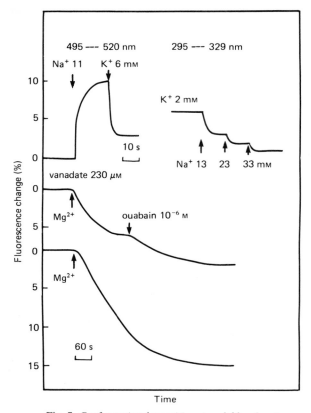

Fig. 7. *Conformational transitions in soluble αβ-units*

Changes in intensity of fluorescence from fluorescein (495–520 nm) or tryptophan (295–328 nm) accompanying transitions between E_2K and E_1Na or from E_1–Tris to E_2–vanadate and E_2–vanadate–ouabain forms of soluble Na,K-ATPase in $C_{12}E_8$. Soluble Na,K-ATPase (20–50 μg) prepared as in Fig. 1, or fluorescein-labelled Na,K-ATPase (Karlish, 1980) was suspended in 50 mM-Tris–HCl, 0.1 mM-EDTA, pH 7.5, in a continously stirred cuvette at 10 °C. Fluorescence was recorded in a Perkin–Elmer MPF 44A fluorimeter. NaCl, KCl, MgCl$_2$, vanadate or ouabain was added from Hamilton syringes.

The phosphorylation experiment in Fig. 8 was performed to examine if soluble Na,K-ATPase could undergo transition between E_1P and E_2P conformations. It is seen that the amount of ADP-sensitive E_1P as a fraction of total phosphoenzyme was smaller for soluble than for membrane-bound enzyme. K-dependent dephosphorylation of soluble Na,K-ATPase is demonstrated in the lower part of Fig. 8. The soluble phosphoenzyme was more sensitive to K^+ than the membrane-bound enzyme. Addition of 100 μM-KCl increased the rate constant for dephosphorylation of soluble enzyme 2.8-fold, as compared with an 1.8-fold increase in the rate constant for dephosphorylation for membrane-bound Na,K-ATPase in agreement with previous data (Jørgensen et al., 1978). These data show that the soluble Na,K-ATPase may exist both in E_1P and E_2P forms. The data furthermore suggest that the equilibrium between phospho

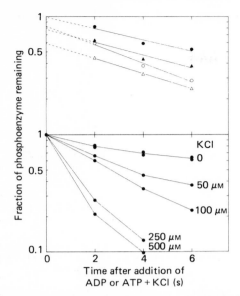

Fig. 8. *ADP-dependent and K-dependent dephosphorylation of soluble Na,K-ATPase*

Upper part: ADP-dependent dephosphorylation of soluble Na,K-ATPase (●, ▲) or membrane-bound enzyme (○, △) at 150 mM-NaCl (●, ▲) or 300 mM-NaCl (○, △). After phosphorylation for 6 s, 2.5 mM-ADP was added and the amount of phosphoenzyme was determined after 2,4 or 6 s as described in the text. Lower part: effect of K⁺ on the rate of dephosphorylation of soluble Na,K-ATPase. After phosphorylation for 6 s, 1 mM-ATP was added without or with K⁺ in the range 50–500 μM and the amount of phosphoenzyme was determined after 2,4, or 6 s. Zero time values were determined by adding perchloric acid before ADP or ATP plus K⁺.

forms of soluble enzyme is poised in direction of E_2 forms as compared with the membrane-bound enzyme.

The conclusion is therefore that both dephospho and phospho forms of soluble $\alpha\beta$-units can undergo the E_1–E_2 transitions. From this point of view the $\alpha\beta$-unit is competent as a Na,K-transport system and the α–α subunit interaction is not an obligatory element in the E_1–E_2 conformational transition.

Localization of E_1–E_2 Transition Within the α-Subunit

We will now return to the membrane-bound Na,K-ATPase. The path of the α-subunit in the membrane has been traced by determining the sidedness of cytoplasmic proteolytic splits and the localization of membrane embedded segments relative to residues involved in formation of sites for ouabain and ATP at the two membrane surfaces (Karlish *et al.*, 1977; Jørgensen, 1974, 1982). In the model presented in Fig. 11 it is illustrated how the catalytic functions are executed in cytoplasmic domains that are separated by intramembrane segments.

Several attempts have been made to determine the amino acid sequence of α-subunit and β-subunit. Progress has been surprisingly slow, and only short sequences around the phosphorylation site and ATP binding area and the

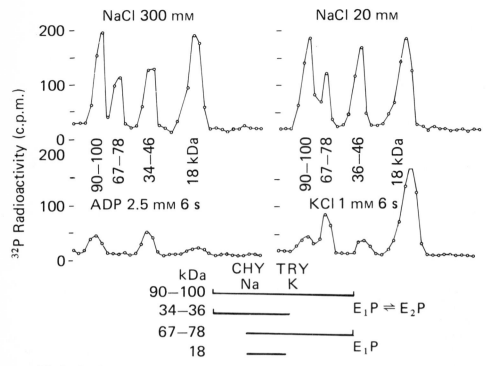

Fig. 9. *Phosphorylation and dephosphorylation of intact α-subunit and fragments of α-subunit obtained by sequential cleavage with chymotrypsin in NaCl and trypsin in KCl medium*

Chymotryptic cleavage at 37 °C was started by adding α-chymotrypsin (CHY) to 25 μg/ml, to 12 ml containing membrane-bound Na,K-ATPase (0.5 mg protein/ml) in 15 mM-Tris–HCl/10 mM-NaCl, pH 7.5. After 15 min at 37 °C the reaction was stopped by adding KCl to 150 mM and cooling to 0 °C. The mixture was centrifuged for 90 min at 50000 rev./min. The pellet was resuspended and diluted to 0.5 mg of protein/ml in 150 mM-KCl/imidazole (25 mM), pH 7.5. Trypsin (TRY) was added to 30 μg/ml. After 7 min at 37 °C the reaction was stopped by adding soyabean trypsin inhibitor to 90 μg/ml. The mixture was centrifuged for 90 min at 50000 rev./min and resuspended in 25 mM-Tris–HCl/150 mM-NaCl and washed twice to remove KCl. The pellet was resuspended in 25 mM-Tris–HCl/150 mM-NaCl, pH 7.5. For phosphorylation 300 μg of protein was suspended in 3 ml of 25 mM-Tris–HCl/3 mM-MgCl$_2$/25 μM-[^{32}P]ATP with 20 mM-NaCl (right) or 300 mM-NaCl (left). For the samples in the upper half of the graph the reaction was stopped after 6 s with 3 ml of 8% perchloric acid (PCA). For the samples in the lower half of the graph either 2.5 mM-ADP (left) or 1 mM-ATP plus 1 mM-KCl (right) were added 6 s after start of phosphorylation. At 2 s later the reaction was stopped with 3 ml of 8% PCA. After sedimentation and wash in 4% PCA the sediment was suspended in 100 μl of 1% NaDodSO$_4$/10 mM-dithiothreitol/50 mM-sodium phosphate, pH 2.4, and a sample was layered on 5–15 T% gradient gels in a 13 cm × 0.5 cm tube prepared by the method of Avruch & Fairbanks (1972).

N-terminus have been determined as yet (Jørgensen, 1982). The problem is that hydrophobic peptide chains precipitate when the protein is split for sequence analysis. In this situation the low resolution model obtained by combining proteolytic cleavage and chemical labelling is the only available framework. The proteolytic splits form precise points of reference in a linear map of α-subunit, and conformational transitions in the protein can be described in terms of differences in spatial arrangement of bonds exposed to proteolysis.

In a series of experiments we found that sequential cleavage, first with chymotrypsin in NaCl and then with trypsin in KCl (Jørgensen, 1975; Jørgensen *et al.*, 1982) can be exploited to isolate an important fragment of the α-subunit and to study structure–function relationships of the protein. In the experiment in Fig. 9, chymotryptic cleavage of bond 3 (CHY Na) in NaCl is allowed to proceed until about half the α-subunits have been split to 78K fragments. Cleavage is stopped by cooling and addition of KCl to 150 mM. The membranes are washed by centrifugation. Trypsin is then added and the α-subunit is cleaved at bond 1 (TRY K) to a 46K fragment and the 78K fragment to a 18K fragment. In the experiment in Fig. 9, cleavage with chymotrypsin and trypsin were adjusted to give almost equal proportions of four different peptide fragments that can incorporate ^{32}P from γ-[^{32}P]ATP. This allows direct comparison of the reactivity to ADP or K^+ of four different phosphopeptides, the intact α-subunit, 78K fragment, 46K fragment and 18K fragment in the same preparation of membrane-bound enzyme. At high NaCl, all phosphopeptides are sensitive to ADP, but dephosphorylation of 78K and 18K fragments is more complete than for the 46K fragment and the intact α-subunit. In contrast only intact α-subunit and 46K fragment are sensitive to K^+, and the 78K and 18K fragments are resistant to K^+. Thus, after cleavage of the chymotrypsin sensitive bond, the 78K and 18K fragments can accept phosphate from ATP to form E_1P, but they are unable to undergo transition to E_2P. In contrast tryptic cleavage of bond 1 in KCl does not interfere with the transition from E_1P to E_2P.

This technique provides us with two opportunities. One is the isolation of an important fragment of the α-subunit, the 18K fragment, by chromatography in NaDodSO$_4$ on Sephacryl 300 S columns. Motion of this peptide relative to other parts of α-subunit must be an important element of the E_1–E_2 transition since its *N*-terminus is exposed to cleavage in E_1 (bond 3) and its *C*-terminus is exposed in E_2 (bond 1).

Effects of Proteolytic Cleavage on Cation Binding and Translocation

The other opportunity is examination of the relationship between the conformational change and cation binding and transport. Cleavage of bond 3 with chymotrypsin in NaCl stabilizes the phosphoenzyme in E_1P whereas E_1P–E_2P transitions are allowed after cleavage of bond 1 with trypsin in KCl. Selective cleavage is therefore a powerful tool to examine the relationship of E_1–E_2 transitions to cation exchange. Examination of ATP–ADP-dependent Na/Na-exchange and ATP–P$_i$-dependent K/K-exchange showed that both types of exchange remain uneffected after cleavage with trypsin in KCl. This is the situation where E_1P–E_2P transitions are preserved, as judged by analysis of fluorescence or reactivity of phosphopeptides. In contrast, all cation exchange is blocked after chymotryptic cleavage, demonstrating that cation exchange reactions are tightly coupled to E_1P–E_2P transitions (Jørgensen *et al.*, 1982).

As expected, this interference of chymotryptic cleavage with the conformational transition is not limited to the phospho forms. Fig. 10 shows that Na/K-dependent fluorescence changes are abolished by chymotryptic cleavage. The K–nucleotide antagonism is also lost. This explains the drastic change in nucleotide binding

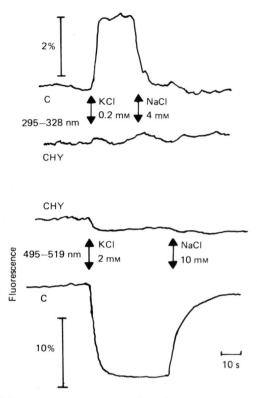

Fig. 10. *Effect of chymotryptic cleavage on conformational transitions monitored by fluorescence from membrane-bound Na,K-ATPase*

Cation induced changes in emission from tryptophan (295–328 nm) or fluorescein (495–519 nm) attached to the α-subunit (Karlish, 1980) in untreated Na,K-ATPase (C) or after chymotryptic cleavage in NaCl (CHY). The cuvettes contained 2.5 ml of 25 mM-Tris–HCl/0.1 mM-EDTA, pH 7.5, and 25 μg of Na,K-ATPase protein. KCl or NaCl was added from Hamilton syringes. Fluorescence measurement was as in Fig. 7.

behaviour after chymotryptic cleavage. We found that chymotrypsin enzyme binds ATP with high affinity both in 150 mM-KCl and in 150 mM-NaCl, whereas control enzyme, as usual, binds ATP with low affinity in KCl medium. Similarly, the dissociation constant for Rb^+ binding to cleaved enzyme was only reduced 1.3-fold by the addition of 94 μM-ATP , and this addition of ATP reduced the dissociation constant for Rb^+ binding to native enzyme by 4.1-fold (not shown). ^{86}Rb binding is the same after chymotryptic cleavage as in native Na,K-ATPase, about $2Rb^+$ per α-subunit with $K_d = 7$–9 μM, but vanadate binding is abolished, in agreement with the observation that transition to E_2 is blocked.

Na-binding sites are preserved after chymotryptic cleavage since Na^+ stimulates transfer of γ-phosphate from ATP to the protein with higher apparent affinity than in control (Jørgensen *et al.*, 1982). The high rate of ADP–ATP-exchange and the ADP-sensitivity of the phosphointermediate, together with the dissappearance of Na–Na-exchange, show that the cleaved phosphoenzyme is stabilized in the E_1P form. This phosphoenzyme incorporates 3 Na^+ ions in the occluded state per phosphyorylation site (Glynn et al., 1984, 1985).

Although the chymotrypsin enzyme is stabilized in E_1 forms with respect to configuration of ATP-binding site and fluorescence emission, the cleaved enzyme can still bind K^+ (or Rb^+) with high affinity and it can occlude K^+ (or Rb^+) in a process that is unaffected by ATP (Glynn et al., 1985). In this occluded form, $E_1(K_n)$, of the cleaved enzyme, the complex is unaffected by ATP or ADP.

Chymotryptic cleavage results in an altered structure of the protein that prevents the motion of the segments in the α-subunit that is associated with E_1–E_2 transition in the native enzyme. Nevertheless, the dephospho form of the cleaved enzyme occludes K^+ (or Rb^+) and the phosphoform occludes Na^+ (Glynn et al., 1984). To explain this one can assume that the pump has one set of cation transport sites that are exposed to the cytoplasm in the E_1 conformation. They bind $K^+ +$ (or Rb^+) in the dephospho form and Na^+ in the phospho form. These cation sites can exist in two forms, an open and a closed or occluded configuration. Transition from open to closed form is assumed to involve only a limited change in conformation of co-ordinating carbonyl residues and substitution of solvent molecules from the inner sphere of the cation. This is a local structural change which is distinct from the E_1–E_2 transition.

Model for the E_1–E_2 Conformational Change in α-Subunit

Fig. 11 summarizes the observations made in the two series of experiments. The conformational transition between E_1 and E_2 consists of movements within one α-subunit and need not involve α–α-subunit interaction. We have also demonstrated that E_1P–E_2P transitions are tightly coupled to cation exchange. In this model the transition involves movement of the 18K segment of the α-subunit from relatively hydrophilic to relatively hydrophobic micro-environments during the transition from E_1 to E_2.

From this model there are two possibilities for explaining that chymotryptic cleavage of bond 3 abolishes the motion of the protein. One is that the cleavage interrupts the co-operation between ionizable groups in the N-terminal sequence and the phosphorylated segment. Removal of an ionizable group at the N-terminal end of the α-subunit by cleavage of bond 2 (Jørgensen et al., 1982) leaves the enzyme as an 'invalid', with reduced Na,K-ATPase activity, due to a shift in conformational equilibrium in direction of the E_1 forms (Jørgensen & Karlish, 1980). An ionizable group with pK in the range 6.5–7 is proposed to be located near the N-terminal end because cleavage of bond 2 with removal of the N-terminus reduces the sensitivity of the activation curve for Na^+ to changes in pH.

The other possible explanation is that cleavage of bond 3 interrupts the association between the aspartyl phosphate residue and a structure involved in cation binding and translocation, which may be formed by the part of the α-subunit between bond 3 and bond 2. The data do not, however, provide direct evidence for location of cation binding sites, since it is uncertain to what extent the defective cation exchange is secondary to effects on the conformational equilibrium between E_1 and E_2.

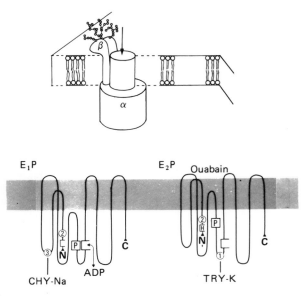

Fig. 11. *Models for independent function of αβ-unit and for arrangement of the α-subunit in E₁ and E₂ forms of Na,K-ATPase*

The encircled numbers mark the sites of primary tryptic cleavage in KCl (1 and 2) or NaCl (3) and chymotryptic cleavage in NaCl (3). In the E_1P conformation, bonds 2 and 3 are exposed to cleavage and bond 1 is protected. Transition to E_2P is accompanined by protonation of an ionizable group close to the N-terminus. In the E_2P conformation, bond 3 is protected and bond 1 is exposed to trypsin; the position of bond 2 is such that it is cleaved secondary to cleavage of bond 1 within the same α-subunit. It is proposed that transition from E_1P to E_2P is accompanined by movement by a part of the segment containing the aspartyl phosphate, from a relatively hydrophilic to a more hydrophobic environment. The segements between bonds 2 and 3 is proposed to engage in cation binding and formation of the pathway across the membrane. For further explanations see the text.

Relation of E_1–E_2 Transition to Cation Exchange

The notion that the rather large change in the structure of α-subunit is coupled to cation translocation was proposed in 1975–1978, when these changes in protein structure were first described. The E_1–E_2 transitions, as detected by tryptic digestion (Jørgensen *et al.*, 1978) or fluorescence changes (Karlish & Yates, 1978), were considered to be coupled to the actual movement of cation across the membrane, Na⁺ extrusion to E_1P–E_2P and K⁺ uptake to E_2–E_1 transition, as illustrated in the scheme in Fig. 12. The dephosphoenzyme exchanges K⁺ for Na⁺ at the cytoplasmic surface and the phosphoenzyme exchanges Na⁺ for K⁺ at the extracellular surface, but these exchange reactions involve only minor perturbations of protein structure. In addition to the evidence discussed above, these views gain strong support in a series of experiments on partial cation exchange reactions, cation binding and occlusion, and pre-steady state ion fluxes.

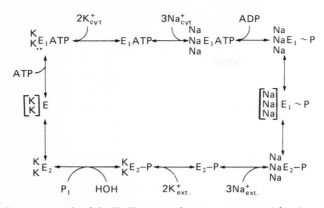

Fig. 12. E_1–E_2 reaction cycle of the Na,K-pump with ping-pong, sequential cation translocation

Combination with ATP and release of ADP and P_i take place on the cytoplasmic surface. Cation sites are exposed to the cytoplasmic surface in E_1 forms. Binding of Na^+ is a condition for transfer of γ-phosphate from ATP to the aspartyl group in α-subunit. Phosphorylation is followed by occlusion of 3 Na^+ per phosphate group. In the E_2 form Na^+ cation sites are exposed to the extracellular medium where Na^+ is released and binding of $K^+_{ext.}$ is coupled to hydrolysis of the phosphoenzyme. Dephosphorylation is followed by occlusion of 2 K^+. Uptake of K^+ is associated with E_2–E_1 transition and accelerated by ATP. Na^+ and K^+ within brackets denote occluded ions.

Cation binding and Occlusion in the Reaction Cycle

The relationships between ion movements, phosphoryl transfer and E_1–E_2 transition are shown in the E_1–E_2 scheme in Fig. 12, a modification of previous schemes (cf. Cantley, 1981; Jørgensen, 1982; Karlish & Stein, 1982; Glynn et al., 1984). In the normal clockwise operation, the major structural transitions between E_1 and E_2 forms of the α-subunit constitute the cation translocation steps, Na-extrusion involving the E_1P–E_2P transitions and K-uptake the E_2K–E_1ATP transition. Na^+ and K^+ are transported in sequence, Na^+ first and then K^+, in a ping-pong, sequential reaction.

In absence of other ligands Rb^+ or K^+ binding to cytoplasmic sites of Na,K-ATPase enters an occluded state within the molecule, where they are unable to exchange with ions in the medium (Glynn & Richards, 1982). Also in the course of the working cycle the Na,K-pump passes states in which Na^+ or K^+ is occluded. Rb^+ catalysing hydrolysis of the phosphoenzyme enters the occluded state from the extracellular surface as originally described by Post. Release of the occluded Rb^+ is regulated by ATP and P_i in the sense that they control the rate and the route by which Rb^+ leaves (Karlish & Stein, 1982). The rate in the forward direction of the cycle is increased by ATP and reversal is increased by P_i. This mechanism ensures tight coupling of cation flux to ATP hydrolysis and the rate and direction of pumping is determined by the relative concentrations of the ligands ATP and P_i.

The E_1P form of the protein can occlude 3 Na^+ ions per phosphorylation site (Glynn et al., 1984). Demonstration of the occluded form requires that the α-subunit is stabilized in the E_1P form by cleavage of bond 3 with chymotrypsin in NaCl medium (Jørgensen et al., 1982). The rate constants involved in occlusion and release of Na^+ from the occluded state are consistent with the view that occlusion of intracellular Na^+ in an $(3Na)E_1P$ form of phosphoenzyme and its release to the exterior after transition to the $3NaE_2P$ form are responsible for translocation of Na^+ through the pump, as shown in Fig. 12.

As alternatives to the E_1–E_2 scheme in Fig. 12, Skou (1975) and Plesner *et al.* (1981) have proposed reaction cycles where E_1P and E_2P are not intermediates in the Na,K-dependent reactions. They question whether transition rates of E_1P to E_2P are fast enough to account for release of P_i during turnover. The alternatives include models in which Na,K-exchange is simultaneous, in contrast to the ping-pong, sequential reaction. In contrast to this it is concluded from recent data that phosphorylation of the α-subunit is a necessary condition for hydrolysis of ATP by Na,K-ATPase, both in the presence of Na^+ alone and with Na^+ plus K^+ in the medium (Hobbs *et al.*, 1983; Fortes & Lee, 1984). In this preparation from eel electroplax the E_1P–E_2P transition rates at 500 s^{-1} are among the fastest reactions of the catalytic cycle (Hobbs *et al.*, 1983).

Pre-steady state fluxes. Another important test of the E_1–E_2 scheme and its coupling to ping-pong ion translocation is the measurement of pre-steady state fluxes of Na^+ and K^+ in single turnover cycles. Forbush (1984) loaded right-side-out vesicles with caged ATP and initiated a single turnover of the pump with a flash of light. Karlish & Kaplan (1985) measured ATP dependent uptake of ^{22}Na in reconstituted vesicles at 0 °C. Both experiments show an initial burst of ^{22}Na uptake, reflecting transfer in the first turnover of the Na,K-pump which is insensitive to K^+ at the extracellular surface. The time of appearance of the burst shows that Na^+ transport involves more steps than formation of E_1P and that Na^+ efflux is an early event in the pump cycle relative to K^+ influx. Thus, the studies of the first turnover of the pump show that the ions are transported in sequence, Na^+ first and then K^+, consistent with the ping pong, sequential mechanism of Na,K-transport in coupling with E_1–E_2 transitions as proposed above.

Work in the author's laboratory is supported by the Danish Medical Research Council and Novo's Foundation.

References

Avruch, J. & Fairbanks, G. (1972) *Proc. Natl. Acad. Sci. U.S.A.* **69**, 1216–1220
Brotherus, J. B., Møller, J. V. & Jørgensen, P. L. (1981) *Biochem. Biophys. Res. Commun.* **100**, 146–154
Brotherus, J. B., Jacobsen, L. & Jørgensen, P. L. (1983) *Biochim. Biophys. Acta* **731**, 290–293
Cantley, L. C. (1981) *Current Topics in Bioenergetics* **11**, 201–237
Castro, J. & Farley, R. A. (1979) *J. Biol. Chem.* **254**, 2221–2228
Deguchi, N., Jørgensen, P. L. & Maunsbach, A. B. (1977) *J. Cell Biol.* **75**, 619–634
Dux, L. & Martonosi, A. (1983) *J. Biol. Chem.* **258**, 2599–2603
Fischer, T. H. (1983) *Biochem. J.* **211**, 771–774
Forbush, B. (1984) *Anal. Biochem.* **140**, 495–505
Fortes, P. A. G. & Lee, J. A. (1984) *J. Biol. Chem.* **259**, 11176–11179
Glynn, I. M. & Karlish, S. J. D. (1975) *Annu. Rev. Physiol.* **37**, 13–53
Glynn, I. M. & Richards, D. E. (1982). *J. Physiol.* **330**, 17–43
Glynn, I. M., Hara, Y. & Richards, D. E. (1984) *J. Physiol.* In press
Glynn, I. M., Richards, E. E. & Hara, Y. (1985) in *The Sodium Pump, 4th International Conference on Na,K-ATPase*, (Glynn, I. M. & Ellory, J. C., eds.), Company of Biologists, Cambridge, in the press

Gresalfi, T. J. & Wallace, B. A. (1984) *J. Biol. Chem.* **259**, 2622–2628
Hayashi, Y., Takagi, T., Kaezawa, S. & Matsui, H. (1983) *Biochim. Biophys. Acta* **748**, 153–167
Hebert, H., Jørgensen, P. L., Skriver, E. & Maunsbach, A. B. (1982) *Biochim. Biophys. Acta* **689**, 571–574
Hobbs, A. S., Albers, R. W. & Froehlich, J. P. (1983) *J. Biol. Chem.* **258**, 8163–8168
Jensen, J. & Ottolenghi, P. (1983) *Biochim. Biophys. Acta* **731**, 282–289
Jørgensen, P. L. (1974) *Methods Enzymol.* **32**, 277–290
Jørgensen, P. L. (1975) *Biochim. Biophys. Acta* **401**, 399–415
Jørgensen, P. L. (1982) *Biochim. Biophys. Acta* **694**, 27–68
Jørgensen, P. L. & Karlish, S. J. D. (1980) *Biochim. Biophys. Acta* **597**, 305–317
Jørgensen, P. L., Klodos, I. & Petersen, J. (1978) *Biochim. Biophys. Acta* **507**, 8–16
Jørgensen, P. L., Skriver, E., Hebert, H. & Maunsbach, A. B. (1982) *Ann. N.Y. Acad. Sci.* **402**, 203–219
Karlish, S. J. D. & Kaplan, J. H. (1985) in *The Sodium Pump, 4th International Conference on Na,K-ATPase* (Glynn, I. M. & Ellory, J. C., eds.), Company of Biologists, Cambridge, in the press
Karlish, S. J. D. & Stein, W. D. (1982) *J. Physiol.* **328**, 317–331
Karlish, S. J. D. & Yates, D. W. (1978) *Biochim. Biophys. Acta* **527**, 115–130
Karlish, S. J. D., Jørgensen, P. L. & Gitler, C. (1977) *Nature (London)* **269**, 715–717
Maunsbach, A. B. & Jørgensen, P. L. (1974) *Proc. Eight Int. Congr. Elect. Microsc. Australia* **2**, 214–215
Michel, H., Oesterhelt, D. & Henderson, R. (1980) *Proc. Natl. Acad. Sci. U.S.A.* **77**, 338–342
Moczydlowski, E. G. & Fortes, P. A. G. (1981) *J. Biol. Chem.* **256**, 2346–2356
Mohraz, M. & Smith, P. R. (1984) *J. Cell Biol.* **98**, 1836–1864
Plesner, I. W., Plesner, L., Nørby, J. G. & Klodos, I. (1981) *Biochim. Biophys. Acta* **83**, 346–356
Sharkey, R. G. (1983) *Biochim. Biophys. Acta* **730**, 327–341
Skou, J. C. (1975) *Quart. Rev. Biophys.* **7**, 401–434
Skriver, E., Maunsbach, A. B. & Jørgensen, P. L. (1980) *J. Cell Biol.* **86**, 746–754
Skriver, E., Maunsbach, A. B. & Jørgensen, P. L. (1981) *FEBS Lett.* **131**, 219–222
Skriver, E., Maunsbach, A. B. & Jørgensen, P. L. (1983) *Current Topics in Membranes and Transport* **19**, 119–122
Zampighi, G., Kyte, J. & Freytag, W. (1984) *J. Cell Biol.* **98**, 1851–1864

Biochem. Soc. Symp. **50**, 81–95
Printed in Great Britain

Calcium Channel Modulators and Calcium Channels

R. TOWART* and M. SCHRAMM**

*Miles Laboratories Limited, Stoke Court, Stoke Poges, Bucks., U.K. and **Institute
of Pharmacology, Bayer A.G., Postfach 101709, D5600 Wuppertal 1, West Germany*

Synopsis

In recent years calcium has become recognized as an important 'second messenger', in that an increase in free intracellular calcium ion concentration is involved in many aspects of cellular activation. In excitable cells such as smooth muscle or cardiac tissue an influx of extracellular calcium ions through voltage sensitive calcium channels plays a major role in increasing the cytoplasmic free calcium concentration. It is now known that the activity of these calcium channels may be inhibited or stimulated by a range of ions, toxins and drugs. This article outlines the biochemistry and pharmacology of these 'calcium modulators', and discusses their present and future role as both biological tools, and as potent and selective drugs.

Introduction

It has been known for over 100 years that calcium ions are necessary for certain types of cellular activity, and there is now incontrovertible evidence that calcium functions as a vital intracellular regulator (for review see Borle, 1981; Campbell, 1983; Rasmussen & Barrett, 1984). To examine the role of intracellular calcium in this context, at least two additional topics must be addressed, namely, the mechanisms for calcium sequestration and efflux, and the mechanisms for calcium influx. It may seem paradoxical to mention efflux of calcium before influx, but it must be remembered that eukaryotic cells have an intracellular free calcium concentration of between 30–300 nM, whereas their extracellular environment has an ionized calcium concentration of approximately 1 mM (see Fig. 1); there is thus a considerable concentration gradient driving calcium into the cell, and active mechanisms are employed to remove calcium from the cytoplasm, and eventually to excrete calcium from the cell. These mechanisms are vital for the viability of the cell, and are discussed fully in several recent reviews (e.g. Penniston, 1983).

Controlled pathways for calcium influx are probably not directly vital for the viability of the cell, but are vital for the survival of the organism. A range of nervous and chemical messages integrate the behaviour of individual cells as part of the functioning organism. The neurotransmitters, circulating and local hormones and tissue factors which constitute these 'first messengers' frequently affect intracellular events by means of 'second messengers', such as calcium ions

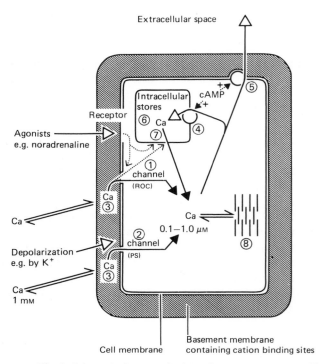

Fig. 1. *Schematic diagram of calcium homoeostasis in excitable cells*

Extracellular free calcium is of the order of 1 mM, whereas the cytoplasmic free calcium concentrations range from 0.01–1 μM, depending on the state of activation. Excitation by means of agonists ('first messengers') often allows release of intracellularly stored calcium (6), possibly by changes in phospholipid 'second messengers' (e.g. inositol triphosphate; see the text). In addition, stimulation may allow calcium influx via 'receptor-operated channels' (1). Electrical activity or depolarization of the cell allows activation of voltage or 'potential sensitive' calcium channels (2). Calcium homoeostasis is maintained in the short term by uptake of calcium into intracellular stores such as sarcoplasmic reticulum or mitochondria (4), and in the long term by extrusion via Ca^{2+}-ATPase pumps (5) or Na/Ca antiporters. cAMP, Cyclic AMP

or cyclic nucleotides. The biochemistry and pharmacology of the influx of calcium ions as second messengers is the subject of this review. It will attempt to integrate some of the findings of the last decade, to explain the biological importance of calcium as a 'second messenger', and to discuss briefly some of the unanswered questions about calcium ion movements.

The biological importance of calcium in unquestioned; apart from its structural role, in the form of bones and teeth in higher forms of life, it plays a vital intracellular role. The evolutionary significance of calcium can be judged from the fact that all known eukaryotic cells maintain, by active means, a very low intracellular concentration of free calcium ions, and that the intracellular calcium binding proteins, of which calmodulin is the best known, have a very highly conserved amino acid structure (see Means *et al.*, 1982). This, and the fact that overload of free calcium ions is cytotoxic (Farber, 1981), indicates the necessity for life of efficient calcium homoeostasis.

A simplified diagram of calcium homoeostasis is shown in Fig. 1. Its key points are that the cell membrane is extremely impermeable to calcium ions, which typically are present extracellularly in a concentration of around 1 mM. Cytoplasm free calcium is maintained at around 0.01–0.1 μM by energy-dependent mechanisms of intracellular sequestration and extrusion to the outside of the cell. These mechanisms, fascinating in their own right, are outwith the subject of this review. The intracellular roles of calcium ions have been extensively reviewed. The excellent monograph of Campbell (1983) describes the field in detail.

Of major importance to the physiologist and the biochemist is the informational role of the free cytoplasmic calcium concentration. Thus many electrical stimuli, transmitters, hormones or growth factors, which act on the cell membrane, cause an increase in intracellular free calcium concentration. Much of this increase is caused by redistribution of calcium already stored in the cell (see Fig. 1) and Berridge among others has reviewed the role of inositol triphosphate as a messenger for the release of intracellular calcium ions (Berridge, 1984). However there are many tissues where at least some of the calcium ions enter the cell from outside; cardiac tissue, smooth muscle and many endocrine and exocrine glands are dependent on calcium influx. The calcium influx 'channels', and their modulators, are discussed in detail below.

Historical Overview

Campbell (1983) has described in depth the increasing awareness of the biological role of intracellular calcium ions over the last century. The concept of calcium influx as a biological messenger is, however, much more recent, and to some extent followed the seminal work of biophysicists like Hodgkin and Huxley on the sodium channel. Cole (1979) has related how the availability of giant squid axons allowed the elucidation of the fundamental mechanisms of the voltage-sensitive sodium channel which is responsible for the propagation of the nerve impulse. The concept of an ion selective membrane channel with distinct activation and inactivation kinetics was put together by these workers (Hodgkin & Huxley, 1952) and remains, with only small alterations, viable (see Lazdunski et al., in this Symposium).

Removal of extracellular sodium ions, of course, prevents the normally observed flow of current through the channels. Fatt & Katz (1953) observed that sodium removal did not completely inhibit responses in crab muscle, and noted that calcium ions were the most important candidates for charge carriers, as the residual response could be increased or decreased by altering the calcium concentration appropriately. They suggested that influx of calcium was responsible for the transport of charge.

By the 1960s it had been realized that the action potential of cardiac tissue was dependent not only on sodium ions, but also upon calcium ions (see Reuter, 1973). The very much slower kinetics of the calcium movements suggested the term 'slow channels' to describe the movement of calcium into the heart. Work in the 1970s consolidated the concept of the influx of calcium ions into cardiac tissue, smooth muscle, and various exocrine and endocrine glands as a second messenger to couple excitation and contraction or secretion.

Pharmacological work in the 1960s by Fleckenstein with cardiac tissue and

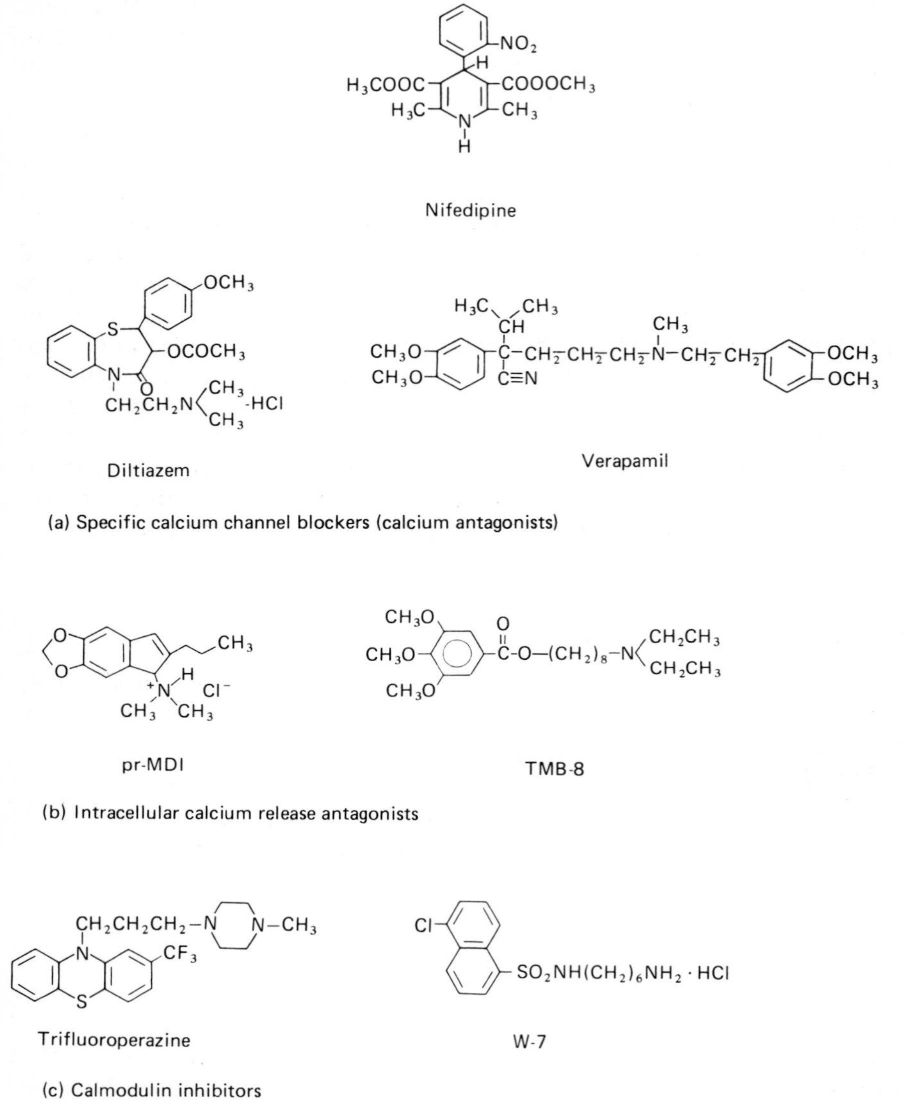

Nifedipine

Diltiazem

Verapamil

(a) Specific calcium channel blockers (calcium antagonists)

pr-MDI

TMB-8

(b) Intracellular calcium release antagonists

Trifluoroperazine

W-7

(c) Calmodulin inhibitors

Fig. 2. *Chemical structures of some compounds which affect calcium homoeostasis or excitation–event coupling*

smooth muscle showed that some metal ions, and some drugs then in development such as verapamil or nifedipine, could selectively inhibit the movement of calcium ions into the cell, and the term 'calcium antagonist' was introduced (see Fleckenstein, 1983). The 1970s saw the introduction into medicine of nifedipine and several other powerful but selective calcium antagonist drugs, and their widespread clinical use as antianginal, antihypertensive and antiarrhythmic agents. Some of these compounds are illustrated in Fig. 2(*a*). Also in the 1970s came the realization that there were several different sorts of calcium channel, characterized by Bolton (1979) as 'voltage sensitive' and 'receptor operated' calcium channels. This is almost certainly an oversimplification, and the topic will be discussed in detail below. However, a characteristic of many voltage sensitive calcium channels is their susceptibility to blockage by the organic calcium antagonists shown in Fig. 2(*a*) (Fleckenstein, 1977).

A major advance in the understanding of calcium channels came in the late 1970s when more potent and selective analogues of the dihydropyridine (DHP) derivative nifedipine became available. The extreme potency of these drugs, and their stereoselectivity, suggested that they could bind specifically to some site in or near the calcium channel (Towart *et al.*, 1981), and the availability in the early 1980s of high specific activity radioactively labelled DHP derivatives has prompted a detailed investigation of the potential binding sites (e.g. Belleman *et al.*, 1981; Glossmann *et al.*, 1982; Triggle & Janis, 1984*a*, *b*). Study of the DHP derivatives also led to the finding that the calcium channels could be activated (Schramm *et al.*, 1983) as well as blocked by these compounds, leading to the concept of calcium channel modulators, and the suspicion (as yet unproven) that these drugs might be mimicking some endogenous modulator of calcium channel activity.

Movement of Calcium Ions Across Membranes

The concept of calcium ion influx as a universal second messenger immediately raises the question: how do hydrophilic calcium ions cross the hydrophobic lipid bilayer of the cell membrane?

Possible methods include:

> passive leak through membrane imperfections
> movement through ion channels specific for Ca^{2+}
> ionophore systems
> vesiculation and endocytosis systems
> cell to cell movement of calcium
> antiporter systems
> pumping systems

The two last-named systems, especially the Na/Ca antiporter, and the Ca^{2+}-ATPase calcium pump, are mainly involved in moving calcium ions out of the cell, against the massive electrochemical gradient, and will be discussed in the next chapter. However, discussion of the possible contributions of leaks, channels, ionophores and endocytosis systems is more than just philosophical; the actual mechanism of calcium ion influx will affect the amounts of ion transported, the time scale and kinetics of the response, the electrogenicity or otherwise of the signal, and the susceptibility of the influx to changes in resting potential, temperature or energy supplies. More importantly, different mechanisms of calcium influx will be differentially susceptible to modulators of influx such as neurotransmitters, hormones, mediator substances, other ions, toxins or drugs. Again space does not permit detailed discussion of some of these points, but relevant reviews will be cited for the interested reader.

Passive Leaks

Careful measurement of calcium influx by using labelled calcium and lanthanum ions (to displace membrane-bound calcium ions) indicate a resting calcium
influx (see Flaim *et al.*, 1984). This resting calcium influx is unaffected by calcium
channel blockers, and is considered to be due to calcium 'leaking' into the cell,
down the steep concentration gradient (see Fig. 1), via membrane imperfections
or via channels incompletely specific for other ions. In normal tissue these
amounts of calcium are presumably quickly sequestered and/or extruded, as
activation does not occur under resting conditions. However, in damaged or
diseased tissues this passive leak of calcium could be of importance; calcium
overload is a consequence of cardiac ischaemia (see Fleckenstein, 1983) and
excess calcium influx may contribute to vascular hyper-reactivity in hypertension
(Bühler *et al.*, 1984).

Little is known about the modulation (if any) of calcium leak, and no drugs
are known which inhibit it.

Influx Channels

Ion channels have been of great interest to biologists for the past 40 years,
and channels selective for sodium, potassium, calcium and chloride ions have
been described. The sodium channel(s) have been the ones most investigated:
they have been solubilized and isolated, and their amino acid structure has been
elucidated; they can be reconstituted as working channels into artificial
membranes. Less is known about the channels which are selective for calcium
ions, partly because of their heterogeneity, but the success of some of the
techniques described below makes it likely that at least one sort of calcium
channel will be fully characterized in the near future (see Lazdunski *et al.*, in
this Symposium).

Basically an ion channel seems to be a proteinaceous macromolecule with a
hydrophilic core through which (probably hydrated) ions may pass. The
macromolecule contains at least one 'selectivity gate', which limits the conductivity to ions of appropriate charge and radius, and activation and inactivation gates which allow it to open, but for a limited time. The activation process
is often voltage-dependent, but may also or instead be ligand controlled. Many
channels appear to be chemically modulated, by phosphorylation (Reuter, 1983),
or other influences. Membrane channels selective for calcium ions have been
demonstrated in a large number of eukaryotic cells, but in contrast to the sodium
channel, whose properties are remarkably similar from tissue to tissue and from
species to species, the calcium channels show a marked heterogeneity, even
within one species. This probably reflects an earlier evolutionary emergence of
calcium channels, and a much more varied biophysical role. Many excellent
reviews of calcium channels have appeared recently, including those of Reuter
(1973), Hagiwara & Byerly (1981), Tsien (1983) and Reuter (1983).

Ionophore Mechanisms

In contrast to channels, ionophores are molecules which transport ions across membranes by virtue of their often hydrophilic interiors and lipophilic exteriors. The distinction between channels and ionophores has been amusingly summarized by Rothstein (1980). Although natural calcium ionophores form part of the calcium ion accumulation process (MacLennan et al., 1980), there is no incontrovertible evidence for participation of an endogenous ionophore in calcium influx mechanisms. The changes in membrane phospholipid turnover on ligand–receptor interaction has, however, given rise to much speculation as to the possible role of phospholipids, especially phosphatidic acid, as ionophores involved in receptor activated calcium influx (e.g. Putney et al., 1980).

Vesiculation or Endocytotic Mechanisms

Theoretically calcium could be 'ingested' by some pinocytotic vesicle mechanism, whereby membrane invaginations enclose and engulf calcium-containing extracellular fluid. No examples are known to the authors where any such mechanism transports 'messenger' calcium into the cell, but a pinocytotic transcellular calcium transport may be involved in the absorption of calcium by intestine or kidney tubules (see Campbell, 1983).

Cell-to-Cell Calcium Movement

Comparatively little is known about movement of calcium ions between cells. Low resistance channels, mainly bidirectional, are present in some tissues, and are responsible for intercellular movement of ions and metabolites, as well as for the rapid spread of electrical impulses in the myocardium (e.g. Lowenstein et al., 1978). Presumably small amounts of calcium ions may flow through these channels if concentration gradients exist. However, these channels are extremely sensitive to intracellular free calcium ion concentration, and close when it rises to $0.1–10\ \mu\mathrm{M}$. This is presumably a protective mechanism which dissociates damaged or dying cells (in which free calcium concentrations rapidly increase) from the healthy ones.

There are therefore several different ways in which calcium may enter cells, either in the resting state, or during stimulation. As mentioned earlier, most is known about voltage sensitive calcium channels, partly because they are increasingly accessible by electrophysiological techniques, and partly because of the range of drugs now available to modulate their activity.

Nomenclature: Calcium 'Channels', 'Gates' and Calcium 'Antagonists'

The nomenclature of the calcium influx modulators described in this review has excited fierce controversy in recent years. The term 'calcium antagonist' was coined independently but more or less simultaneously by Fleckenstein and Godfraind (Fleckenstein et al., 1969; Godfraind & Kaba, 1969; see also Fleckenstein, 1983) to denote ions ('inorganic' calcium antagonists) or drugs ('organic' calcium antagonists, see Fig. 2*a*) that interfere with the influx of calcium ions into excitable cells. Many authors have argued against this usage, preferring terms such as 'calcium channel blocker', 'slow channel inhibitor' etc. In addition some authors (e.g. Rahwan *et al.*, 1981) have extended the term calcium antagonist to include substances such as aminoindines and trimethoxy-benzoate derivatives (see Fig. 2*b*), which interfere with the release of calcium ions from intracellular organelles. Calmodulin inhibitors (see Fig. 2*c*) also affect calcium-dependent events, without affecting calcium influx directly. However, it is our feeling that the term 'calcium antagonists' has become universally understood by researchers and clinicians alike, and despite some sympathy for those who argue to the contrary, we here use the term to refer to the drugs like verapamil and nifedipine which selectively inhibit the influx of calcium ions. Similarly we use the term 'calcium agonist' to refer to drugs like BAY k8644 which selectively increase the influx of calcium ions. We feel that 'agonist' and 'antagonist' are justified by the recently discovered existence of specific binding sites for these compounds (see below).

Similarly we would prefer to keep an open mind, for the present, on the use of the word 'channel'. Recent electrophysiological work, especially the use of patch–clamp techniques, has shown beyond reasonable doubt that many of the phenomena concerning calcium influx and its modulation can be explained by the existence of selective, quantally conducting voltage-sensitive channels in the cell membrane of excitable cells (e.g. Reuter, 1983; Lee & Tsien, 1983). Despite this the so-called 'receptor operated channels', which can be distinguished pharmacologically as pathways of calcium ion influx in various tissues (see below), have not been positively identified as channels after the Hodgkin–Huxley model. Much work has been carried out to examine the role of phosphatidic acid or phosphatidylinositol (PI) turnover in the receptor-induced activation of calcium 'gating' (Putney *et al.*, 1980; Michell, 1982) and it is not inconceivable that some iontophoretic mechanism is involved in the 'receptor-operated' calcium influx.

Measurement of Calcium Influx

The role of calcium influx in cell function can only be examined if reliable methods exist for its measurement. A variety of methods, both direct and indirect, now exist. All have advantages, disadvantages and limitations. Some of the factors which contribute to the difficulties of measurement have been alluded to above: the passive 'leak' of calcium across cell membranes due to the great electrochemical gradient, the very low (30–300 nM) cytoplasmic concentration, despite the large amount of calcium stored in intracellular organelles, which bring the total intracellular concentration to around 1–10 mM (e.g. Campbell, 1983), and the efficient calcium efflux mechanisms. Another problem is that, in many cells, the glycocalyx binds large amounts of calcium ions, making conventional radioactive tracer studies difficult. The problem of interpretation of results is further complicated by the different contribution of intra- and extra-cellular stores of calcium (skeletal muscle contraction is not affected actively by removal of extracellular calcium). Other tissues (e.g. heart) utilize a small influx of extracellular calcium ('activator' calcium) to release much larger amounts of calcium already stored in sarcoplasmic reticulum (Saida & Van Breemen, 1983).

Calcium influx may be investigated by electrophysiological means, especially by the 'patch–clamp' method (see Reuter, 1983; Lee & Tsien, 1983). Radio-labelled calcium may be used to measure influx of calcium, but only when membrane-bound calcium ions have been removed by the use of lanthanum (Van Breemen et al., 1977). In addition a range of fluorescent or luminescent indicators is becoming available which measure intracellular free calcium concentration, and when appropriate precautions are taken, may be used to investigate calcium influx (see Dormer et al., 1985).

Different Calcium Channels

In addition to the passive 'leak' of calcium into cells, which may be inhibited by lanthanum, but is unaffected by organic calcium antagonists (Ress & Flaim, 1982), there is evidence for at least two controlled means of calcium entry into excitable cells (Bolton, 1979; Meisheri et al., 1981). Bolton has called these 'voltage-operated channels' (VOC) and 'receptor-operated channels' (ROC). The existence and importance of VOC is not in doubt. Patch–clamp studies have demonstrated the existence of VOC in heart cell membranes (e.g. Reuter, 1983) and they are affected by calcium antagonists and agonists in the appropriate ways. It is more difficult to show their existence directly in smooth muscle, but ligand binding studies with tritiated DHP derivatives and the effects of calcium antagonists and calcium agonists on contraction and electrical activity in smooth muscle give evidence that similar or identical structures are responsible for the increased calcium influx into depolarized smooth muscle.

The concept of 'receptor-operated calcium channels' presents greater problems. Although the release of sequestered intracellular calcium ions is a feature of the neurotransmitter- or autocoid-induced contraction of many tissues, we are here interested in effects on influx, not intracellular distribution. Exton (1983) and Bolton (1985) have recently summarized the evidence for receptor-operated calcium influx, and proposed mechanisms whereby this may occur. Meisheri et al. (1981) have described effects of amrinone as an inhibitor of receptor-operated calcium channels in vascular smooth muscle, but a selective inhibitor has not been foumd, and to our knowledge there exists no electrophysiological findings that unequivocally demonstrate the existence of receptor-operated calcium channels.

Calcium Channel Modulators

The activity of the voltage-operated calcium channel may be modulated in several ways. In cardiac muscle, but probably not in other systems, cyclic AMP-dependent protein kinase phosphorylates the channels, and increases the probability of activation (see Reuter, 1983). This appears to be the basis for the positive inotropic effects of catecholamines or phosphodiesterase inhibitors.

The channels may be blocked by a range of di- or tri-valent ions, such as cobalt,

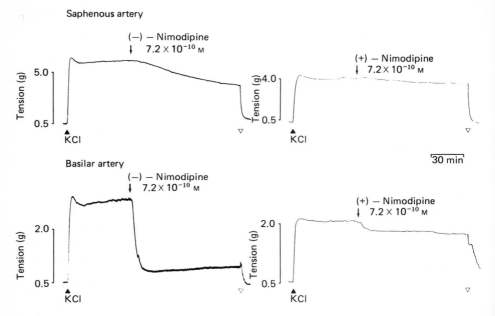

Fig. 3. *Original recording of the effects of the optical isomers of the nifedipine analogue nimodipine on two isolated blood vessels*

Note that the (−)-isomer is very much more potent than the (+)-isomer, and that the cerebral artery (basilar, below) is more affected than the peripheral artery (saphenous artery, above). (From Towart *et al.*, 1982, with permission.)

nickel or lanthanum, which have ionic radii close to that of the calcium ion. These are referred to as 'inorganic' calcium antagonists, but their main interest is as tools to investigate the physiology of the calcium channel (Kohlhardt *et al.*, 1973). Certain toxins affect the calcium channel, especially maitotoxin (Freedman *et al.*, 1984), and a range of diverse drugs, which, in addition to their main mechanism of action, inhibit (Triggle, 1982) or stimulate (El-Fakahany *et al.*, 1984) calcium ion influx through voltage-sensitive channels.

The major interest in recent years has been the development of 'specific' calcium antagonists (and agonists, see below). As outlined above, these compounds are not only highly effective drugs in the treatment of many cardiovascular disorders (see, e.g., Stone & Antman, 1983; Towart, 1985), but also potent and specific blockers of the calcium channel. An original tracing is shown in Fig. 3, which illustrates some of the properties of these drugs. The prototype DHP, nifedipine, is symmetrical (see Fig. 1) but some of its analogues are unsymmetrically substituted, and have been prepared as optically pure stereoisomers. In addition to the extreme potency of these compounds (subnanomolar amounts are able completely to reverse contractions), it can be seen that the (−) isomer of nimodipine is more potent than the (+) isomer (Towart *et al.*, 1982). Also demonstrated in Fig. 3 is an example of 'tissue selectivity'. Here it can be seen that nimodipine is more potent on cerebral vessels than on peripheral vessels. In general the potency of calcium antagonists can

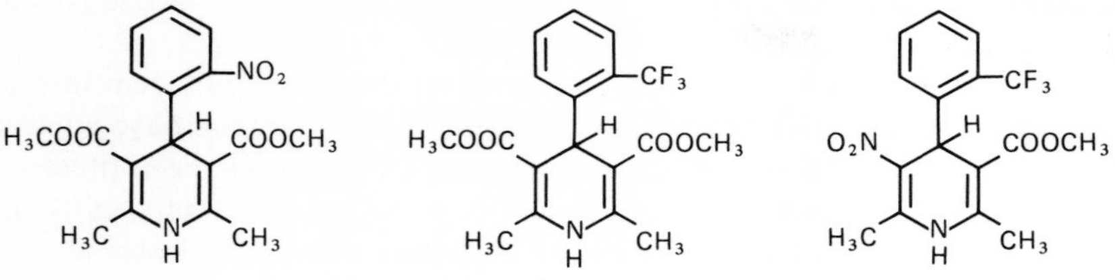

Fig. 4. *Structures of nifedipine (left) and its ortho-CF₃ analogue (middle), two calcium antagonists, and BAY k8644 (right), a prototype calcium agonist*

be expressed as coronary vessels $\doteqdot$ cerebral vessels $>$ peripheral vessels $\doteqdot$ smooth muscle $>$ cardiac muscle $>$ neurones $\doteqdot$ skeletal muscle. The reasons for this selectivity of action are not entirely clear, but may be partly explained by the different sources of calcium ions required for activation of different tissues (Fleckenstein, 1977) and also by the effects of membrane potential on the binding affinity of the calcium antagonists to their receptors in the calcium channel (Bean, 1984). It is this selectivity of action which provides the calcium antagonists with their usefulness as cardiovascular drugs: selective effects on the coronary circulation and on the peripheral resistance vessels explain their action as antianginal agents, for example. In addition it is becoming apparent that different calcium antagonists possess different profiles of tissue selectivity. Thus drugs are presently being developed which may preferentially affect cerebral vessels (e.g. nimodipine, Scriabine *et al.*, 1985) or peripheral resistance vessels (e.g. nitrendipine, Scriabine *et al.*, 1984).

A major development in the study of calcium channel modulators has been the discovery that slight alterations to the molecular structure of the classical calcium antagonist nifedipine produces compounds with opposite effects (see Fig. 4). Thus BAY k8644 and its analogues (Franckowiak *et al.*, 1983) increase calcium influx through voltage-sensitive channels, and have positive inotropic and hypertensive effects (Schramm *et al.*, 1983). These drugs bind to the DHP receptor in the calcium channel (see below), and provide the first evidence that one molecular structure may stimulate or inhibit calcium channel activity.

Ligand Binding Studies

As outlined above, the extreme potency and stereoselectivity of the DHP calcium antagonists had suggested that specific binding sites for calcium antagonists existed in or near the voltage-sensitive calcium channel. A variety of groups (see Belleman *et al.*, 1981; Glossmann *et al.*, 1982; Triggle & Janis, 1984*a*; see also Lazdunski, in this Symposium) quickly established that this was in fact the case, and that specific, high affinity sites for DHP derivatives were present in tissues sensitive to the drugs. Surprisingly there were also large numbers of these sites in tissues such as brain or skeletal muscle, which are not normally affected by calcium antagonists (see Janis & Triggle, 1984, for discussion). Indeed these sites may also be detected in non-excitable cells (e.g. liver; Gopalakrishnan & Triggle,

1984) and even in plant membranes (Hetherington & Trewavas, 1984), suggesting a very wide biological distribution of these sites.

The non-DHP calcium antagonists such as verapamil or diltiazem interact allosterically with the DHP binding site, suggesting that they too have adjacent but distinct binding sites in the calcium channel. Evidence has been presented for a single non-DHP site (Murphy *et al.*, 1983), but most workers show that verapamil and diltiazem bind to distinct and separate sites (e.g. Glossman *et al.*, 1982). The properties and biochemistry of calcium antagonist binding has been extensively reviewed (e.g. Glossmann *et al.*, 1982; Triggle & Janis, 1984*a*, *b*; Lazdunski, this Symposium). Most importantly, these studies are leading to solubilization, isolation and purification of the proteins involved in binding and channel function. This should lead ultimately to elucidation of the amino acid structure of these proteins, and may give information about channel subtypes and tissue selectivity.

Structure–Activity Relationships of Calcium Modulators

The structure–activity relations (SAR) of the analogues of the three classical calcium antagonists have been examined in detail: verapamil (Manhold *et al.*, 1978; Rodenkirchen *et al.*, 1981); diltiazem (Sato *et al.*, 1971); nifedipine (Loev *et al.*, 1974; Bossert *et al.*, 1979; Rodenkirchen *et al.*, 1979; Triggle & Swamy, 1983; Meyer *et al.*, 1983). In addition some attempts have been made to investigate the structure of the acceptor site of the verapamil molecule (Hoeltje, 1982).

However predictively the present knowledge of SAR may be optimizing the general potency of calcium antagonists, it has so far yielded little information about the molecular structure of their binding sites, or of the potential tissue selectivity of the compounds. Indeed even ligand binding studies give little if any useful information about tissue selectivity (Gould *et al.*, 1984; Bristow *et al.*, 1984). A correlation between antagonist activity and dihydropyridine ring puckering as measured by X-ray crystallography has been found (Langs & Triggle, 1985), but the predictive capacity of this approach is obviously limited.

In the same way little is known about the distinction between antagonists and agonists. Although there are many different DHP substructures which function as calcium agonists, there is no evidence known to us that analogues of verapamil, diltiazem or other calcium antagonistic structures can function as agonists. This, of course, suggests that the DHP binding site, as opposed to the adjacent verapamil and/or diltiazem sites, is the most important for calcium channel modulation, and is consistent with the greater potency of DHP derivatives.

The potent prototype calcium agonist BAY k8644 differs from calcium antagonists in only one substitution of the dihydropyridine ring, the 3 position (see Fig. 4). Many more DHP substructures are known with agonistic properties (Franckowiak *et al.*, 1983; Truog *et al.*, 1985) but predictive generalizations of the calcium agonist structure have not yet appeared. Crystallographic findings by Triggle's group have implicated lack of ring puckering

Table 1. *Possible new applications for calcium antagonistic drugs**

(1) **Relaxation of smooth muscle**	(2) **Prevention of calcium overload etc.**
Cerebral vasospasm	Cardioprotective effects
Cerebral insufficiency	Hypertrophic cardiomyopathy
Migraine	Cardiac preservation
Peripheral vascular disease	Cerebro-protective actions
Asthma	Prevention of atherosclerosis
Oesophageal and intestinal spasm	Hepato-protective actions
Unstable bladder	Prevention of arterial calcinosis
Uterine hyper-reactivity	and diabetic complications
	Antimetastic effects
	Insulinoma

* Adapted from Towart (1985), where relevant references are given.

as a possibly important factor with BAY k8644 (see Langs & Triggle, 1985), but again predictive capacity is limited.

Clinical Pharmacology of Calcium Modulators

The detailed clinical pharmacology of the calcium channel modulators is outwith the scope of this review, but present and future indications are summarized in Table 1. The interested reader is referred to any of the numerous reviews of the past few years for detailed accounts and bibliography (e.g. Stone & Antman, 1983; Towart, 1985). The different clinical spectra of the 'classical' calcium antagonists verapamil, nifedipine and diltiazem prompted the search for more 'indication specific' drugs in the late 1970s, and several calcium antagonists have been developed which show predilective effects on particular tissues. Thus nitrendipine is especially potent as an antihypertensive agent (Scriabine *et al.*, 1984), and nimodipine is proving successful in combatting a range of cerebrovascular diseases from cerebral vascular spasm (Allen *et al.*, 1983) to migraine (Meyer & Hardenberg, 1983). This trend will undoubtedly continue as more is learned both of the structure–activity relationships of the calcium antagonists themselves, and of the pathophysiology of the diseases in question.

The specific calcium agonists such as BAY k8644 have been discovered only recently, and although they are increasingly being employed as pharmacological tools to investigate the working of the calcium channel, have not yet been developed for human medicine. Although it is a potent hypertensive and positive inotropic agent, BAY k8644 is also a potent constrictor of coronary arteries, and as such is unsuited for human use. However, the discovery of this novel class of pharmacological agents will undoubtedly lead to the development of newer, more selective agents which can be applied to diseases such as hypotension or cardiac insufficiency without the undesirable side effects of the present prototype drugs.

Conclusions and Perspectives

The underlying theme of this review is that the vital intercellular role of calcium is regulated to a large extent by control of calcium influx, especially in excitable and secretory cells. A range of drugs and toxins is being used to investigate the biology of the calcium influx, and many of the drugs are already widely used clinically. The next few years should see further investigation of the role of alterations in calcium channel numbers or properties in disease states, and possibly the identification of endogenous or environmental factors which control the functioning of the calcium channel.

We thank many colleagues at Bayer A.G. and at Miles Laboratories, especially Dr Kazda, Dr Wehinger and Dr. Sturton for frequent helpful discussions. Especially we are grateful to Miss Lynne Smith for typing the manuscript.

References

Allen, G. S. and others (1983) *New Engl. J. Med.* **308**, 619–624
Bean, B. P. (1984) *Proc. Natl. Acad. Sci. U.S.A.* **81**, 6388–6392
Belleman, P., Ferry, D., Luebbbecke, F. & Glossmann, H. (1981) *Arzneim-Forsch.* (*Drug Res*). **31**, 2064–2067
Berridge, M. J. (1984) *Biochem. J.* **220**, 345–360
Bolton, T. B. (1979) *Physiol. Rev.* **59**, 607–718
Bolton, T. B. (1985) in: *Control and Manipulation of Calcium Movement* (Parratt, J. R., ed.), pp. 147–168, Raven Press, New York
Borle, A. B. (1981) *Rev. Physiol. Biochem. Pharmacol.* **90**, 13–153.
Bossert, F., Horstmann, H., Meyer, H. & Vater, W. (1979) *Arzneim-Forsch.* (*Drug Res*). **29**, 226–229
Bristow, M. R., Ginsberg, R., Laser, J. A., Mcanley, B. J. & Minobe, W. (1984) *Br. J. Pharmacol.* **82**, 309–320
Bühler, F. R., Bolli, P. & Hulther, V. L. (1984) in: *Calcium Antagonists and Cardiovascular Disease* (Opie, L. H., ed.), pp. 313–322, Raven Press, New York
Campbell, A. K. (1983) *Intracellular Calcium: its Universal Role as Regulator*, John Wiley and Sons, New York.
Cole, K. S. (1979) *Annu. Rev. Physiol.* **41**, 1–24
Dormer, R. L., Hallett, M. B. & Campbell, A. K. (1985) in: *Control and Manipulation of Calcium Movement* (Parratt, J. R., ed.), pp. 1–27, Raven Press, New York
El-Fakahany, E., Eldefrawi, A. T., Murphy, D. L., Aguayo, L. G., Triggle, D. J., Albuquerque, E. X. & Eldefrawi, M. E. (1984) *Mol. Pharmacol.* **25**, 369–378
Exton, J. F. (1983) in: *Calcium and Cell Function* (Cheung, W. Y., ed.), vol. IV, pp. 64–97, Academic Press, New York
Farber, J. L. (1981) *Life Sci.* **29**, 1289–1295
Fatt, P. & Katz, B. (1953) *J. Physiol.* (*London*) **120**, 171–204
Flaim, S. F., Ratz, P. H. & Ress, R. J. (1984) in: *Hypertension: Physiological Basis and Treatment* (Ong, H. H. & Lewis, J. C., ed.), pp. 269–299, Academic Press, New York
Fleckenstein, A. (1977) *Annu. Rev. Pharmacol. Toxicol.* **17**, 49–66
Fleckenstein, A. (1983) *Calcium Antagonism in Heart and Smooth Muscle.* John Wiley and Sons, New York
Fleckenstein, A., Tritthart, H., Fleckenstein, B., Herbet, A. & Gruen, G. (1969) *Pflugers Arch. ges Physiol.* **307**, R25
Franckowiak, G., Boeshagen, H., Bossert, F., Goldmann, S., Meyer, H., Wehinger, E., Schramm, M., Thomas, G. & Towart, R. (1983) *German Patent DE* 3,130,041
Freedman, S., Miller, R. J., Miller, D. M. & Tindall, D. R. (1984) *Proc. Natl. Acad. Sci. U.S.A.* **81**, 4582–4585
Glossmann, H., Ferry, D. R., Luebbecke, F., Mewes, R. & Hoffmann, F. (1982) *Trends Pharmacol. Sci.* **3**, 431–437
Godfraind, T. & Kaba, A. (1969) *Br. J. Pharmacol.* **36**, 549–560
Gopalakrishnan, V. & Triggle, C. R. (1984) *Can. J. Physiol. Pharmacol.* **62**, 1249–1253
Gould, R. J., Murphy, K. M. M. & Snyder, S. H. (1984) *Mol. Pharmacol.* **25**, 236–241

Hagiwara, S. & Byerly, L. (1981) *Annu. Rev. Neurosci.* **4**, 69–125
Hetherington, A. M. & Trewavas, A. J. (1984) *Plant Sci. Lett.* **35**, 109–113
Hodgkin, A. L. & Huxley, A. F. (1952) *J. Physiol. (London)* **117**, 500–544
Hoeltje, H. D. (1982) *Arch. Pharm. (Weinheim)* **315**, 317–323
Janis, R. A. & Triggle, D. J. (1984) *Drug Development Res.* **4**, 257–274
Kohlhardt, M., Bower, B., Krause, H. & Fleckenstein, A. (1973) *Pfluegers Arch. ges Physiol.* **338**, 115–123
Langs, D. & Triggle, D. J. (1985) in: *Proceedings of IX International Congress of Pharmacology* (Paton, W., Mitchell, J. & Turner, P., eds.), vol. II, pp. 323–328, Macmillan, London
Lee, K. S. & Tsien, R. W. (1983) *Nature (London)* **302**, 790–794
Loev, B., Goodman, M. M., Snador, K. M., Tedesci, R. & Macko, E. (1974) *J. Med. Chem.* **17**, 956–965
Lowenstein, W. R., Kanno, Y. & Socolar, S. J. (1978) *Fed. Proc. Amer. Soc. Cell Biol.* **37**, 2645
MacLennan, D. H., Reitheimer, R. A. F., Shoshan, V., Campbell, K. P., LeBel, D., Herrmann, T. R. & Shamoo, A. E. (1980) *Ann. N.Y. Acad. Sci.* **358**, 138–148
Manhold, R., Steiner, R., Haas, W. & Kaufmann, R. (1978) *N.S. Arch. Pharmacol.* **320**, 217–226
Means, A. R., Tash, J. S. & Chafouleas, J. G. (1982) *Physiol. Rev.* **62**, 1–39
Meisheri, H. D., Hwang, O. & Van Breemen, L. (1981) *J. Memb. Biol.* **59**, 19–25
Meyer, H., Kazda, S. & Bellemann, P. (1983) *Annu. Rep. Med. Chem.* **18**, 79–88
Meyer, J. S. & Hardenberg, J. (1983) *Headache* **23**, 266–277
Michell, R. H. (1982) *Cell Calcium* **3**, 429–440
Murphy, K. M. M., Gould, R. J., Largent, B. L. & Snyder, S. H. (1983) *Proc. Natl. Acad. Sci. U.S.A.* **80**, 860–864
Penniston, J. T. (1983) in: *Calcium and Cell Function*, (Cheung, W. Y., ed.), vol. IV, pp. 100–149. Academic Press, New York
Putney, J. W., Weiss, S. J., van de Walle, C. M. & Haddas, R. A. (1980) *Nature (London)* **284**, 345–347
Rahwan, R. G., Witiak, D. T. & Muir, W. W. (1981) *Annu. Rep. Med. Chem.* **16**, 257–268
Rasmussen, H. & Barrett, P. Q. (1984) *Physiol. Rev.* **64**, 938–984
Ress, R. J. & Flaim, S. F. (1982) *Circulation* **66** (Suppl. II), 141 (Abst.)
Reuter, H. (1973) *Prog. Biophys. Mol. Biol.* **26**, 1–43
Reuter, H. (1983) *Nature (London)* **301**, 569–574
Rodenkirchen, R., Bayer, R., Steiner, R., Bossert, F., Meyer, H. & Moeller, E. (1979) *N.S. Arch. Pharmacol.* **310**, 69–78
Rodenkirchen, R., Bayer, R. & Manhold, R. (1981) *Arzneim-Forsch. (Drug Res.)* **31**, 773–780
Rothstein, A. (1980) *Ann. N.Y. Acad. Sci.* **358**, 96–102
Saida, K. & Van Breemen, C. (1983) *Pfluegers Arch.* **397**, 166–167
Sato, M., Nagao, T. & Yamaguchi, I. (1971) *Arzneim-Forsch. (Drug Res.)* **21**, 1338–1343
Schramm, M., Thomas, G., Towart, R. & Franckowiak, G. (1983) *Nature (London)* **303**, 535–537
Scriabine, A., Vanov, S. & Deck, K. (eds.) (1984) *Nitrendipine*, Urban & Schwarzenberg, Baltimore
Scriabine, A., Battye, R., Hoffmeister, F., Kazda, S., Towart, R., Garthoff, B., Schluter, G., Rämsch, K. D. & Scherling, D. (1985) in: *New Drugs Annual*, vol. 3, in the press
Stone, P. H. & Antman, E. M. (eds.) (1983) *Calcium Channel Blocking Agents in the Treatment of Cardiovascular Disorders*, Futura, New York
Towart, R. (1985) in: *Control and Manipulation of Calcium Movement* (Parratt, J. R., ed.), pp. 169–187, Raven Press, New York
Towart, R. & Wehinger, E. & Meyer, H. (1981) *N.S. Arch. Pharmacol.* **317**, 183–185
Towart, R., Wehinger, E., Meyer, H. & Kazda, S. (1982) *Arzneim-Forsch. (Drug Res.)* **32**, 338–346
Triggle, D. J. (1982) in: *Calcium Blockers: Mechanisms of Action and Clinical Applications* (Flaim, S. F. & Zelis, R., eds.), pp. 121–134, Urban and Schwarzenberg, Baltimore
Triggle, D. J. & Janis, R. A. (1984*a*) in: *Modern Methods in Pharmacology* (Spector, S. & Back, N., eds.), pp. 1–28, Alan R. Liss Inc., New York
Triggle, D. J. & Janis, R. A. (1984*b*) *J. Card. Pharmacol.* **6**, S949–S955
Triggle, D. J. & Swamy, V. C. (1983) *Circ. Res.* **52** (Suppl. 1), 17–28
Truog, A. G., Branner, H., Criscione, T., Falbert, M., Kuhnis, H., Meier, M. & Rogg, H. (1985) in: *Calcium in Biological Systems* (Rubin, R. P. *et al.*, eds.), in the press
Tsien, R. W. (1983) *Annu. Rev. Physiol.* **45**, 341–358
Van Breemen, C., Hwang, O. & Siegel, B. (1977) in: *Excitation–Contraction Coupling in Smooth Muscle* (Casteels, R. *et al.*, eds.), pp. 243–252, Elsevier, Amsterdam

Biochem. Soc. Symp. **50**, 97–125
Printed in Great Britain

Role of Water in Processes of Energy Transduction: Ca^{2+}-Transport ATPase and Inorganic Pyrophosphatase

LEOPOLDO DE MEIS

*Instituto de Ciências Biomédicas, Departamento de Bioquímica, Centro de Ciências da Saúde,
Universidade Federal do Rio de Janeiro, Cidade Universitária, Ilha do Fundão – Rio de Janeiro,
CEP 21910, Brasil*

Synopsis

After the proposal of the chemiosmotic theory by Mitchell (1966, 1979) it has been recognized that different membrane-bound enzymes are able to use the energy derived from ionic gradients for the synthesis of ATP. These include the F_1-ATPases of mitochondria and chloroplasts, the Ca^{2+}-dependent ATPase of sarcoplasmic reticulum and the (Na$^+$,K$^+$)-ATPase of plasma membrane. In these systems the process of energy transduction is fully reversible. The enzyme can use the energy derived from the hydrolysis of ATP to build up a concentration gradient of ions across the membrane and, in the reverse process, use the energy derived from the gradient to synthesize ATP. Another interesting system in which these forms of energy are interconverted is found in photo-synthetic bacteria. In chromatophores of *Rhodospirillum rubrum* there is a membrane-bound pyrophosphatase that, like the transport ATPases, catalyses the synthesis of pyrophosphate from P_i when a light-dependent proton gradient is formed across the chromatophore membrane. Like F_1-ATPase, this enzyme is also able to generate an electrochemical potential gradient of protons at the expense of pyrophosphate hydrolysis. The mechanism by which the energy derived from a gradient is used by membrane-bound enzymes to catalyse the synthesis of high-energy phosphate compounds is still far from understood. Among the different enzymes studied, Ca^{2+}-dependent ATPase is probably the system in which most is known about the mechanism of energy transduction. We now know of experimental conditions which allow us to move the different intermediary steps of the catalytic cycle of the enzyme in the direction of ATP synthesis. Thus, ATP synthesis can be attained after a single catalytic cycle in the absence of a transmembrane Ca^{2+} gradient. The net synthesis of ATP can be promoted by a variety of pertubations, including Ca^{2+}, pH and water activity. These experiments indicate that during the catalytic cycle different forms of energy are interconverted by the Ca^{2+}-dependent ATPase. The ultimate step of the cycle seems to be a change of water activity within the catalytic site of the ATPase.

A common feature of all membrane-bound enzymes mentioned above is that during the catalytic cycle there are steps in which the hydrolysis of a phosphate compound (ATP, pyrophosphate or an acyl phosphate residue) is accompanied by only a small change in free energy. In conditions similar to those found in the cytosol, the hydrolysis of these phosphate compounds is accompanied by a much larger change in free energy. At present it is not clear why the free energy of hydrolysis of an acyl phosphate residue, ATP or pyrophosphate changes so much, depending on whether they are in solution or on the enzyme surface. This phenomenon appears to be crucial to the mechanism of energy transduction in the living cell. Recent experiments performed in our laboratory show that a discrete change of water activity in the medium leads to a significant change of the ΔG^0 of pyrophosphate hydrolysis. In fact, when the water activity of the medium is decreased by the addition of organic solvents, soluble inorganic pyrophosphatase can catalyse the synthesis of pyrophosphate in amounts similar to that attained with chromatophores of *R. rubrum* under illumination. In this paper the mechanism by which Ca^{2+}-dependent ATPase and inorganic pyrophosphatase use the energy derived from ionic gradients for the synthesis of high-energy phosphate compounds is discussed. The possibility is raised that a change of water activity in the catalytic site might represent a common step in the process of energy transduction of all membrane-bound enzymes.

The Sarcoplasmic Reticulum

This is a membrane-bound compartment arranged as a network of tubules and cisternae which surround the myofibrils of skeletal muscle. The membrane of the sarcoplasmic reticulum is highly specialized, inasmuch as 60–80% of the total protein of the membrane is Ca^{2+}-dependent ATPase. In the living cell the sarcoplasmic reticulum controls the cytoplasmic Ca^{2+} concentration, thereby controlling muscle contraction and relaxation. Vesicles derived from the sarcoplasmic reticulum can be isolated from muscle homogenates by differential centrifugation. The vesicles are right-side out; the outer surface of the vesicle's membrane corresponds to the surface of the sarcoplasmic reticulum that faces the cytoplasm. The Ca^{2+}-dependent ATPase is asymmetrically embedded in the membrane of the vesicles. One part of the protein protrudes from the outer (cytoplasmic) surface of the membrane and the other part is buried in the lipid layer. The ATPase is a single polypeptide chain with a molecular weight in the range 100000–120000 Da; it can be solubilized by a large variety of detergents and can be purified as a lipoprotein complex (for reviews see Hasselbach, 1978; Tada *et al.*, 1978; de Meis & Vianna, 1979; de Meis, 1981; de Meis & Inesi, 1982).

Active Transport of Ca^{2+}

What follows is a short account of the process of ATP hydrolysis and Ca^{2+} transport in vesicles derived from the sarcoplasmic reticulum. This was shown by Ebashi & Lipmann (1962) and by Hasselbach & Makinose (1961). When suspended in a medium containing ATP and Mg^{2+}, these vesicles can reduce the Ca^{2+} concentration of the medium from 0.1 mM to less than 0.10 μM. In this process the ATPase uses the chemical energy derived from the hydrolysis of ATP to pump Ca^{2+} into the vesicles and to maintain a steep Ca^{2+} concentration gradient, which is formed across the vesicle membrane. Two calcium ions are

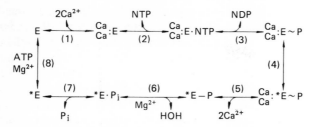

Fig. 1. *Reaction sequence of ATP hydrolysis and Ca²⁺ transport*

pumped inside the vesicles for each ATP molecule hydrolysed. The true substrate of the Ca^{2+}-dependent ATPase is the complex Mg^{2+}–ATP. The catalytic cycle of the enzyme is initiated by the binding of Ca^{2+} to a high-affinity site (K_s about 1 μM at pH 7) located on the part of the ATPase molecule facing the outer surface of the vesicle. At pH 7.0, two Ca^{2+} bind to each enzyme molecule in a co-operative process. After the binding of Ca^{2+}, a γ-carboxyl group of an aspartyl residue located in the catalytic site of the enzyme is phosphorylated by ATP, forming an acyl phosphate residue. After phosphorylation, the enzyme undergoes a conformational change and the Ca^{2+} binding site which was facing the outer surface of the vesicle now faces the vesicle lumen. This is associated with an increase of the K_s from 1 μM to 1 mM, which permits the release of the bound Ca^{2+} in the vesicle lumen. After dissociation of Ca^{2+} from the enzyme, the acyl phosphate residue in the catalytic site is hydrolysed and the enzyme returns to its first conformation to initiate a new cycle of Ca^{2+} transport.

The apparent K_m of the ATPase for Mg^{2+}–ATP is in the range 1–3 μM. At higher concentrations (apparent K_m 50–20 μM) Mg^{2+}–ATP increases the rate of phosphoenzyme hydrolysis and the rate of enzyme return to the conformation having high affinity for Ca^{2+}. Thus, Mg^{2+}–ATP is both a substrate and a regulator of the Ca^{2+}-dependent ATPase. Besides ATP, other nucleoside triphosphates, acetyl phosphate, *p*-nitrophenyl phosphate, carbamoyl phosphate and other artificial energy-yielding substrates can support the active transport of Ca^{2+}. The affinity of the enzyme for Mg^{2+}–ATP is one or more orders of magnitude greater than for the other substrates.

The reaction sequence shown in Fig. 1 describes the steps involved in the process of ATP hydrolysis and Ca^{2+} transport (Carvalho *et al.*, 1976; de Meis & Vianna, 1979; de Meis, 1981). In this sequence, E and *E represent two distinct conformations of the enzyme. The Ca^{2+} binding site in the E form faces the outer surface of the vesicle and has an apparent K_s for Ca^{2+} of 1 μM at pH 7.0. After binding of Ca^{2+} this form is phosphorylated by different nucleoside triphosphate but not by P_i. The Ca^{2+} binding site in the *E form faces the inner surface of the vesicle and has an apparent K_s for Ca^{2+} in the range 1–2 mM at pH 7.0. This form is phosphorylated by P_i but, under physiological conditions, is not phosphorylated by ATP. In presence of acetyl phosphate or low concentrations of nucleotide the conversion of *E into E is the rate-limiting step of the catalytic cycle. Increasing concentrations of nucleotide increases the rate of interconversion, with ITP, GTP and other energy-yielding substrates being much less effective than ATP. The data summarized above were discussed in detail in previous reviews (Hasselbach, 1978; de Meis & Inesi, 1982). Some of the experimental evidence supporting the sequence shown in Fig. 1 will be presented in subsequent sections of this paper.

Reversibility of the Calcium Pump

It was shown by Makinose & Hasselbach (1971) that the entire process of Ca^{2+} transport can be reversed. A steep Ca^{2+} concentration gradient is formed across the membrane when vesicles previously loaded with Ca^{2+} are incubated in a medium containing EGTA, a Ca^{2+} chelating compound. In this condition Ca^{2+} flows out of the vesicles at slow rate, owing to the low permeability of the membrane. The Ca^{2+} efflux is sharply increased when ADP, P_i and Mg^{2+} are added to the incubation medium (Barlogie et al., 1971). The increment of Ca^{2+} efflux is coupled to the synthesis of ATP (Makinose & Hasselbach, 1971). For every two calcium ions released from the vesicles, one molecule of ATP is synthesized. The synthesis of ATP is initiated by phosphorylation of the enzyme by P_i (Makinose, 1972; Yamada & Tonomura, 1972; Yamada et al., 1972). The acyl phosphate residue thus formed can be measured by incubating vesicles previously loaded with Ca^{2+} in a medium at pH 7 containing EGTA, Mg^{2+} and $^{32}P_i$. Under optimal conditions, $2–4 \mu mol$ of phosphoenzyme is formed/g of protein. The phosphorylation of the enzyme by P_i is not accompanied by an increment of Ca^{2+} efflux. The subsequent addition of ADP to the medium leads to a decrease of the steady-state level of phosphoenzyme, synthesis of ATP and enhancement of the rate of Ca^{2+} efflux. The following minimal reaction sequence was therefore proposed (Makinose, 1972).

$$E + P = E \sim P + HOH \tag{1}$$

$$E \sim P + ADP = E + ATP \tag{2}$$

Another index of the reversal of the Ca^{2+} pump is the $ATP \rightleftharpoons P_i$ exchange reaction. Makinose (1971) observed that when vesicles were incubated in a medium containing all the reactants required for the pump to work forward and backwards, i.e. $0.2 \text{ mM-}CaCl_2$, Mg^{2+}, ATP, ADP and $^{32}P_i$, Ca^{2+} was first accumulated by the vesicles until a steady state was reached, in which the Ca^{2+} efflux was balanced by the ATP-driven influx. As soon as net uptake ceased, a steady state of exchange between $^{32}P_i$ and the $(\gamma$-P)ATP began. During the initial phase of Ca^{2+} accumulation, the vesicles catalyse the hydrolysis of ATP but not the incorporation of $^{32}P_i$ into the ATP pool. The $ATP \rightleftharpoons P_i$ exchange was abolished when the membranes of the vesicles were rendered leaky by various procedures that did not affect the transport ATPase (Makinose, 1971; de Meis & Carvalho, 1974). $ATP \rightleftharpoons P_i$ exchange indicates that the enzyme operates simultaneously forwards (ATP hydrolysis) and backwards (ATP synthesis from ADP and $^{32}P_i$). From these data it was inferred that when the steady state between Ca^{2+} influx and Ca^{2+} efflux was reached, the energy derived from the hydrolysis of ATP was used to maintain the Ca^{2+} gradient, and the energy derived from the gradient was used for the synthesis of ATP (Makinose, 1971; de Meis & Carvalho, 1974).

When the reversal of the Ca^{2+} pump was first described, it was proposed that the energy required for the synthesis of ATP was derived from the osmotic energy of the Ca^{2+} gradient (Barlogie et al., 1971; Makinose, 1971, 1972; Yamada et al., 1972). This interpretation was particularly attractive in the light of the chemiosmotic hypothesis proposed by Mitchell (1966, 1979) to account for the synthesis of ATP in mitochondria. The osmotic energy of the Ca^{2+} gradient was regarded as driving the first step of ATP synthesis, the phosphorylation of the ATPase by P_i. The finding that the protein phosphate was an energy-rich acyl phosphate contributed to this conclusion. The ΔG^0 values calculated from the ratio of Ca^{2+} concentration in the vesicles to that in the medium were compatible with those required for the synthesis of ATP or for the formation of an acyl phosphate residue (Makinose, 1971, 1972; Yamada & Tonomura, 1972; Yamada et al., 1972).

Phosphoenzyme of Low and High Energy: Phosphorylation by P_i in the Absence of a Ca^{2+} Gradient

Beginning in 1972, work in my laboratory was conducted aiming at the understanding of the mechanism by which the sarcoplasmic reticulum ATPase catalyses the synthesis of ATP. The first question we raised was whether the synthesis of ATP was related to the unidirectional movement of calcium ions forced through the enzyme by the gradient (osmotic energy) or whether it was simply related to the enzyme being exposed to two media of different composition. The enzyme expands through the vesicle membrane (Blasie et al., 1982; Hasselbach et al., 1983). When a Ca^{2+} concentration gradient is formed, the part of the enzyme which faces the outer surface of the vesicle is exposed to a medium without Ca^{2+} (EGTA) and the part of the enzyme which faces the lumen is exposed to a medium containing a high Ca^{2+} concentration (10–30 mM).

In order to explore this possibility, leaky vesicles were prepared by using different procedures (de Meis & Carvalho, 1974; Masuda & de Meis, 1973). In leaky vesicles, a Ca^{2+} gradient cannot be formed. The membrane is disrupted and, although the enzyme remains fully active, the vesicles are no longer able to accumulate Ca^{2+}. We found that the ATPase is phosphorylated by P_i in the absence of a transmembrane Ca^{2+} gradient (Masuda & de Meis, 1973). This was attained by incubating leaky vesicles in a medium without Ca^{2+} (excess EGTA) containing P_i and Mg^{2+} (Fig. 2). As in the presence of a Ca^{2+} gradient, the phosphoenzyme formed with leaky vesicles is an acyl phosphate (Masuda & de Meis, 1973; de Meis & Masuda, 1974). This finding was confirmed in different laboratories (Kanazawa, 1975; Knowles & Racker, 1975; Rauch et al., 1977; Beil et al., 1977; Hasselbach, 1978; Punzengruber et al., 1978; Prager et al., 1979; Lacapere et al., 1981; Martin & Tanford, 1981; Inesi et al., 1982; Guillain et al., 1984) and clearly shows that the energy derived from the Ca^{2+} gradient is not necessary for the first step of the reversal of the Ca^{2+} pump. The apparent K_m for P_i varies significantly with the pH of the medium and according to whether leaky vesicles or vesicles previously loaded with Ca^{2+} are used (de Meis, 1976). Table 1 shows that the presence of Ca^{2+} gradient across the membrane promotes an increase of the apparent affinity of the enzyme for P_i. This is more

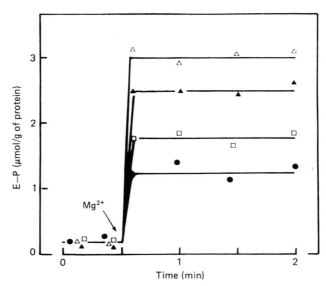

Fig. 2. *Time course of E–P formation and its dependence on P_i concentration and Mg^{2+}*

The assay medium composition was 10 mM-Tris–maleate buffer (pH 6.0), 1 mM-EGTA and 0.5 mM- (●), 1 mM- (□), 2 mM- (▲) or 4 mM- (△) $^{32}P_i$. The reaction was started by the addition of leaky vesicles. The arrow points to the addition of $MgCl_2$ with a final concentration of 10 mM. For further details see Masuda & de Meis (1973).

Table 1. *P_i affinity for E–P formation*

Free Ca^{2+} concentration in the vesicle lumen is higher in vesicles loaded with calcium phosphate than in vesicles loaded with calcium oxalate. For details see de Meis (1976).

Vesicles	K_{P_i} for E–P formation (mM)	
	pH 6.1	pH 7.0
Empty or leaky	3.5 ± 0.5 (8)	25.3 ± 4.2 (3)
Loaded with calcium oxalate	3.1 ± 0.4 (3)	8.7 ± 0.9 (3)
Loaded with calcium phosphate	0.7 ± 0.1 (7)	2.5 ± 0.4 (4)

pronounced at pH 7 than at pH 6. The low affinity for P_i at pH 7.0 probably accounts for earlier failures to measure phosphorylation by P_i in the absence of a transmembrane Ca^{2+} gradient (Makinose, 1972; Yamada et al., 1972; Kanazawa & Boyer, 1973). At 30 °C and pH 6.0, $K_{eq.}$ for phosphoenzyme hydrolysis in the absence of a gradient is in the range 0.1–1.5 ($\Delta G° + 5.7$ to -0.83 kJ/mol). This was determined by measuring the rate constant of phosphoenzyme formation and phosphoenzyme hydrolysis in transient kinetics experiments (Guimarães-Motta & de Meis, 1980; de Meis et al., 1982).

The phosphorylation of leaky vesicles by P_i is inhibited by Ca^{2+}. The inhibition is reversible (Fig. 3). The enzyme is rephosphorylated by P_i if the added Ca^{2+} is chelated with an excess of EGTA. Inhibition by Ca^{2+} is observed when either leaky vesicles or vesicles previously loaded with Ca^{2+} are used

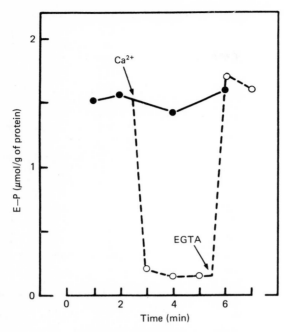

Fig. 3. *Reversible effect of Ca²⁺*

The assay medium composition was 10 mM-Tris–maleate buffer (pH 6.0), 0.1 mM-EGTA, 10 mM-MgCl$_2$ and 4 mM-^{32}P$_i$. To a separate aliquot of this medium, CaCl$_2$ to a final concentration of 0.3 mM and EGTA to a final concentration of 2 mM were added as shown by arrows. For further details see Masuda & de Meis (1973).

(Barlogie *et al.*, 1971; de Meis, 1976). The Ca^{2+} concentration required for half-maximal inhibition of phosphorylation by P$_i$ both in the presence and the absence of a gradient is about 1 μM at pH 7, which is in the same range of Ca^{2+} concentrations required for half-maximal activation of ATP hydrolysis.

In a previous report Yamada *et al.* (1972) observed an inhibition of the enzyme phosphorylation by P$_i$ when Ca^{2+} was added to a medium containing intact vesicles loaded with Ca^{2+}. The authors attributed the inhibition to a decrease of the Ca^{2+} gradient. The finding that Ca^{2+} inhibits phosphorylation by P$_i$ of both leaky vesicles and vesicles loaded with Ca^{2+} shows that the inhibition is related to the binding of Ca^{2+} to the outer surface of the membrane and not to a modification of the Ca^{2+} gradient.

The data described above led to the conclusion that ATP and P$_i$ interact with a common active site of the enzyme which undergoes a conformational change depending on the binding of Ca^{2+}. In one conformational state the ATPase would be phosphorylated by ATP. Thus, the binding of Ca^{2+} would determine the choice of substrate for the phosphorylation reaction (Masuda & de Meis, 1973; Carvalho *et al.*, 1976; de Meis & Vianna, 1979). It was already known that the enzyme has a Ca^{2+} binding site of high affinity, K_s about 1 μM at pH 7 (Carvalho & Leo, 1967; Carvalho, 1968; Inesi *et al.*, 1980). Evidence that the ATPase undergoes a conformational change during the catalytic cycle had

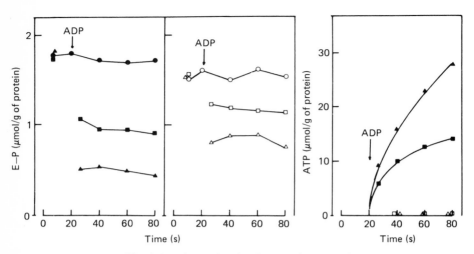

Fig. 4. *Membrane phosphorylation and ATP synthesis*

The assay medium composition was 10 mM-Tris–maleate buffer (pH 6.5), 30 mM-MgCl₂, 15 mM-EGTA and 2 mM-³²Pᵢ (left). For each experiment three sets of tubes were prepared. One served as a control, to which no ADP was added (●, ○). To the other two sets of tubes, ADP (as shown by the arrows) was added to a final concentration of 50 (■, □) or 100 (▲, △) μM. The reaction was started by the addition of sarcoplasmic reticulum vesicles (0.7 mg/ml), which were empty (○, □, △) or loaded with calcium phosphate (●, ■, ▲). After different incubation intervals at 30 °C, samples were taken for the determination of E–P (left) and [γ-³²P]ATP synthesis (right). For further details see de Meis (1976).

already been presented by Landgraf & Inesi (1969) and subsequently confirmed by several authors (Coan & Inesi, 1977; Thorley-Lawson & Green, 1977; Dupont & Leigh, 1978). The data described above permitted characterization of the properties of the two forms of the enzyme: one form is phosphorylated by P_i and the other by ATP. Moreover, the binding of Ca^{2+} would trigger the conversion of one form into the other (Masuda & de Meis, 1973). These intermediary steps of the catalytic cycle are represented by reactions 6, 7, 8 and 1 in the sequence shown in Fig. 1. The time constant of the different intermediary reactions has been measured by means of transient kinetic experiments (for reviews see de Meis, 1981; de Meis & Inesi, 1982). In the absence of ATP, by far the slower reaction is the conversion of *E into 2Ca·E (reactions 8 and 1 in Fig. 1). At 30 °C, the time for half-maximal conversion of *E into 2Ca·E is in the range 400–800 ms whereas that of all other reactions varies between 5 and 80 ms (Rauch et al., 1977; de Meis & Tume, 1977; Guimarães-Motta & de Meis, 1980).

After phosphorylation by P_i, an acyl phosphate residue is formed in the catalytic site of the ATPase both in the presence and the absence of transmembrane Ca^{2+} gradient (de Meis, 1976). However, only the phosphoenzyme formed in presence of a Ca^{2+} gradient is able to transfer its phosphate to ADP (Fig. 4). This indicates that the energy derived from the Ca^{2+} gradient is used not for the phosphorylation of the enzyme by P_i, but for the transfer of phosphate to ADP. Therefore, the phosphoenzyme formed in absence of a Ca^{2+} gradient is

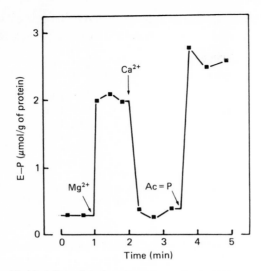

Fig. 5. *Effect of Ca²⁺ and acetyl phosphate on E–P formation*

The assay medium composition was 10 mM-Tris–maleate buffer (pH 6.5), 4 mM $^{32}P_i$, 0.05 mM-EGTA and 0.6 mg of SRV protein/ml. The reaction was started by the addition of SRV. After different incubation intervals at 37 °C, aliquots of 2.5 ml of this medium were removed for E–P determination. Arrows point to the addition of MgCl₂, final concentration in the assay medium 10 mM; CaCl₂, 0.25 mM; acetyl phosphate (AcP), 2 mM. For details see de Meis & Masuda (1974).

referred to as 'low energy' because it is formed spontaneously ($\Delta G^0 + 5.7$ to -0.83 kJ/mol) and is not able to transfer its phosphate to ADP, and that formed in the presence of a Ca²⁺ gradient as 'high energy' because its phosphate can be transferred to ADP, leading to the synthesis of ATP. In this view the capture of the osmotic energy of the Ca²⁺ gradient would permit the conversion of the 'low-energy' phosphoenzyme into the 'high-energy' form. Contrasting with this reasoning is the finding that the phosphoryl group of the phosphoenzyme formed both in the presence and in the absence of a Ca²⁺ gradient is an acyl phosphate. In water, the ΔG^0 of the hydrolysis of both an acyl phosphate residue and ATP is in the range -35 to -40 kJ/mol (Alberty, 1968; George *et al.*, 1970; Haynes *et al.*, 1978).

Simultaneous Phosphorylation of ATPase by Nucleoside Triphosphate and P_i

Phosphorylation of the enzyme of P_i is inhibited by Ca²⁺. We have found (de Meis & Masuda, 1974; de Meis & Sorenson, 1975; de Meis & Boyer, 1978; Vieyra *et al.*, 1979) that this inhibition is overcome by the addition to the medium of an energy-yielding substrate such as acetyl phosphate, ITP, GTP or ATP (Fig. 5 and Table 2). The Ca²⁺ present in the medium activates hydrolysis of the substrate, and phosphorylation by P_i can be detected as long as the substrate is present in the medium. This finding indicates that the enzyme form *E is accumulated and back-phosphorylated by P_i in the steady state established during the hydrolysis of the nucleotide (Fig. 1). The proportion of

Table 2. *Phosphorylation of leaky vesicles by P_i in the presence of Ca^{2+}*

The assay medium composition was 40 mM-Tris–maleate buffer (pH 6.7), 20 mM-MgCl$_2$, 8 mM-^{32}P$_i$, 0.2 mM-CaCl$_2$ and 0.2 mg of leaky vesicle protein/ml.

Additions to assay medium	E–^{32}P (μmol/g of protein)
None	0.10 ± 0.05
1 mM-Acetyl phosphate	1.94 ± 0.18
1 mM-GTP	1.76 ± 0.26
1 mM-ITP	1.89 ± 0.17
1 mM-ATP	0.20 ± 0.08
0.05 mM-ATP*	1.20 ± 0.13

* Contained in addition 1 mM-phosphoenolpyruvate and 50 μg of pyruvate kinase/ml. Data from de Meis & Masuda (1974), Carvalho *et al.* (1976) and de Meis & Boyer (1978).

*E accumulated during the steady state depends on the nucleoside triphosphate used and its concentration in the medium. Table 2 shows that at a concentration of 1 mM, acetyl phosphate, ITP and GTP are equally effective in eliciting phosphorylation by P_i. A small portion of the enzyme is phosphorylated by P_i in the presence of 1 mM-ATP. To obtain a high level of phosphorylation by P_i, it is necessary to decrease the ATP concentration to the micromolar range (Souza & de Meis, 1976). In previous reports it was shown that, besides being a substrate, ATP, but not GTP, ITP and acetyl phosphate, can regulate the catalytic cycle of the enzyme. Inesi *et al.* (1967), measured the rate of hydrolysis of leaky vesicles and found that two different K_m values for ATP can be distinguished on double reciprocal plots; 1–3 μM and 50–200 μM. Later we observed that, in the micromolar concentration range ATP binds only to the catalytic site of the ATPase and at higher concentrations it binds also to a regulatory site which accelerates the velocity of substrate hydrolysis, GTP, ITP and acetyl phosphate being much less effective than ATP (de Meis & de Mello, 1973; Scofano *et al.*, 1979; Scofano & de Meis, 1981). The regulatory role of ATP has now been well documented in different laboratories (Dupont, 1977; Froehlich & Taylor, 1975, 1976; Takisawa & Tonomura, 1978; McIntosh & Boyer, 1983). Taken together these data indicate that, in the steady state established during the hydrolysis of ITP or GTP, a significant portion of the enzyme is in the form *E, which is phosphorylated by P_i (Table 2). ATP accelerates the conversion of *E into E. Thus, in the presence of 1 mM-ATP most of the enzyme would be in the form phosphorylated by the nucleotide and only a small fraction would be in the form phosphorylated by P_i (Table 2).

The discovery of the regulatory role of ATP played a key part in the understanding of the mechanism of energy conservation of the system.

ATP $\rightleftharpoons$ P$_i$ Exchange in the Absence of Ca^{2+} Gradient

When a gradient is formed, the Ca^{2+} concentration required inside the vesicles for the synthesis of ATP is the range 1–10 mM (Makinose & Hasselbach, 1971; Makinose, 1971; Yamada *et al.*, 1972). The finding that a small portion of the

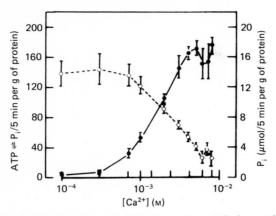

Fig. 6. *Ca²⁺ dependence of ATP ⇌ Pᵢ exchange in leaky vesicles*

The assay medium composition was 30 mM-Tris–maleate buffer (pH 6.8), 10 mM-ATP, 6 mM-³²Pᵢ, 20 mM-MgCl₂ and 0.3 mg of SRV protein/ml. For the ATPase activity (○---○), [γ-³²P]ATP was used. For ATP ⇌ Pᵢ exchange (●—●), ³²Pᵢ was used. The reaction was performed at 37 °C. The values shown in the Figure represent the average ±S.E. of five experiments. For further details see de Meis & Carvalho (1974).

enzyme is phosphorylated by P_i during the hydrolysis of ATP led us to question whether the ATP $\rightleftharpoons P_i$ exchange is activated only when a Ca^{2+} gradient is formed across a vesicle membrane or whether a high Ca^{2+} concentration, similar to that found inside the vesicles when a gradient is formed, would be sufficient by itself to activate this reaction. In other words, whether the energy derived from the gradient or solely a high Ca^{2+} concentration would allow the phosphoenzyme formed from P_i to be converted from 'low energy' into 'high energy'.

In order to untangle these possibilities, leaky vesicles or ATPase solubilized with the detergent X-100 were incubated at pH 7 in media containing different Ca^{2+} concentrations, and the rates of ATP hydrolysis and of ATP synthesis from ADP and P_i were measured simultaneously (de Meis & Carvalho, 1974; Plank et al., 1979; Oliva et al., 1983; McIntosh & Davidson, 1984). When the Ca^{2+} concentration in the medium is sufficient to saturate only the high-affinity binding site (up to 0.1 mM), the enzyme catalyses only the hydrolysis of ATP (Fig. 6). Increasing the Ca^{2+} concentration in the medium to the millimolar range results in simultaneous inhibition of the ATPase activity and activation of ATP synthesis. The lack of a requirement for a transmembrane Ca^{2+} gradient for activation of the ATP $\rightleftharpoons P_i$ exchange reaction is documented further in Fig. 7. In this experiment, intact vesicles were incubated in a medium containing 0.1 mM-CaCl₂ and the rate of ATP synthesis was measured after most of the Ca^{2+} of the medium had been removed by the vesicles. Under these conditions a steep Ca^{2+} concentration gradient is formed across the vesicle membrane. In a parallel experiment, after the vesicle has been loaded, the Ca^{2+} concentration in the medium was increased to 8 mM to approximately equalize the Ca^{2+} concentration on the two sides of the membrane (de Meis & Carvalho, 1974). In both cases, ATP $\rightleftharpoons P_i$ exchange was detected. However, the rate of ATP synthesis is severalfold higher in the presence of a Ca^{2+} gradient than in its absence. The

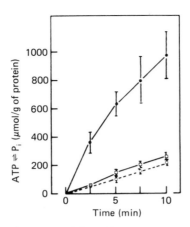

Fig. 7. *Role of the Ca²⁺ concentration gradient*

The basic assay medium composition was 30 mM-Tris–maleate buffer (pH 6.8), 20 mM-MgCl₂, 6 mM-³²Pᵢ and 10 mM-ATP. The reaction was performed at 37 °C: ●—●, 0.3 mg of vesicle protein/ml was incubated in the basic assay medium, containing in addition 0.1 mM-CaCl₂; the rate of ATP ⇌ Pᵢ exchange was measured after 98 % of the Ca²⁺ of the assay medium had been removed by the vesicles; ○---○, same as ●—●, but after the vesicles had been loaded the Ca²⁺ concentration of the medium was raised to 8 mM; □—□, 0.3 mg of leaky vesicle protein/ml was incubated in the basic assay medium, containing in addition 8 mM-CaCl₂. The values shown in the Figure represent the average ± s.e. of three experiments. For further details see de Meis & Carvalho (1974).

rate of ATP ⇌ Pᵢ exchange varies with the Pᵢ concentration in the medium, both in the presence and in the absence of a transmembrane Ca²⁺ gradient. However, the apparent K_m for Pᵢ differs in these two conditions, being in the range 40–60 mM when measured in leaky vesicles and in the range 2–4 mM when a gradient is formed (de Meis & Carvalho, 1974). Thus the variation of the enzyme affinity for Pᵢ seems to be a recurring theme. However, at this stage the possibility was raised that the apparent K_m for Pᵢ might vary owing to a kinetic parameter (the regulatory effect of ATP) and not because of a thermodynamic barrier overcome by the energy derived from the Ca²⁺ gradient. This in fact proved to be the case (Table 3). Note that in the experiments of Figs. 6 and 7 a high ATP concentration was used. Both in the presence and in the absence of a Ca²⁺ gradient, essentially the same apparent K_m for Pᵢ and for a given Pᵢ concentration, essentially the same rate of exchange was attained when (a) ATP was replaced by ITP (Carvalho *et al.*, 1976; Plank *et al.*, 1979), (b) the ATP concentration was decreased to the micromolar range (Scofano & de Meis, 1981; Oliva *et al.*, 1983) or (c) when in the presence of a high ATP concentration, the regulatory effect of ATP was impaired with the use of silver (de Meis & Sorenson, 1975).

The finding that the ATPase is able to drive an ATP ⇌ Pᵢ exchange in the absence of a transmembrane Ca²⁺ gradient indicates that the system must be able to conserve some of the energy released from ATP hydrolysis in a form which permits the resynthesis of a new ATP molecule from ADP and Pᵢ. The degree of energy conservation can be estimated from the ratio between the rates of hydrolysis and synthesis. Table 3 shows that, depending on the experimental

Table 3. *P_i dependence for nucleotide triphosphate = P_i exchange*

Data shown are from de Meis & Carvalho (1974), de Meis & Sorenson (1975), Carvalho *et al.* (1976) & Scofano *et al.* (1979).

Conditions	Substrates	K_m of P_i (mM)	Ratio ATP hydrolysis/ATP synthesis
Intact vesicles,	ATP (5–10 mM)	3–4	2–4
Ca²⁺ gradient	ITP (2–5 mM)	2–4	—
Leaky vesicles plus	ATP (5–10 mM)	40–60	20–50
8 mM-CaCl₂	ATP (0.05 mM)	2–4	2–3
	ITP (2 to 5 mM)	2–4	4–6
Leaky vesicles plus	ATP (5 mM)	6–8	2
Ag⁺ and 8 mM-CaCl₂			

conditions used, the same degree of energy conservation can be attained in the presence and the absence of a transmembrane Ca^{2+} gradient and that, as with the apparent K_m for P_i, this depends on a kinetic parameter related to the regulatory effect of ATP and not to a thermodynamic barrier overcome by the energy derived from the gradient. $ATP \rightleftharpoons P_i$ or $ITP \rightleftharpoons P_i$ exchange involves reactions 2–7 in the sequence shown in Fig. 1. For each P_i molecule released during hydrolysis of nucleotide triphosphate (reactions forward 2–7), one molecule of *E becomes available to participate in the reversal reaction, leading to synthesis of a nucleotide triphosphate molecule (reactions 7–2 backward). The ratio of hydrolysis to synthesis is higher than one to the extent that some of the enzyme units *E are converted into 2Ca·E instead of being driven in the reverse direction through phosphorylation by P_i. Therefore, the faster the forward rate of reactions 8 and 1, the faster the rate of hydrolysis and the smaller the ratio of hydrolysis to synthesis of ATP.

Net Synthesis of ATP: Ca²⁺ Jump

Knowles & Racker (1975) demonstrated that leaky vesicles can synthesize a small amount of ATP in the absence of a Ca^{2+} gradient. This was achieved by a two step procedure where, initially, the enzyme was phosphorylated by P_i at pH 6.3 in the presence of EGTA. Subsequently, upon the addition of ADP and 10 mM-CaCl₂ (Ca²⁺ jump), it was found that the phosphate of the phosphoenzyme was transferred to ADP, forming ATP. Fig. 8 shows a millisecond mixing and quenching experiment confirming the observation of Knowles & Racker (1975). This experiment shows that after the addition of Ca^{2+} and ADP, the disappearance of phosphoenzyme and synthesis of ATP are synchronous, the time for half-maximal ATP synthesis and phosphoenzyme dephosphorylation being in the range 20–40 ms.

The Ca^{2+} concentration added in the second step of the Ca^{2+} jump procedure is sufficient to saturate both Ca^{2+} binding sites of low and high affinity. After binding of Ca^{2+} to the high-affinity site, the phosphoenzyme is hydrolysed (Fig. 3). In transient kinetic experiments it was found that the time for half-maximal conversion of *E–P into 2Ca·E through reactions 6, 7, 8 and 1 shown in the

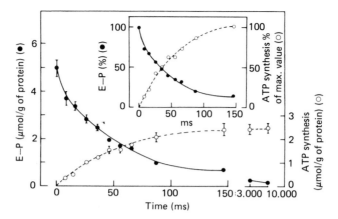

Fig. 8. *Dephosphorylation of E–P and ATP formation*

Syringe A contained 20 mM-Tris–maleate, 2 mM-EGTA, 20 mM-MgCl₂, 8 mM-³²Pᵢ (4 × 10⁷ c.p.m./μmol) and 1 mg of Ca²⁺ + Mg²⁺-dependent ATPase. The pH of this phosphorylating solution was 6.0. Syringe B contained 40 mM-CaCl₂, 4 mM-ADP, 10 mM-glucose and 4 units of hexokinase. Equal volumes (about 1 ml) of A and B were force-mixed together through a capillary tube. After quenching in 250 mM-perchloric acid, the content of phosphoenzyme and amount of [γ-³²P]ATP synthesized were assessed. Each point represents the mean ± s.e. of three experiments. For details see de Meis & Tume (1977).

sequence of Fig. 1 is in the range 400–800 ms (de Meis & Tume, 1977). Thus, in Fig. 8 net synthesis of ATP was observed because the rate at which the phosphoenzyme transfers its phosphate to ADP (reactions 5, 4 and 3 backwards) is much faster than the rate of hydrolysis of the phosphoenzyme promoted by binding of Ca²⁺ to the high affinity sites (reactions 6, 7, 8 and 1 forward). The amount of ATP synthesized by the Ca²⁺-jump procedure is proportional to and never exceeds the number of enzyme sites phosphorylated by Pᵢ (Knowles & Racker, 1975). In other words, the Ca²⁺ jump permits the enzyme to complete only one catalytic cycle, after which the system reaches equilibrium and no more ATP is synthesized. The ATP synthesized after the Ca²⁺ jump does not remain bound to the enzyme, being found free in solution.

Proton Gradient and Synthesis of ATP

The binding of Ca²⁺ to the high- and low-affinity sites of the ATPase is altered by changing the pH of the medium (Figs. 9 and 10). Both sites exhibit an increased affinity for Ca²⁺ with high pH and reduced affinity at low pH values, the relative difference of the two sites remaining unchanged in the range pH 6–8 (Verjovski-Almeida & de Meis, 1977). Based on this finding the rationale was developed that no ATP synthesis will be observed when the same Ca²⁺ concentration exists on both sides of the membrane and at levels sufficient to saturate the high-affinity site but not sufficient to allow binding to the low-affinity site. However, the system could synthesize ATP if the affinity of the two Ca²⁺ binding sites is modified by varying the pH on the two sides of the membrane, making it acidic in the assay medium and alkaline inside the vesicles. In such

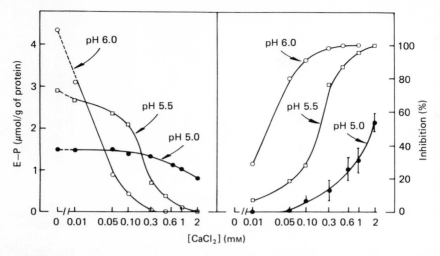

Fig. 9. *Effect of Ca²⁺ concentration and pH on the phosphorylation of the Ca²⁺-dependent ATPase*

The reaction medium contained 20 mM-Tris–maleate buffer, 8 mM-$^{32}P_i$ (10^7 c.p.m./μmol), 20 mM-MgCl$_2$, 1 mg of Ca²⁺-dependent ATPase protein and the CaCl$_2$ concentrations shown in the Figure. For zero Ca²⁺ concentration, CaCl$_2$ was omitted and 2.5 mM-EGTA was added. The final volume was 1 ml, and the pH of the mixture was 6.0 (○), 5.5 (□) or 5.0 (●). The reaction was performed at 25 °C; it was started by the addition of the enzyme and was arrested after 30 s of incubation by the addition of 3 ml of an ice-cold solution of perchloric acid (125 mM) containing 2 mM-orthophosphate. The values represent either the average of two experiments (○, □) or the average ± s.e. of five experiments (●). For details see de Meis & Tume (1977).

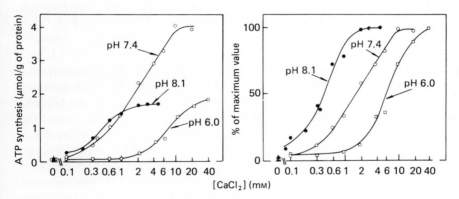

Fig. 10. *Ca²⁺ dependence*

The reaction medium (0.5 ml) contained 20 mM-MgCl$_2$, 8 mM-$^{32}P_i$, 2 units of hexokinase and 1 mg of Ca²⁺-dependent ATPase. The pH of the mixture was 6.0. The reaction was started by the addition of Ca²⁺-dependent ATPase and after 30 s it was added to 0.5 ml of a solution containing 10 mM-glucose, 4 mM-ADP, various concentrations of CaCl$_2$ to give the final concentrations shown in the Figure, and KOH in a concentration sufficient to increase the pH of the final mixture from 6.0 to either 7.4 (○) or 8.1 (●). KOH was not added to those incubations which were maintained at pH 6.0 (□). The reaction was stopped 1 min later by the addition of 0.1 ml of 100% trichloroacetic acid. For details see de Meis & Tume (1977).

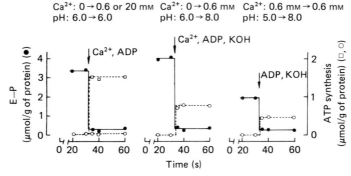

Fig. 11. *pH and Ca²⁺ jump*

Left: the reaction medium (0.5 ml) contained 20 mM-$MgCl_2$, 8 mM-$^{32}P_i$ (2.5×10^7 c.p.m./μmol), 2 units of hexokinase and 1 mg of Ca^{2+}-dependent ATPase protein. The pH of the medium was 6.0. The reaction was performed at 25 °C; it was started by the addition of the enzyme, and after 30 s of incubation (arrow) 0.5 ml of a solution containing 4 mM-ADP, 10 mM-glucose and either 1.2 ($\bigcirc$) or 40 mM-($\square$) $CaCl_2$ was added. Note that the concentrations of these reagents were halved after mixing. The pH of the assay medium remained constant (6.0) after this addition. The reaction was terminated at different incubation intervals as shown in the Figure by the additions of 0.1 ml of 100% (w/v) trichloroacetic acid. Aliquots of the deproteinized supernatant and the protein precipitates were analysed for [γ-^{32}P]ATP and E–^{32}P formation respectively. Middle: the experimental conditions and assay medium were the same as above, except that the second solution, in addition to 4 mM-ADP, 10 mM-glucose and 1.2 mM-$CaCl_2$, also contained 23 mM-KOH. Thus, the pH of the medium increased from 6.0 to 8.0 after mixing. Right: the reaction medium (0.8 ml) contained 9 mM-maleic acid, 20 mM-$MgCl_2$, 8 mM-$^{32}P_i$, 0.6 mM-$CaCl_2$, 2 units of hexokinase and 1 mg of Ca^{2+}-independent ATPase protein. The pH was 5.0. The reaction was started by the addition of the enzyme and 30 s later 0.2 ml of a mixture of ADP, glucose, $CaCl_2$ and KOH was added to give final concentrations of 2 mM-ADP, 5 mM-glucose and 40 mM-KOH. The $CaCl_2$ concentration remained constant (0.6 mM). After this addition, the pH of the medium increased from 5.0 to 8.0. Other experimental conditions were as described in the text: $\bigcirc$, $\square$, ATP synthesis; $\bullet$, E–P. For details see de Meis & Tume (1977).

a situation the phosphorylation by P_i will no longer be inhibited because the Ca^{2+} concentration will no longer be sufficient to saturate the external Ca^{2+} binding site of high affinity, and the Ca^{2+} concentration inside the vesicle could become sufficient to allow the transfer of the phosphate to ADP because of the increased affinity of the internal Ca^{2+} binding site of low affinity. We tested this hypothesis using leaky vesicles and the technique described for the Ca^{2+} jump, except that the Ca^{2+} concentration was maintained constant and the affinity of the two Ca^{2+} binding sites was varied by a sudden change of pH of the medium (de Meis & Tume, 1977; de Meis & Inesi, 1982). Fig. 9 shows that, in the presence of EGTA, the level of phosphoenzyme decreases when the pH value of the medium is decreased from 6 to 5. However, this leads also to a marked reduction of the inhibitory effect of Ca^{2+}. At pH 6, 0.6 mM-$CaCl_2$ completely inhibited phosphorylation by P_i, whereas at pH 5 the same $CaCl_2$ concentration produced only 20–30% inhibition.

In order to test the effect of pH on the transfer of phosphate to ADP, leaky vesicles were first phosphorylated by P_i at pH 6 and then the medium was diluted in solution containing different Ca^{2+} concentrations adjusted at different pH values. Fig. 10 shows that at pH 6 there is no synthesis of ATP in the presence of 0.6 mM-CaCl$_2$ whereas at pH 8.1 the same Ca^{2+} concentration triggers the transfer of most of the phosphoenzyme phosphate to ADP. Fig. 11 shows that after phosphorylation at pH 5, in the presence of 0.6 mM-CaCl$_2$, net synthesis of ATP is measured after the addition to the medium of ADP and KOH in an amount sufficient to raise the pH of the mixture from 5 to 8. Synthesis of ATP is not detected when Ca^{2+} is omitted from the assay medium. When extended to sealed vesicles, these results indicate that the sarcoplasmic reticulum ATPase can catalyse the synthesis of ATP when there are equal Ca^{2+} concentrations but different pH values on the two sides of the membrane. In such a situation then, there is a transmembrane H$^+$ gradient and not a Ca^{2+} gradient, and the synthesis of ATP is not due to the H$^+$ gradient. The difference in pH is required only to modify the affinity of the two Ca^{2+} binding sites of the enzyme. Recently, evidence obtained in different laboratories indicate that a small H$^+$ gradient is formed across the vesicle membrane during Ca^{2+} transport (Madeira, 1979, 1980; Chiesi & Inesi, 1980). Maintenance of large H$^+$ gradients does not seem possible because of the high H$^+$ permeability of the sarcoplasmic reticulum membrane. Besides its implication on the mechanism of Ca^{2+} transport, the data described above raise the possibility that, in membrane systems other than the sarcoplasmic reticulum, when there is a transmembrane H$^+$ gradient, different transport ATPases can be mobilized for the synthesis of ATP, besides those that specifically transport H$^+$ (de Meis & Tume, 1977).

Binding Energy

From the data presented it is concluded that each of the individual reactions involved in the hydrolysis and synthesis of ATP is reversible and can flow forward and backward, without using the osmotic energy that might be derived from the concentration gradient of Ca^{2+} across the membrane. For the synthesis of ATP in sealed vesicles, the large difference of Ca^{2+} concentration on the two sides of the membrane is needed only to meet the difference in affinity of the external and internal Ca^{2+} binding sites. In this view, the synthesis of ATP is promoted by the asymmetric binding of Ca^{2+} on the two sides of the membrane. For the enzyme to be phosphorylated by P$_i$, it is necessary that Ca^{2+} not be bound to the high-affinity site located on the outer surface of the membrane (Figs. 3 and 9). In order to transfer the phosphate from the phosphoenzyme to ADP, Ca^{2+} must bind to the low-affinity site located on the inner surface of the vesicle membrane (Figs. 6, 8, 10 and 11). According to the concept of binding energy, first proposed by Weber (1972a, b, 1974), the energy derived from the binding of Ca^{2+} to the enzyme is used for the synthesis of ATP. The standard free energy, derived from the binding of Ca^{2+} to the low-affinity site, can be calculated by using the equation

$$\Delta G^0 = RT \ln K_a^{in}$$

where K_a^{in} is the association constant for Ca^{2+}. For two Ca^{2+} bound for each enzyme unit at 30 °C, this gives a value of $\Delta G^0 = -35$ kJ/mol. After binding of Ca^{2+}, the phosphoenzyme undergoes a conformational change and the binding site is converted into high-affinity (reaction 4). For this transition

$\Delta G^0 = -RT \ln K_2^{\text{out}}/K_a^{\text{in}}$ where K_a^{out} is the association constant of the high-affinity binding site, which at pH 7 is 10^6 M. For two calcium ions bound for each enzyme unit at 30 °C this gives a value of $\Delta G^0 = -35$ kJ/mol. Therefore, for the overall process, binding of Ca^{2+} to the low-affinity site and conversion of the low- into the high-affinity site the ΔG^0 is -70 kJmol. This is sufficient for the synthesis of one mol of ATP for each mol of enzyme. In order to initiate a new catalytic cycle it is necessary to dissociate the Ca^{2+} bound to the high-affinity site. During reversal of the Ca^{2+} pump, continuous synthesis of ATP is measured because, at the end of each cycle, Ca^{2+} dissociates from this site due to the very low Ca^{2+} concentration found in the medium. Therefore continuous synthesis of ATP is observed because the enzyme cycle is between two media of different composition, one in which the activity of Ca^{2+} is high (vesicle lumen) and another where the activity of Ca^{2+} is very low (assay medium). The energy derived from the large difference of Ca^{2+} activity found in these two compartments is converted into binding energy. This is used for the conversion of the phosphoenzyme from low into high energy (conformational energy) and synthesis of ATP (chemical energy). Thus, during reversal of the Ca^{2+} pump, different forms of energy are interconverted.

Role of Water in the Conversion of Phosphoenzyme from Low into High Energy

The concept of high-energy compounds has been analysed primarily from a theoretical viewpoint. Until 1969 it was thought that intramolecular effects such as opposing resonance, electrostatic repulsions and electron distribution along the P–O–P backbone were the dominant factors contributing to the large negative free energies of hydrolysis of pyrophosphate and other high-energy phosphate compounds (Kalckar, 1941; Hill & Morales, 1951; Pullman & Pullman, 1963; Boyd & Lipscomb, 1969). In these theoretical formulations, water was ignored or regarded solely as a continuous dielectric for the purpose of calculating repulsion energies. George *et al.* (1970) and, later, Haynes *et al.* (1978) proposed that interactions of reactants and product with the solvent might play a more important role than intramolecular effects. Thus the energy of hydrolysis of a phosphate compound would be determined by the differences in solvation energies of reactants and products (George *et al.*, 1970). Haynes *et al.* (1978) calculated the energy of hydrolysis of several phosphate compounds in gas phase, including pyrophosphate and acetyl phosphate and compared these values with those reported in water (Table 4). Note that, in the gas phase, acetyl phosphate is not a high-energy compound. On the contrary, the large positive ΔH of hydrolysis indicates that when reactant and products are not solvated, acetyl phosphate is more stable than its products of hydrolysis. Acetyl phosphate and pyrophosphate have similar energies of hydrolysis in water, but there is a large difference between the corresponding energies of hydrolysis when the reactants are not solvated. Thus, in the gas phase an acyl phosphate cannot transfer its phosphate to pyrophosphate or a similar compound such as ADP.

Based on this observation we raised the possibility that, in the absence of a Ca^{2+} gradient, the catalytic site of the enzyme is hydrophobic so that P_i and

Table 4. ΔE of hydrolysis in gas phase and in water

The values shown are from Haynes *et al.* (1978).

Reaction	ΔE (kJ/mol)	
	Gas phase	Soln. in water
$H_3P_2O_7^- + H_2O \rightarrow$ $H_3PO_4 + H_2PO_4$	-3.9	-30.5
$H_2P_2O_7^{2-} + H_2O \rightarrow 2H_2PO_4^-$	-314.8	-28.5
$CH_3COOPO_3H^- + H_2O \rightarrow$ $CH_3COOH + H_2PO_4^-$	$+22.6$	-36.0
$CH_3COOPO_3^{2-} + H_2O \rightarrow$ $CH_3COOH + HPO_4^{2-}$	$+135.9$	

the aspartic acid would react as if in a gas phase. In this case, the major thermodynamic barrier for the phosphorylation of the enzyme by P_i would be the entry of P_i into the catalytic site, i.e. the partition of P_i from a hydrophilic environment (assay medium) into a hydrophobic environment (catalytic site). Factors facilitating this partition should also facilitate the phosphorylation of the enzyme by P_i. In the absence of a transmembrane Ca^{2+} gradient, the phosphoenzyme formed from P_i would not be able to transfer its phosphate to ADP because of the larger difference of ΔH of hydrolysis of the acyl phosphate and pyrophosphate residues of ATP (Table 4). The conformational change of the enzyme observed after the binding of Ca^{2+} to the low-affinity site would permit the entry of water into the catalytic site, with subsequent solvation of both acyl phosphate residues and ADP. As a result, the ΔH of hydrolysis of the acyl phosphate and pyrophosphate residues would become equal, and the synthesis of ATP would proceed spontaneously.

Experimental conditions which impair the entry of water into the catalytic site should also impair the synthesis of ATP. According to this hypothesis, the concept of 'high-energy' and 'low-energy' forms of phosphoenzyme would be related solely to the water activity in the catalytic site (de Meis *et al.*, 1980). This hypothesis was tested by measuring the phosphorylation of the enzyme by P_i and the synthesis of ATP in the presence of various water–organic solvent mixtures, namely dimethyl sulphoxide (DMSO), glycerol and NN-dimethyl-formamide (DMFA). In the presence of these solvents the phosphorylation of leaky vesicles by P_i is largely facilitated (de Meis *et al.*, 1980, 1982; Otero & de Meis, 1982; Kosk-Kosicka *et al.*, 1983; Chiesi *et al.*, 1984). The apparent K_m for P_i no longer varies with the pH of the medium. In addition, there is a significant increase of the apparent affinity of the enzyme for P_i. In the presence of 40% DMSO, the apparent K_m for P_i decreases three (pH 6) to five (pH 8) orders of magnitude (Table 5). The decrease of the apparent K_m for P_i seems to be related to an increase of the hydrophobicity of the medium promoted by the addition of organic solvents.

If the catalytic site of the enzyme is hydrophobic, then the partition of P_i from the assay medium into the catalytic site should be facilitated when the difference

Table 5. *Apparent K_m for P_i at different pH values*

DMFA, *NN*-Dimethylformamide; DMSO, dimethyl sulphoxide. For further details see de Meis *et al.* (1980).

Organic solvent added (v/v)	Apparent K_m (M)		
	pH 6.0	pH 7.0	pH 8.0
None	1.5×10^{-3}	10^{-2}	$\gg 10^{-2}$
Glycerol (20%)	0.8×10^{-3}		1.5×10^{-3}
Glycerol (40%)	0.5×10^{-3}	10^{-3}	10^{-3}
DMFA (13%)	0.5×10^{-3}		1.1×10^{-3}
DMFA (26%)			0.7×10^{-3}
DMSO (20%)	7.0×10^{-5}	2.5×10^{-5}	4.0×10^{-5}
DMSO (40%)	7.0×10^{-6}	7.0×10^{-6}	2.0×10^{-6}

Table 6. *Partition of P_i between media of different hydrophobicity*

The assay medium composition was 50 mM-Tris–maleate buffer, 0.5 mM-EGTA, 10 mM-MgCl$_2$, 2 mM-^{32}P$_i$ (4×10^8 c.p.m./μmol), and the Me$_2$SO or glycerol concentrations shown. To 1 ml of the assay medium was added 1 ml of benzene or benzene/isobutyl alcohol. The tube was vigorously stirred for 60 s. After phase separation an aliquot of the organic phase was counted for ^{32}P in a scintillation of P$_i$ which can be detected in the organic phase 5×10^{-10}. The partition coefficient was calculated by dividing the concentration of P$_i$ in the organic phase by its concentration in the aqueous phase. nd, Non-detectable. DMSO, Dimethyl sulphoxide. For details see de Meis *et al.* (1980).

Addition to assay medium (v/v)	Organic–aqueous phase partition coefficient		
	Benzene	Benzene/isobutyl alcohol, (v/v)	Benzene/isobutyl alcohol (55:45, v/v)
None	nd	nd	nd
DMSO (40%)	nd	2.0×10^{-5}	1.6×10^{-4}
Glycerol (40%)	nd	nd	10^{-5}

of hydrophobicity between these compartments is decreased. To test this hypothesis, we measured the partition of P$_i$ between the water phase (phosphorylation medium) and different mixtures of benzene–isobutyl alcohol (organic phase). The hydrophobicity of the organic phase increased with an increase in the benzene/isobutyl alcohol ratio. In accordance with the hypothesis raised above, it was found that P$_i$ entered in significant amounts into the organic layer only when DMSO or glycerol was included in the medium (Table 6). The amount of P$_i$ which entered the organic layer increased as the benzene/isobutyl alcohol ratio decreased. DMSO was more effective than glycerol in promoting the entry of P$_i$.

The synthesis of ATP, measured by the Ca^{2+} jump procedure, is abolished in the presence of 40% DMSO (Fig. 12). After the addition of ADP and Ca^{2+}, the phosphoenzyme is slowly hydrolysed. This is not accompanied by the synthesis of ATP. The inhibition of the ATP synthesis seems to be related to the decrease of water activity due to the DMSO in the medium. A rapid cleavage

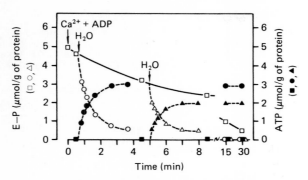

Fig. 12. *Inhibition of ATP synthesis by Me₂SO*

The enzyme (7 mg of protein/ml) was phosphorylated at 0 °C in 1.7 ml of a medium containing 30 mM-Tris–maleate buffer (pH 8.0), 0.05 mM-EGTA, 5 mM-MgCl₂, 1 mM-³²P$_i$ and 40% (v/v) dimethyl sulphoxide; 15 min after the addition of enzyme 0.017 ml of ADP (20 mM) and 0.02 ml of CaCl₂ (20 mM) were simultaneously added. The final concentration of Ca²⁺ and ADP were 0.15 and 0.40 mM, respectively. The reaction was arrested at different incubation intervals after the addition of ADP and Ca²⁺. In a parallel set of tubes, after the addition of ADP and CaCl₂, 0.1 ml of the mixture was diluted in 1.9 ml of a solution of identical composition at 0 °C except that ³²P$_i$ and dimethyl suphoxide were omitted. In the Figure this is represented as (↓H₂O). Radioactive phosphoenzyme (open symbols) and ATP synthesis (closed symbols) were measured as described in the text: □, ■, 40% (v/v) dimethyl sulphoxide; ○, ●, △, ▲, after dilution of dimethyl sulphoxide from 40 to 2% (v/v). For details see de Meis *et al.* (1980).

of the phosphoenzyme coupled with a stoichiometric synthesis of ATP was observed if, after the addition of ADP and Ca²⁺, the DMSO concentration was suddenly decreased from 40% to 2% (de Meis *et al.*, 1980; de Meis & Inesi, 1982; Kosk-Kosicka *et al.*, 1983; Chiesi *et al.*, 1984). The hypothesis proposed states that 'high'- and 'low'-energy forms of the phosphoenzyme are correlated with the availability of water at the catalytic site of the enzyme. The experimental results are consistent with this hypothesis. As predicted, phosphorylation of the enzyme by P$_i$ is facilitated when the hydrophobicity of the medium is increased by the addition of organic solvents. Finally, after the addition of Ca²⁺, the phosphoenzyme is able to transfer its phosphate to ADP only if the water activity of the medium is increased by dilution of the organic solvent. In a previous section of this review it was mentioned that during reversal of the Ca²⁺ pump different forms of energy are interconverted by the Ca²⁺-dependent ATPase. The data obtained with the use of organic solvents indicate that conformation energy and solvation energy are interconverted. Thus, the following sequence of energy transduction is proposed:

$$\Delta Ca^{2+} \text{ activity} \rightleftharpoons \text{binding energy} \rightleftharpoons \text{conformational energy} \rightleftharpoons \text{solvation}$$
$$\text{energy} \rightleftharpoons \text{chemical energy}$$

where ΔCa²⁺ activity refers to the difference of Ca²⁺ concentration found in the vesicle lumen and in the assay medium when the Ca²⁺ pump is reversed. Additional evidence that in the catalytic site there is a hydrophobic–hydrophilic transition during the catalytic cycle has been recently reported by Dupont & Pougeois (1983) and Nakamoto & Inesi (1984).

Table 7. *Variability of the $K_{eq.}$ of phosphate compounds*

References: (a) de Meis & Vianna (1979), de Meis (1981), de Meis & Masuda (1974), Masuda & de Meis (1973), Lacapere *et al.* (1981), de Meis *et al.* (1982), Guillain *et al.* (1984). (b) Boyer *et al.* (1973, 1982), Cross & Boyer (1975). (c) Janson *et al.* (1979), Springs *et al.* (1981), Cooperman (1982). (d) Alberty (1968), George *et al.* (1970), Haynes *et al.* (1978). (e) Flodgaard & Fleron (1974), de Meis (1984).

| | | $K_{eq.}$ of hydrolysis (M) | |
Enzyme	Phosphate compound	At the catalytic site	In solution
Ca^{2+}-dependent ATPase of sarcoplasmic reticulum and (Na$^+$+K$^+$)-ATPase of plasma membrane	Acyl phosphate residue	0.1–1.5[a]	10^5–10^6 [d]
F$_1$-ATPase of mitochondria and chloroplasts, myosin	ATP	1–10 [b]	10^5–10^6 [d]
Yeast inorganic pyrophosphatase	PP$_i$	4.5 [c]	10^3–10^4 [e]

Phosphate Compounds of Low and High Energy

A common feature of the enzymes shown in Table 7 is that phosphate compounds which have a high energy of hydrolysis in water can be spontaneously formed in the catalytic site without an apparent need of an energy input.

The (Na$^+$, K$^+$)-ATPase of plasma membrane catalyses a continuous synthesis of ATP when a Na$^+$ gradient is formed across the membrane (Garrahan & Glynn, 1966, 1967). In the absence of an ionic gradient this enzyme is also spontaneously phosphorylated by P$_i$, forming an acyl phosphate residue of low energy (Post *et al.*, 1975; Taniguchi & Post, 1975). After the binding of Na$^+$ to a low-affinity site of the enzyme (K_s about 300 mM at pH 7.4) the phosphoenzyme is converted into a high-energy form and becomes able to transfer its phosphate to ADP, leading to the synthesis of ATP. Like the Ca^{2+}-dependent ATPase, in the absence of ionic gradients the (Na$^+$, K$^+$)-ATPase is also able to catalyse a continuous ATP = P$_i$ exchange in the presence of a high Na$^+$ concentration (Taniguchi & Post, 1975; Moraes & de Meis, 1982).

The elegant water–phosphate exchange experiments performed in Boyer's laboratory (Boyer *et al.*, 1973, 1982; Cross & Boyer, 1975; Hutton & Boyer, 1979) revealed that ATP is spontaneously formed in the catalytic site of F$_1$-ATPase of mitochondria and chloroplasts without the need of energy which may be derived from a H$^+$ gradient. This ATP remains tightly bound to the enzyme and does not dissociate into the assay medium unless a H$^+$ gradient is formed across the membrane of the organele.

Recently it has been shown that soluble yeast inorganic pyrophosphatase retains tightly bound pyrophosphate (Janson *et al.*, 1979; Springs *et al.*, 1981; Cooperman, 1982). In the conditions prevailing in the cytosol, the observed equilibrium constant ($K_{obs.}$) of pyrophosphate hydrolysis is in the range 10^3–10^4 (Flodgaard & Fleron, 1974; de Meis, 1984). In contrast, the $K_{obs.}$ of the tightly bound pyrophosphate is 5 (Janson *et al.*, 1979; Cooperman, 1982). In analogy with the F$_1$-ATPase, the membrane-bound pyrophosphatase found in *Rhodo-*

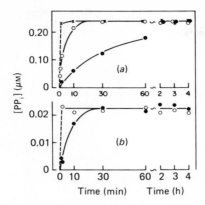

Fig. 13. *Rate of pyrophosphate formation*

The assay medium consisted of 4 mM-^{32}P (pH 6.6) and 100 mM- (*a*) or 10 mM- (*b*) MgCl$_2$. The reaction was performed at 35 °C. It was started by the addition of yeast inorganic pyrophosphate to a final concentration of 0.100 μg/ml (×), 0.010 μg/ml (○) or 0.001 μg/ml (●). For further details see de Meis (1984).

spirillum rubrum catalyses the synthesis of pyrophosphate when a H$^+$ gradient is formed across the chromatophore membrane (Baltscheffsky *et al.*, 1966, 1982; Lahti, 1983; Nyren & Baltscheffsky, 1983).

For the Ca^{2+}-dependent ATPase the data obtained with organic solvents indicate that an acyl phosphate residue can be formed spontaneously owing to the low water activity found in the catalytic site of the enzyme. Recently it has been shown that there is an increase in the amount of tightly bound ATP formed in the catalytic site of F$_1$-ATPase when the water activity of the medium is decreased by the addition of DMSO (Sakamoto & Tonomura, 1983; Yoshida, 1983; Cross *et al.*, 1984). Therefore we raised the possibility that a low water activity could be a common feature permitting the spontaneous synthesis of acyl phosphate residues and tightly bound ATP and pyrophosphate at the catalytic site of the different enzymes shown in Table 7. To explore this possibility, a direct approach is to measure the $K_{eq.}$ of hydrolysis of a phosphate compound in solutions with different water activities. For this purpose, inorganic pyrophosphate was chosen. Yeast inorganic pyrophosphatase is a soluble enzyme which catalyses the synthesis of pyrophosphate (Flodgaard & Fleron, 1974; de Meis, 1984) at a velocity which depends on the concentration of enzyme used (Fig. 13). With different enzyme concentrations, essentially the same amounts of pyrophosphate were found after long incubation intervals. In all subsequent experiments it was assumed that equilibrium between synthesis and hydrolysis of pyrophosphate had been attained when the concentration of radioactive pyrophosphate in the medium was the same in the presence of two different enzyme concentrations after two different incubation intervals (de Meis, 1984). With the use of very high enzyme concentrations (0.5–1.0 mg of protein/ml), Janson *et al.* (1979) and Springs *et al.* (1981) have shown the formation of enzyme-bound pyrophosphate. The amount of tightly bound pyrophosphate formed was proportional to the amount of enzyme present and, on a molar basis,

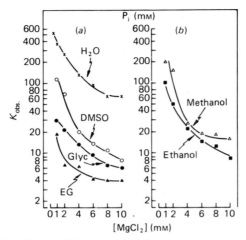

Fig. 14. *Effect of different organic solvents in the presence of* 1 mM-$^{32}P_i$

The assay medium consisted of 50 mM-Tris–HCl buffer (pH 7.8), 1 mM-$^{32}P_i$ without organic solvent
($\times$) or with 30% dimethyl sulphoxide (DMSO) ($\bigcirc$), 60% glycerol (Glyc) ($\bullet$), 60% ethylene glycol
(EG) ($\blacktriangle$), 30% methanol ($\triangle$), or 30% ethanol ($\blacksquare$). The enzyme concentrations were either 1 μg
and 10 μg/ml ($\bigcirc$, $\bullet$, $\blacktriangle$) or 0.1 and 1.0 μg/ml ($\times$, $\triangle$, $\blacksquare$). Essentially the same results were obtained
after 2 h and 4 h incubation at 35 °C. The values shown in the Figure are averages of the values
obtained with these incubation intervals and enzyme concentrations. For further details see de Meis
(1984).

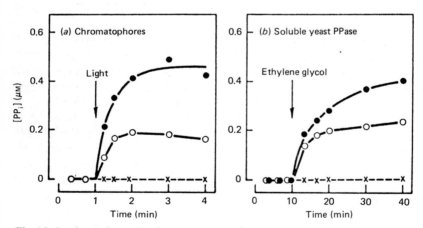

Fig. 15. *Synthesis of pyrophosphate by chromatophores of Rhodospirillum rubrum and by soluble*
yeast inorganic pyrophosphate

The assay medium composition was 50 mM-Tris–HCl, pH 7.8, 10 mM-MgCl$_2$ and either 1 mM-$^{32}P_i$
($\bigcirc$, $\times$) or 2 mM-$^{32}P_i$ ($\bullet$, $\times$). In (a), with chromatophores (total of 0.1 mg of bacteriochlorophyll/ml)
in the absence ($\bigcirc$, $\bullet$) and the presence ($\times$) of 5 μM of the H$^+$ ionophore carbonyl cyanide
p-trifluoromethoxyphenylhydrazone. The arrow indicates when the light was turned on (20 W/m²).
In (b) with soluble yeast inorganic pyrophosphatase (5 μg/ml). Conditions ($\bigcirc$, $\bullet$) were as in (a)
except that, instead of chromatophores, soluble yeast inorganic pyrophosphatase (5 μg/ml) was
used. The arrow indicates the simultaneous addition of ethylene glycol, Tris–HCl (pH 7.8), $^{32}P_i$,
MgCl$_2$ and enzyme in amounts sufficient to attain a final concentration of ethylene glycol of 60%
(v/v) and to maintain the concentration of the reagents the same as before addition of organic
solvent. $\times$, Without ethylene glycol.

Table 8. *Pyrophosphate of high and low energy*

The assay medium contained 50 mM-Tris–HCl buffer (pH 8.0), 1 mM-$^{32}P_i$ and 0.9 mM-$MgCl_2$. The concentrations of yeast inorganic pyrophosphatase were 0.001 and 0.005 μg/ml in totally aqueous media, 0.050 and 0.025 μg/ml in the presence of ethylene glycol and 1 and 5 μg/ml in the presence of polyethylene glycol 8000. Essentially the same results were obtained in each case with both enzyme concentrations after incubation for 4 and 7 h at 30 °C. The pyrophosphate concentrations are the average $\pm$ S.E. of the number of experiments shown in parentheses.

Additions	$[PP_i]$ (μM)	$K_{obs.}$ (M)	$\Delta G^0_{obs.}$ (kJ/mol)
None	0.002 ± 0.001 (5)	500.0	-15.5
Ethylene glycol (60%, v/v)	0.063 ± 0.011 (4)	15.9	-7.1
Polyethylene glycol 8000 (50%, w/v)	8.305 ± 0.136 (11)	0.1	$+5.4$

corresponded to about 10% of the total enzyme used. The concentrations of enzyme used in our experiments (Fig. 13–15) were much smaller than those required to measure enzyme-bound pyrophosphate. In the presence of 1 μg of enzyme protein/ml, the concentration of enzyme-bound pyrophosphate present in the mixture should be about 0.003 μM (pyrophosphatase M_r is 63000–71000; Lahti, 1983), i.e. much smaller than that measured in our experimental conditions. Thus, the measured pyrophosphate corresponds to the equilibrium concentration in solution and the error derived from enzyme-bound pyrophosphate is not significant.

In totally aqueous medium (Fig. 14), increasing Mg^{2+} concentration leads to a decrease of $K_{obs.}$ for pyrophosphate hydrolysis (Flodgaard & Fleron, 1974). However, in water the value of $K_{obs.}$ never reaches values as low as that reported for the tightly bound pyrophosphate, regardless of the pH value or the Mg^{2+} and P_i concentration in the medium. A more pronounced decrease of $K_{obs.}$ is observed when the water activity of the medium is decreased by the addition of different organic solvents or of polymers of ethylene glycol (de Meis, 1984). In the presence of a low Mg^{2+} concentration (0.9 mM) there is a remarkable decrease of $K_{obs.}$ when the water activity of the solution is decreased with polymers of ethylene glycol. The ΔG^0 of pyrophosphate hydrolysis changes from a negative value to a positive value after the addition to the medium of 50 g of polyethylene glycol 8000/100 ml (Table 8). The pyrophosphate synthesized in the presence of polyethylene glycol is readily hydrolysed when the concentration of the polymer in the medium is suddenly decreased.

The values of $K_{obs.}$ measured in Fig. 13, Table 8 and Fig. 14 represent the contribution of the $K_{eq.}$ of different ionic species. In the presence of ethylene glycol there is a decrease of $K_{eq.}$ of all ionic reactions calculated.

Springs *et al.* (1981) reported a value of 4.5 for the $K_{obs.}$ of the tightly bound pyrophosphate. In the range pH 7.2–8.0 values of $K_{obs.}$ between 5.0 and 0.1 were obtained in the presence of Mg^{2+} and different organic solvents. Therefore it may be that both Mg^{2+} and water structure are responsible for the low $K_{obs.}$ of pyrophosphate hydrolysis of the tightly bound pyrophosphate found on the catalytic site of the enzyme. In physiological conditions, magnesium enters the catalytic site both as Mg^{2+}–pyrophosphate and as Mg^{2+}–P_i complexes (Cooperman, 1982; Lahti, 1983).

Synthesis of Pyrophosphate in the Absence of H^+ Gradient

Coupled to the electron transport chain in the chromatophore of *Rhodospirillum rubrum* there is a membrane-bound pyrophosphatase that catalyses the synthesis of pyrophosphate in light (Baltscheffsky *et al.*, 1966, 1982; Lahti, 1983; Nyren & Baltscheffksy, 1983). Synthesis is impaired when the electrochemical proton gradient formed across the chromatophore membrane by light is collapsed by the addition of a proton ionophore to the medium. When the water activity of the medium is decreased by the addition of organic solvent, soluble yeast pyrophosphatase catalyses the synthesis of pyrophosphate in amounts similar to those attained with the use of chromatophores under illumination (Fig. 15). For a given P_i concentration, the maximal amount of pyrophosphate synthesized by chromatophores varies, depending on the pH and $MgCl_2$ concentration in the medium. A similar pH and $MgCl_2$ dependence is observed for the synthesis of pyrophosphate catalysed by soluble pyrophosphatase in the presence of organic solvent. A comparison of Figs. 15 and Table 8 shows that, in the presence of polyethylene glycol 8000, soluble pyrophosphatase can catalyse the synthesis of a larger amount of pyrophosphate than that attained with chromatophores under illumination.

Conclusions

For theoretical consideration of the mechanism of energy transduction, F_1-ATPase, Ca^{2+}-dependent ATPase and (Na^+, K^+)-ATPase have been analysed together under the assumption that the mechanism of ATP synthesis is similar in all of them (Boyer *et al.*, 1977; Tanford, 1983). This generalization is derived from the fact that all of these ATPases are membrane-bound enzymes, all are able to use the energy derived from a transmembrane ionic gradient for the synthesis of ATP, all undergo a conformational change and in the catalytic site of all ATPases a high-energy phosphate compound is spontaneously formed. Inorganic pyrophosphatase may also be included in this generalization. At present it is not known whether or not this enzyme undergoes a conformational change during the catalytic cycle. However, the tightly bound pyrophosphate is comparable with the tightly bound ATP of F_1-ATPases and in chromatophores of *Rhodospirillum rubrum* the energy derived from the H^+ gradient is used for the synthesis of pyrophosphate. The data obtained with the Ca^{2+}-dependent ATPase and with inorganic pyrophosphatase suggests that solvent structure at the catalytic site is involved in the process of energy transduction. It may be that for the different enzymes shown in Table 7, the energy derived from ionic gradients is used to create media with a particular composition and structure of the solvent. In one medium, the phosphate compound would be of low energy and could be formed spontaneously (e.g. a high concentration of Mg^{2+} and lower water activity). In a second medium, the molecule would become of high energy (e.g. low Mg^{2+} concentration and high water activity). Media of different composition can be found in two compartments of the cell, as, for instance, in the cytosol and in the lumen of the sarcoplasmic reticulum (de Meis, 1981) or, at the molecular level, in the catalytic site of an enzyme. Energy would be used by the enzyme to change its conformation, and two different microenvironments would be found in the catalytic site, one before and another after the conformational change (de Meis, 1982, 1984; de Meis *et al.*, 1980). According to the mechanism proposed, synthesis of a compound and its change from low energy to high energy can occur in the presence of a large excess of water because it is not dependent on the water concentration in the medium but on the manner in which water is organized around the molecule (George *et al.*, 1970). Notice that in Figs. 13 and 14 and Table 8 the water concentration in the different mixtures of solvents used varied between 22 and 39 M.

This work was supported by grants from FINEP (Financiadora de Estudos e Projetos) and CNPqC Conselho Nacional de Desenvolvimento Científico e Tecnológico).

References

Alberty, R. A. (1968) *J. Biol. Chem.* **243**, 1337–1343
Baltscheffsky, M., Von Stedingt, L. V., Heldt, H. W. & Klingenberg, M. (1966) *Science* **153**, 1120–1124
Baltscheffsky, M., Baltscheffsky, H. & Borrk, J. (1982) *Topics in Photosynthesis* **4**, 249–271
Barlogie, B., Hasselbach, W. & Makinose, M. (1971) *FEBS Lett.* **12**, 267–268
Beil, F. U., Chak, D. & Hasselbach, W. (1977) *Eur. J. Biochem.* **81**, 151–164
Blasie, J. K., Herbette, L., Pierce, D., Pascolini, D., Scarpa, A. & Fleischer, S. (1982) *Ann. N.Y. Acad. Sci.* **402**, 478–484
Boyd, D. B. & Lipscomb, W. N. (1969) *J. Theor. Biol.* **25**, 403–420
Boyer, P. D., Cross, R. L. & Momsen, W. (1973) *Proc. Natl. Acad. Sci. U.S.A.* **70**, 2837–2839
Boyer, P. D., Chance, B., Ernster, L., Mitchell, P., Racker, E. & Slatter, E. C. (1977) *Annu. Rev. Biochem.* **46**, 955–1026
Boyer, P. D., Kohlbrenner, W. F., Melntosh, D. B., Smith, L. T. & O'Neal, C. C. (1982) *Ann. N.Y. Acad. Sci.* **402**, 65–83
Carvalho, A. P. (1968) *J. Gen. Physiol.* **51**, 427–442
Carvalho, A. P. & Leo, B. (1967) *J. Gen. Physiol.* **50**, 1327–1352
Carvalho, M. G. C., Souza, D. O. & de Meis, L. (1976) *J. Biol. Chem.* **251**, 3629–3636
Chiesi, M. & Inesi, G. (1980) *Biochemistry* **19**, 2912–2918
Chiesi, M., Zurini & Carafoli, E. (1984) *Biochemistry* **23**, 2595–2600
Coan, C. R. & Inesi, G. (1977) *J. Biol. Chem.* **252**, 3044–3049
Cooperman, B. S. (1982) *Methods Enzymol.* **87**, 526–548
Cross, R. L. & Boyer, P. D. (1975) *Biochemistry* **14**, 392–398
Cross, R. L., Cunninghan, D. & Tamura, J. K. (1984) *Current Topics in Cellular Regulation* **24**, 335–344
de Meis, L. (1976) *J. Biol. Chem.* **251**, 2055–2062
de Meis, L. (1981) *The Sarcoplasmic Reticulum, Transport and Energy Transduction*, The Wiley Series: Transport in the Life Sciences (Bittar, E. E., ed.), vol. 2. John Wiley and Sons, New York
de Meis, L. (1982) *Annals N.Y. Acad. Sci.* **402**, 535–548
de Meis, L. (1984) *J. Biol. Chem.* **259**, 6090–6097
de Meis, L. & Boyer, P. D. (1978) *J. Biol. Chem.* **253**, 1556–1559
de Meis, L. & Carvalho, M.G.C. (1974) *Biochemistry* **13**, 5032–5038
de Meis, L. & de Mello, M. C. F. (1973) *J. Biol. Chem.* **248**, 3691–3701
de Meis, L. & Inesi, G. (1982) *J. Biol. Chem.* **257**, 1289–1294
de Meis, L. & Inesi, G. (1982) in *The Transport of Calcium by Sarcoplasmic Reticulum and Various Microsomal Preparations in Membrane Transport of Calcium* (Carafoli, E., ed.), pp. 141–186, Academic Press, New York
de Meis, L. & Masuda, H. (1974) *Biochemistry* **13**, 2057–2062
de Meis, L. & Sorenson, M. M. (1975) *Biochemistry* **14**, 2739–2744
de Meis, L. & Tume, R. K. (1977) *Biochemistry* **16**, 4455–4563

de Meis, L. & Vianna, A. L. (1979) *Annu. Rev. Biochem.* **48**, 275–292
de Meis, L., Carvalho, M. G. C. & Sorenson, M. M. (1979) in *Concepts on Membranes in Regulation and Excitation* (Rocha e Silva, M. & Soarez-Kurtz, G., eds.), pp. 7–19, Raven Press, New York
de Meis, L., Martins, O. B. & Alves, E. W. (1980) *Biochemistry* **19**, 4252–4261
de Meis, L., Otero, A. S., Martins, O. B., Alves, E. W., Inesi, G. & Makamoto, R. (1982) *J. Biol. Chem.* **257**, 4993–4998
Dupont, Y. (1977) *Eur. J. Biochem.* **72**, 357–363
Dupont, Y. & Leigh, J. R. (1978) *Nature (London)* **273**, 396–398
Dupont, Y. & Pougeois, R. (1983) *FEBS Letts.* **156**, 93–98
Ebashi, S. & Lipmann, F. (1962) *J. Cell Biol.* **14**, 389–400
Flodgaard, H. & Fleron, P. (1974) *J. Biol. Chem.* **249**, 3465–3474
Froehlich, J. P. & Taylor, E. W. (1975) *J. Biol. Chem.* **250**, 2013–2021
Froehlich, J. P. & Taylor, E. W. (1976) *J. Biol. Chem.* **251**, 2307–2315
Garrahan, P. J. & Glynn, I. M. (1966) *Nature (London)* **211**, 1414–1415
Garrahan, P. J. & Glynn, I. M. (1967) *J. Physiol. (London)* **192**, 237–256
George, P., Witonsky, R. J., Trachtman, M., Wu, C., Dorwost, W., Richman, L., Richaman, W., Shurayh, F. & Lentz, B. (1970) *Biochim. Biophys. Acta* **223**, 1–15
Guillain, F., Champeil, P. & Boyer, P. D. (1984) *Biochemistry* **23**, 4754–4761
Guimarães-Motta, H. & de Meis, L. (1980) *Arch. Biochem. Biophys.* **203**, 395–403
Hasselbach, W. (1978) *Biochim. Biophys. Acta* **515**, 23–53
Hasselbach, W. & Makinose, M. (1961) *Biochem. Z.* **333**, 518–528
Hasselbach, W., Medda, P., Migala, A. & Agostini, B. (1983) *Z. Naturforsch.* **38c**, 1015
Haynes, D. M., Kenyon, G. L. & Kollman, P. A. (1978) *J. Am. Chem. Soc.* **100**, 4331–4340
Hill, T. L. & Morales, M. H. (1951) *J. Am. Chem. Soc.* **73**, 1656–1660
Hutton, R. L. & Boyer, P. D. (1979) *J. Biol. Chem.* **254**, 9990–9993
Inesi, G., Goodman, J. J. & Watanabe, S. (1967) *J. Biol. Chem.* **242**, 4637–4643
Inesi, G., Kurzmack, M., Coan, C. & Lewis, D. E. (1980) *J. Biol. Chem.* **255**, 3025–3031
Inesi, G., Watanabe, T., Coan, C. & Murphy, A. (1982) *Ann. N.Y. Acad. Sci.* **402**, 515–534
Janson, C. A., Degani, C. & Boyer, P. D. (1979) *J. Biol. Chem.* **254**, 3743–3749
Kalckar, H. M. (1941) *Chem. Rev.* **28**, 71–142
Kanazawa, T. (1975) *J. Biol. Chem.* **250**, 113–119
Kanazawa, T. & Boyer, P. D. (1973) *J. Biol. Chem.* **248**, 3163–3172
Knowles, A. F. & Racker, E. (1975) *J. Biol. Chem.* **250**, 1949–1951
Kosk-Kosicka, D., Kurzmack, M. & Inesi, G. (1983) *Biochemistry* **22**, 2559–2567
Lacapere, J. J., Gingold, M. P., Champeil, P. & Guillan, F. (1981) *J. Biol. Chem.* **256**, 2302–2306
Lahti, R. (1983) *Microbiol. Rev.* **47**, 169–179
Landgraf, W. C. & Inesi, G. (1969) *Arch. Biochem. Biophys.* **130**, 111–118
Madeira, V. M. (1979) *Arch. Biochem. Biophys.* **193**, 22–27
Madeira, V. M. (1980) *Arch. Biochem. Biophys.* **200**, 319–325
Makinose, M. (1971) *FEBS Lett.* **12**, 269–270
Makinose, M. (1972) *FEBS Lett.* **25**, 113–115
Makinose, M. & Hasselbach, W. (1971) *FEBS Lett.* **12**, 271–272
Martin, W. M. & Tanford, C. (1981) *Biochemistry* **20**, 4597–4603
Masuda, H. & de Meis, L. (1973) *Biochemistry* **12**, 4581–4585
McIntosh, D. B. & Boyer, P. D. (1983) *Biochemistry* **22**, 2867–2875
McIntosh, D. P. & Davidson, G. A. (1984) *Biochemistry* **23**, 1959–1965
Mitchell, P. (1966) *Chemiosmotic Coupling in Oxidative and Photosynthetic Phosphorylation*, Glynn Research, Bodmin, U.K.
Mitchell, P. (1979) *Eur. J. Biochem.* **95**, 1–20
Moraes, V. L. G. & de Meis, L. (1982) *Biochim. Biophys. Acta* **688**, 131–137
Nakamoto, R. K. & Inesi, G. (1984) *J. Biol. Chem.* **259**, 2961–2970
Nyren, P. & Baltscheffsky, M. (1983) *FEBS Lett.* **155**, 125–130
Oliva, J. M., de Meis, L. & Inesi, G. (1983) *Biochemistry* **22**, 5822–5825
Otero, A. S. & de Meis, L. (1982) *Z. Naturforsch.* **37**, 527–531
Plank, B., Hellman, G., Punzengruber, C. & Suko, J. (1979) *Biochim. Biophys. Acta* **550**, 259–268
Post, R. L., Toda, G. & Rogers, F. N. (1975) *J. Biol. Chem.* **250**, 691–701
Prager, R., Punzengruber, C., Kolassa, N., Winker, F. & Suko, J. (1979) *Eur. J. Biochem.* **97**, 239–250
Pullman, A. & Pullman, B. (1963) *Quantum Biochemistry*, Interscience, New York
Punzengruber, C., Prager, R., Kolassa, N., Winkler, F. & Suko, J. (1978) *Eur. J. Biochem.* **92**, 349–359
Rauch, B., Chak, D. & Hasselbach, W. (1977) *Z. Naturforsch.* **32c**, 828–834
Sakamoto, J. & Tonomura, Y. (1983) *J. Biochem. (Tokyo)* **93**, 1601–1614

Scofano, H. M. & de Meis, L. (1981) *J. Biol. Chem.* **256**, 4282–4285
Scofano, H. M., Vieyra, A. & de Meis, L. (1979) *J. Biol. Chem.* **254**, 10227–10231
Souza, D. O. & de Meis, L. (1976) *J. Biol. Chem.* **251**, 6355–6359
Springs, B., Welsh, K. M. & Cooperman, B. S. (1981) *Biochemistry* **20**, 6384–6391
Tada, M., Yamamoto, T. & Tonomura, Y. (1978) *Physiol. Rev.* **58**, 1–79
Takisawa, H. & Tonomura, Y. (1978) *J. Biochem. (Tokyo)* **83**, 1275–1284
Tanford, C. (1983) *Annu. Rev. Biochem.* **52**, 379–410
Taniguchi, K. & Post, R. L. (1975) *J. Biol. Chem.* **250**, 3010–3018
Thorley-Lawson, D. A. & Green, N. M. (1977) *Biochem. J.* **167**, 739–748
Verjovski-Almeida, S. & de Meis, L. (1977) *Biochemistry* **16**, 329–334
Vieyra, A., Scofano, H. M., Guimarẽs-Motta, H., Tume, R. K. & de Meis, L. (1979) *Biochim. Biophys. Acta* **568**, 437–445
Weber, G. (1972a) *Biochemistry* **11**, 864–878
Weber, G. (1972b) *Proc. Natl. Acad. Sci. U.S.A.* **69**, 3000–3003
Weber, G. (1974) *Ann. N.Y. Acad. Sci.* **227**, 486–496
Yamada, S. & Tonomura, Y. (1972) *J. Biochem. (Tokyo)* **72**, 417–425
Yamada, S., Sumida, M. & Tonomura, Y. (1972) *J. Biochem. (Tokyo)* **72**, 1537–1548
Yoshida, M. (1983) *Biochem. Biophys. Res. Commun.* **114**, 907–912

Biochem. Soc. Symp. **50**, 127–149
Printed in Great Britain

Integral Membrane Protein Translocations in the Mechanism of Insulin Action

SAMUEL W. CUSHMAN and IAN A. SIMPSON

Experimental Diabetes, Metabolism and Nutrition Section, Molecular, Cellular and Nutritional Endocrinology Branch, National Institute of Arthritis, Diabetes, and Digestive and Kidney Diseases, National Institutes of Health, Bethesda, Maryland 20205 U.S.A.

Synopsis

The subcellular distributions of insulin and insulin-like growth factor type II (IGF-II) receptors, and glucose transporters, have been examined in basal and insulin-stimulated rat adipose cells. Plasma membranes (PM), high-density microsomes (HDM) and low-density microsomes (LDM) were prepared by differential ultracentrifugation. Insulin receptors were quantified by ^{125}I-insulin binding or lactoperoxidase ^{125}I-iodination and immunoprecipitation, IGF-II receptors by ^{125}I-IGF-II binding, and glucose transporters by specific D-glucose-inhibitable [^{3}H]cytochalasin B binding. In the basal state, more than 90% of the cells' insulin receptors are localized to PM, and $\sim 90\%$ of the cells' glucose transporters and IGF-II receptors are associated with LDM. In the maximally insulin-stimulated state, the number of insulin receptors in PM is decreased by $\sim 30\%$, of which approximately half are recovered in LDM and the remainder in HDM in an inverted configuration. Concomitantly, the numbers of glucose transporters and IGF-II receptors in LDM are decreased by $\sim 60\%$ and $\sim 22\%$, respectively, with stoichiometric numbers appearing in PM. All three redistribution processes are rapid ($t_{\frac{1}{2}} = 2$–3 min), achieving new steady states in 5–10 min. The redistributions of glucose transporters and IGF-II receptors are half-maximal at ~ 0.1 nM-insulin, whereas insulin receptor redistribution correlates with receptor occupancy ($1/2_{\max} \cong 3$ nM). Thus, insulin stimulates the rapid and simultaneous subcellular translocations of its own receptors and, in the opposite direction, IGF-II receptors and glucose transporters.

Introduction

A fundamental action of insulin in regulating glucose metabolism in such peripheral tissues as muscle and adipose tissue is its stimulatory effect on glucose transport (Levine & Goldstein, 1955; Park *et al.*, 1959; Crofford & Renold, 1965*a, b*). However, although kinetic studies by several investigators have clearly established that this stimulatory effect of insulin occurs through an increase in the maximum transport velocity ($V_{\max}$), such kinetic analyses by themselves are unable to differentiate among an increase in glucose transporter intrinsic activity, an increase in the number of glucose transporters, and/or some combination of both (Narahara & Ozand, 1963; Vinten et al., 1976; Taylor & Holman, 1981). Thus, a technique for assaying glucose transporter concentration in the plasma membrane fraction of cells was developed in our laboratory employing cytochalasin B, a relatively high affinity competitive inhibitor of glucose transport in a large number of cell types (Wardzala, 1979; Wardzala et al., 1978). In this technique, the number of glucose transporters is determined by Scatchard analysis of specific D-glucose-inhibitable equilibrium [³H]cytochalasin B binding in the presence of cytochalasin E; the latter inhibits cytochalasin B binding to sites other than glucose transporters without influencing the characteristics of cytochalasin B binding to the glucose transporter itself. This technique was then used to demonstrate that insulin stimulates glucose transport in the isolated rat adipose cell primarily through an increase in the plasma membrane concentration of functional glucose transporters (Wardzala et al., 1978; Wardzala, 1979).

More recently, this same technique has been used in our laboratory to examine the subcellular distribution of glucose transporters among membrane fractions prepared from rat adipose cells by differential ultracentrifugation (Cushman & Wardzala, 1980; Karnieli et al., 1981; Wheeler et al., 1982; Simpson et al., 1983; Horuk et al., 1983; Smith et al., 1984). The results have provided evidence for a novel concept of insulin's stimulatory action on glucose transport, namely, the translocation of glucose transporters to the plasma membrane from a large intracellular pool (Cushman et al., 1984). Such a translocation hypothesis was simultaneously and independently proposed by Kono and coworkers (Suzuki & Kono, 1980; Kono et al., 1981, 1982; Ezaki & Kono, 1982; Smith et al., 1984; Kono, 1984), who used sucrose gradient procedures to prepare rat adipose cell subcellular membrane fractions and a reconstitution technique (Robinson et al., 1982) to assess glucose transport activity. Additional supporting evidence for this hypothesis has been reported by Lienhard and coworkers (Lienhard et al., 1982; Gorga & Lienhard, 1984), who used rat adipose cells and either a rabbit antiserum prepared against the purified human erythrocyte glucose transporter or a one glucose transporter per vesicle reconstitution technique, and by Czech and coworkers (Pessin et al., 1984; Oka & Czech, 1984) by directly photolabelling the rat adipose cell glucose transporter with cytochalasin B. Further studies have now (1) extended this hypothesis to the stimulation of glucose transport by insulin in isolated guinea pig (Horuk et al., 1983) and human (unpublished results) adipose cells, and rat diaphragm (Wardzala & Jeanrenaud, 1981, 1983) and heart (Watanobe et al., 1984), and (2) demonstrated that the insulin resistant glucose transport observed in adipose cells from the streptozotocin diabetic rat (Karnieli et al., 1981), the high fat-fed rat (Hissin et al., 1982a), and the aged, obese male rat (Hissin et al., 1982b) appears to be explained by a depletion of glucose transporters from the intracellular pool in the basal state and a corresponding reduction in the translocation of glucose transporters in response to insulin.

Most recently, reports from our laboratory (Wardzala et al., 1984) and by Oka et al. (1984) suggest that insulin increases insulin-like growth factor type II (IGF-II) binding to isolated rat adipose cells through a similar translocation process. In addition, further reports from our laboratory demonstrate that

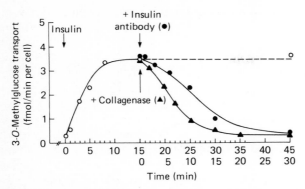

Fig. 1. *Stimulation of glucose transport activity by insulin at 37 °C in the isolated rat adipose cell*
Time course (○) in response to 0.67 nM-(100 μunits/ml) insulin and reversal (●, ▲), by using a
300-fold excess of the IgG fraction of an anti-insulin antiserum or by removing insulin with 3 mg
of crude collagenase/ml (in part from Karnieli *et al.*, 1981).

during the insulin-induced translocations of glucose transporters and IGF-II
receptors from intracellular pools to the plasma membrane, insulin concomitantly
induces the internalization of its own receptor from the plasma membrane into
two separate and distinguishable intracellular compartments (Wang *et al.*, 1983;
Simpson *et al.*, 1984; Sonne & Simpson, 1984; Hedo & Simpson, 1984; Simpson
& Hedo, 1984). The present report briefly reviews the evidence primarily from
our own laboratory for these translocations of integral membrane proteins in
the mechanism of insulin action.

Glucose Transporter Translocation

The stimulatory action of insulin on glucose transport in isolated rat adipose
cells is rapid, fully reversible, and insulin concentration-dependent. The time
courses of the response to a maximal but physiological concentration of insulin
(0.68 nM, 100 μunits/ml) and its reversal at 37 °C are illustrated in Fig. 1
(Karnieli *et al.*, 1981). Here and elsewhere in this report, glucose transport
activity is assessed by measuring the uptake of 0.1 mM-3-*O*-[^{14}C]methylglucose,
a non-metabolizable glucose analogue (Gliemann *et al.*, 1972; Foley *et al.*, 1978;
Gliemann & Rees, 1983). After the addition of insulin at 0 min, glucose transport
activity rapidly increases ($t_{\frac{1}{2}} \cong 4.0$ min) until a new steady state is achieved by
~ 10 min; the latter is then maintained for up to several hours in the presence
of insulin at a level 20–40-fold that observed in the basal state. The subsequent
addition of a 300-fold excess of anti-insulin antibody, on the other hand, fully
restores the cells to the basal state over a 30–45 min period with a $t_{\frac{1}{2}}$ of ~ 9.0 min.
An almost twofold more rapid reversal of the response to insulin can be achieved
by rapidly destroying the insulin present with a crude collagenase preparation
(Kono *et al.*, 1981, 1982); however, half-times less than 5 min are not observed
even under these conditions. The response to insulin is also dependent on the
concentration of hormone added, with half-maximal and maximal steady state
effects observed at ~ 0.11 nM and ~ 0.35 nM, respectively (Karnieli *et al.*, 1981).

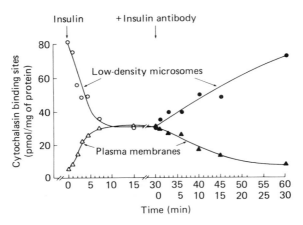

Fig. 2. *Stimulation of the subcellular redistribution of glucose transporters by insulin at 37 °C in the isolated rat adipose cell*

Time course (○, △) in response to 0.67 nM-(100 μunits/ml) insulin and reversal (●, ▲), with a 300-fold excess of the IgG fraction of an anti-insulin antiserum (from Karnieli *et al.*, 1981).

The corresponding time courses of the effects of insulin and anti-insulin antibody on the subcellular distribution of glucose transporters in these same preparations of adipose cells at 37 °C are illustrated in Fig. 2. Here, glucose transporters are quantified by using the specific D-glucose-inhibitable cytochalasin B binding assay developed in our laboratory (Wardzala *et al.*, 1978; Wardzala, 1979; Cushman & Wardzala, 1980); plasma membranes and low-density microsomes are prepared by homogenizing the cells at the indicated times and subjecting the homogenates to a differential ultracentrifugation procedure (Cushman & Wardzala, 1980; Simpson *et al.*, 1983). Similar results have been obtained by Kono *et al.* (1982) using a reconstitution technique to assess the subcellular distribution of glucose transport activity and sucrose gradients to prepare the equivalent membrane fractions.

As described above for glucose transport activity itself, addition of insulin to the intact adipose cells at 0 min is associated with a rapid ($t_{\frac{1}{2}} \cong 2.5$ min) increase in the concentration (per mg of membrane protein) of glucose transporters in the plasma membranes and a concomitant decrease in their concentration in the low-density microsomes. Again, a new and stable steady state is achieved by ~ 10 min at levels $\sim$ sixfold greater in the plasma membranes and $\sim 60\%$ lower in the low-density microsomes than the corresponding levels observed in the basal state. Furthermore, the concentrations of glucose transporters in both membrane fractions are fully restored to their basal levels 30–45 min after the addition of anti-insulin antibody to the intact cells ($t_{\frac{1}{2}} \cong 9.0$ min).

The ~ 1.5 min lag between the appearance of glucose transporters in the plasma membranes (Fig. 2) and the detection of increased glucose transport activity in the intact cells (Fig. 1) in response to insulin is highly reproducible and suggests either that the insertion of glucose transporters into the plasma membrane is a multi-step process and/or that glucose transporters must be activated during the insertion process in order to become functional. The latter

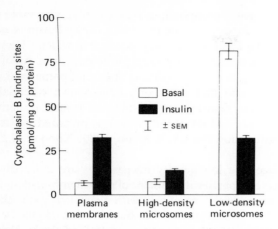

Fig. 3. *Steady-state subcellular distribution of glucose transporters at 37 °C in the isolated rat adipose cell*

Distributions are shown for the basal state and in response to 0.67 nM-(100 μunits/ml) insulin (from Simpson *et al.*, 1983).

possibility appears less likely than the former in view of a recent report describing the high degree of similarity in the transport characteristics of glucose transporters reconstituted from the plasma membranes of insulin-stimulated cells and the low-density microsomes of basal cells (Smith *et al.*, 1984), but cannot be excluded. In contrast, the $t_{\frac{1}{2}}$ values for the reversal of insulin's effects on glucose transporter concentration in the plasma membranes and on glucose transport activity in the intact cells by anti-insulin antibody are coincident, thus suggesting that the response to insulin and its reversal occur by distinctly different mechanisms. The effect of insulin on the subcellular distribution of glucose transporters is also hormone concentration-dependent with half maximal and maximal steady-state responses observed at insulin concentrations which are virtually identical to those producing the corresponding effects on glucose transport activity in the intact cells (Karnieli *et al.*, 1981).

Fig. 3 demonstrates the steady-state subcellular distribution of glucose transporters at 37 °C among the three major membrane fractions prepared in our laboratory from basal and maximally insulin-stimulated adipose cells: plasma membranes, high-density microsomes, and low-density microsomes (Cushman & Wardzala, 1980; Simpson *et al.*, 1983). Glucose transporters are assayed as specific D-glucose-inhibitable cytochalasin B binding sites and the results are expressed as concentrations per mg of membrane protein. Similar distributions have been observed: (1) in our laboratory by the Western blot technique using an affinity-purified rabbit IgG prepared against the purified human erythrocyte glucose transporter (Wheeler *et al.*, 1982), by a photolabelling procedure in which [³H]cytochalasin B is cross-linked to the glucose transporter using a photoactivated cross-linking agent (Horuk *et al.*, 1983), and by reconstitution of glucose transport activity (Cushman *et al.*, 1984); (2) by Lienhard *et al.* (1982) using either the Western blot technique and their own rabbit anti-human erythrocyte glucose transporter antiserum or a reconstitution technique in which a given phospholipid vesicle contains no more than a single glucose transporter (Gorga & Lienhard, 1984); (3) by Pessin *et al.* (1984) and Oka *et al.* (1984) using a photolabelling procedure in which [³H]cytochalasin B is linked to the glucose transporter by direct photoactivation. All of these studies were carried out using similar membrane fractions prepared by differential ultracentrifugation. Kono and coworkers have shown essentially the same steady-state subcellular distributions of glucose transport activity using a reconstitution technique and membrane fractions prepared by sucrose density gradient ultracentrifugation (Suzuki & Kono, 1980; Kono *et al.*, 1982; Kono, 1984).

In the basal state, the low-density microsomes are markedly enriched in glucose transporters compared with both the plasma membranes and high-density microsomes. Treatment of the intact cells with a maximally stimulating concentration of insulin, however, increases the concentration of glucose transporters ~ fivefold in the plasma membranes and ~ twofold in the high-density microsomes, and concomitantly decreases that in the low-density microsomes by ~ 60% compared with the basal state. Glucose transporter concentrations in the mitochondrial/nuclear fraction are below the level of detection in both the basal and insulin-stimulated states (not illustrated).

A qualitative assessment of the membrane composition of these subcellular fractions has been carried out by using marker enzyme activities both in our own laboratory (Cushman & Wardzala, 1980; Simpson *et al.*, 1983) and by Kono *et al.* (1982). The plasma membrane fraction is relatively enriched in both 5′-nucleotidase and isoprenaline-stimulated adenylate cyclase activities, marker enzymes characteristic of plasma membranes; the high-density microsomes are relatively enriched in both rotenone-insensitive NADH–cytochrome *c* reductase and glucose-6-phosphate phosphatase activities, marker enzymes characteristic of endoplasmic reticulum; and the low-density microsomes are enriched in both UDP-galactose:*N*-acetylglucosamine galactosyltransferase and CMP-*N*-acetylneuraminate:asialofetuin *N*-acetylneuraminyltransferase (sialyltransferase), marker enzymes characteristic of the Golgi apparatus. The mitochondrial/nuclear fraction is relatively enriched in the mitochondrial marker enzyme activity citrate synthase. The lysosomal marker enzyme activity acid phosphatase is relatively enriched in the high-density microsomes (Sonne & Simpson, 1984). Electron microscopic studies of these membrane fractions are fully consistent with the relative distributions of marker enzyme activities and demonstrate the presence of primarily sealed membrane vesicles of decreasing size in the plasma membranes, and high- and low-density microsomes (Simpson *et al.*, 1983).

Insulin treatment of intact adipose cells does not detectably influence any of the general characteristics of these membrane fractions including the recoveries of membrane protein from the original homogenates, the relative distributions of marker enzyme activities, their electron microscopic appearance, and their gross protein composition as assessed by sodium dodecylsulphate/polyacrylamide gel electrophoresis (Simpson *et al.*, 1983). Furthermore, a comparison of the relative distributions of glucose transporters in both the basal and insulin-stimulated states with those of the marker enzyme activities demonstrate that the presence of glucose transporters in the high-density microsomes is probably due to contamination of this intermediate fraction with plasma membranes and low-density microsomes (Simpson *et al.*, 1983). Finally, the marked relative enrichment of glucose transporters in the basal state and marked reduction of glucose transporters in the insulin-stimulated state in the low-density microsomes clearly identity this membrane fraction as the intracellular pool from which glucose transporters are translocated to the plasma membrane in response to insulin. However, while the low-density microsomes are also relatively enriched in membranes of the Golgi apparatus, the lack of correlation between the distributions of glucose transporters and Golgi marker enzyme activities over all subcellular fractions strongly suggests that the intracellular pool is actually associated with either a highly specialized subpopulation of Golgi membrane vesicles or a unique membrane species for which no specific marker enzyme activity has yet been identified (Simpson *et al.*, 1983). The apparent intracellular localization of this large pool of glucose transporters is confirmed by Oka & Czech (1984) by direct photolabelling with [³H]cytochalasin B in the intact cells; these investigators demonstrate that high extracellular concentrations of glucose analogues that are only slowly transported into the cells inhibit labelling of the glucose transporters in the subsequently prepared plasma membranes, but not those in the low-density microsomes.

A determination of the stoichiometry of the translocation process requires expression of the numbers of glucose transporters on a per cell basis. However, because of the adipose cell's large droplet of stored triglyceride and the hydrophobic nature of cytochalasin B, the numbers of glucose transporters per cell cannot be directly determined and an indirect assessment must be carried out by using membrane protein and marker enzyme activity recoveries. When the recoveries of plasma membrane and intracellular glucose transporters are assumed to parallel those of the 5′-nucleotidase and galactosyltransferase activities, respectively, a total number of $\sim 3.7 \times 10^6$ glucose transporters per adipose cell from the epididymal fat pads of ~ 180 g Sprague-Dawley rats can be estimated whether or not the intact cells have been exposed to insulin (Simpson *et al.*, 1983). Furthermore, $\sim 90\%$ of this total number is associated with the intracellular pool in the basal state and $\sim 53\%$ of this intracellular pool is translocated to the plasma membrane in response to insulin.

Thus, insulin appears to stimulate glucose transport in the isolated rat adipose cell through a rapid, reversible, and insulin concentration-dependent exocytic-/endocytic-like translocation of glucose transporters from a large intracellular pool to the plasma membrane. Further studies not described here provide evidence for a similar translocation mechanism in guinea pig (Horuk *et al.*, 1983) and human (unpublished observations) adipose cells and rat diaphragm (Wardzala & Jeanrenaud, 1981, 1983) and heart (Watanobe *et al.*, 1984). In addition, the marked insulin resistance observed at the glucose transport level in adipose cells from the streptozotocin diabetic rat (Karnieli *et al.*, 1981), the high fat-fed rat (Hissin *et al.*, 1982*a*), and the aged, obese male rat (Hissin *et al.*, 1982*b*) now appears to be explained by a marked reduction in the intracellular pool of glucose transporters in the basal state and a corresponding decrease in the translocation of glucose transporters in response to insulin. Similarly, the hyper-insulin-responsive glucose transport observed in the adipose cells of young, genetically obese Zucker rats appears to be accounted for by a markedly increased basal intracellular pool of glucose transporters and an enhanced translocation in response to insulin (Guerre-Milo et al., 1985).

While the insulin resistant systemic glucose metabolism associated with diabetes, high fat feeding and obesity is chronic in nature, an acute antagonism to insulin's stimulatory effects on peripheral tissue glucose utilization has recently been shown to accompany catecholamine administration (Diebert & De Fronzo, 1980). Furthermore, when isolated rat adipose cells are treated with adenosine deaminase to remove the adenosine which is invariably present in these cell preparations (adenosine is a potent adenylate cyclase inhibitor), catecholamines (Taylor & Halperin, 1979; Kashiwagi et al., 1983), as well as ACTH (Taylor & Halperin, 1979) and glucagon (Green, 1983), can be shown to acutely and markedly inhibit insulin's stimulatory effect on glucose transport. Initial studies in our own laboratory on the mechanism of these latter effects suggested that isoprenaline, in the absence of adenosine, inhibits both the translocation of glucose transporters in response to insulin and the intrinsic activity of the glucose transporter through a cyclic AMP-mediated process (Smith et al., 1984).

In order to further evaluate the mechanism of these counter-regulatory effects of the lipolytic hormones on insulin-stimulated glucose transport, glucose transport activity and the subcellular distribution of glucose transporters have recently been re-examined under the stringent incubation conditions described by Honnor et al. (1985a) to achieve and maintain steady-state cyclic AMP-dependent protein kinase (A-kinase) activity ratios. These ratios represent an indirect measure of cellular cyclic AMP concentrations and range in value from 0 to 1. The critical elements in these new incubation conditions are designed to prevent the juxtacellular accumulation of non-esterified fatty acids and consequent irreversible inhibition of adenylate cyclase during the stimulation of lipolysis, and include the presence of serum albumin at concentrations of 5% or greater and the use of high shaking speeds (110 cycles/min in our own laboratory). 2.5 mM-Glucose is also present in the incubation medium to prevent cellular ATP depletion and promote fatty acid re-esterification (Smith et al., 1984); this glucose concentration competitively inhibits 0.1 mM-3-O-methyl-glucose transport by only 20%.

The results of these studies have provided two unexpected findings. First, the following observations suggest that the counter-regulatory effects of lipolytic hormones such as isoprenaline, ACTH and glucagon, and antilipolytic hormones such as adenosine, nicotinic acid and PGE_1, on insulin-stimulated glucose transport activity in rat adipose cells are mediated through a cyclic AMP-independent mechanism. (1) In the presence of endogenous or exogenous adenosine, isoprenaline stimulates the protein kinase activity ratio to a level sufficient to produce nearly maximum lipolysis, but does not influence maximally insulin-stimulated glucose transport activity. (2) In the absence of a lipolytic hormone, the removal of adenosine with adenosine deaminase by itself inhibits insulin-stimulated glucose transport activity up to 30% without affecting the basal protein kinase activity ratio. (3) In the absence of adenosine, the exposure of insulin-stimulated cells to increasing isoprenaline concentrations progressively

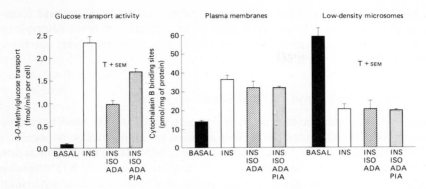

Fig. 4. *Counter-regulation of insulin-stimulated glucose transport activity and the subcellular distribution of glucose transporters by adenylate cyclase stimulators and inhibitors at steady state at 37 °C in the isolated rat adipose cell*

Basal, the response to 6.7 nM-(1000 μunits/ml) insulin (INS), the subsequent response to 200 nM-isoprenaline (ISO) and simultaneous adenosine removal with 1 unit of adenosine deaminase ((ADA)/ml), and the further subsequent response to 1 μM-phenylisopropyladenosine (PIA).

inhibits glucose transport activity to a maximum of 70% and concomitantly stimulates the protein kinase activity ratio to its maximum level of 1.0. The simultaneous addition of phenylisopropyladenosine, a non-metabolizable adenosine analogue, prevents these effects of isoprenaline in a concentration-dependent manner. However, when the phenylisopropyladenosine is added only after achievement of the steady-state response to isoprenaline, it still restores glucose transport activity in a concentration-dependent fashion to the control level, but no longer influences the established protein kinase activity ratio.

Thus, these classic regulators of adenylate cyclase activity appear to influence glucose transport through their own specific stimulatory and inhibitory receptors (R_s and R_i), and presumably their corresponding stimulatory and inhibitory nucleotide regulatory subunits (N_s and N_i), but do so independently of their effects on cellular cyclic AMP concentration. Under similar conditions, Londos and coworkers (Honnor *et al.*, 1985*b*; Londos *et al.*, 1985) have shown that the lipolytic and antilipolytic actions of these agents in the absence of insulin can be fully accounted for by their effects on the protein kinase activity ratio, but that the antilipolytic action of insulin has both protein kinase activity ratio-dependent and -independent components.

Second, as is clearly illustrated in Fig. 4, the marked effects of isoprenaline, adenosine removal with adenosine deaminase, and phenylisopropyladenosine on insulin-stimulated glucose transport activity under these conditions (Fig. 4*a*) occur in the absence of corresponding alterations in the insulin-stimulated subcellular distribution of glucose transporters between the plasma membranes (Fig. 4*b*) and low-density microsomes (Fig. 4*c*). Thus, these agents appear to regulate plasma membrane glucose transporter intrinsic activity and not the translocation process stimulated by insulin. However, a marked reduction in the sensitivity of the glucose transport response to insulin by lipolytic hormones in the absence of adenosine, and its restoration to normal by phenylisopropyladenosine, as noted by others (Green, 1983) and in our own laboratory (not illustrated), may involve both glucose transporter intrinsic activity and trans-location, and possibly the signalling mechanism through which the response to insulin is actually generated. These processes are currently under intensive investigation.

Type II Insulin-like Growth Factor Receptor Recycling

Several laboratories have reported that insulin stimulates the binding of type II insulin-like growth factor (IGF-II) to isolated rat adipose cells (Schoenle et al., 1977; King et al., 1982; Oppenheimer et al., 1983). This stimulatory action of insulin on IGF-II binding appears to be specifically mediated by the insulin receptor and closely parallels the stimulation of glucose oxidation by insulin (King et al., 1982). Furthermore, examination of IGF-II binding at 24 °C by Scatchard plots suggests that insulin increases IGF-II receptor apparent affinity (K_a) in the absence of changes in apparent receptor number (King et al., 1982; Oppenheimer et al., 1983). However, when plasma membranes and low-density microsomes are prepared by a differential ultracentrifugation procedure similar to that used in our own laboratory, Oppenheimer et al. (1983) report that IGF-II receptor concentration per mg of membrane protein is enriched 3.5-fold in the low-density microsomes compared with the plasma membranes from basal cells, and that insulin treatment of the intact cells is accompanied by an $\sim 60\%$ increase in IGF-II receptor concentration in the plasma membranes and a concomitant $\sim 40\%$ decrease in the low-density microsomes. These latter results are consistent with a translocation of IGF-II receptors in response to insulin similar to that postulated for glucose transporters as described above.

Thus, IGF-II binding to rat adipose cells and its stimulation by insulin have been re-examined in our laboratory using KCN to prevent the possible internalization and/or cycling of IGF-II receptors during the binding assay. Kono et al. (1981) have previously shown that KCN rapidly and fully inhibits both the insulin-induced translocation of glucose transporters from the intra-cellular pool to the plasma membrane and their retranslocation in the reverse direction during the reversal of the response to insulin.

Fig. 5 illustrates Scatchard plots of steady-state ^{125}I-IGF-II association in the absence (Fig. 5a) and presence (Fig. 5b) of 1.0 mM-KCN at 24 °C with basal and maximally insulin-stimulated rat adipose cells (Wardzala et al., 1984). In both cases, cells were preincubated for 20 min at 37 °C in the absence or presence of 7.0 nM-insulin; when present for the IGF-II binding assay, KCN was added for the last 5 min of this 37 °C preincubation. The cells were then transferred to 24 °C, IGF-II was added, and the cells were incubated for an additional 40 min. The presence of KCN does not appreciably influence the incubation time required to achieve steady-state binding. The results of a representative experiment carried out in the absence of KCN (Fig. 5a) essentially confirm those reported by King et al. (1982) and Oppenheimer et al. (1983) and suggest that insulin increases IGF-II receptor apparent K_a in the absence of an effect on apparent receptor number.

In contrast, the presence of KCN during the IGF-II binding assay (1) reduces apparent receptor number by more than 95% in the basal cells and almost 80%

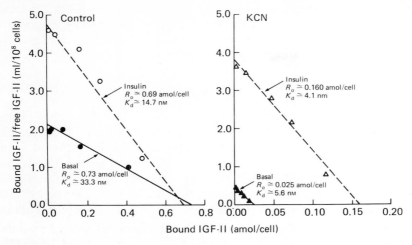

Fig. 5. *Effect of KCN on steady-state IGF-II association at 24 °C with the isolated rat adipose cell* Scatchard plots of the absence or presence of 1 mM-KCN in the basal and maximally insulin-stimulated states (from Wardzala et al., 1984).

in the insulin-stimulated cells, and (2) increases receptor apparent K_a by ~ sevenfold in the basal cells, and ~ threefold in the insulin-stimulated cells (Fig. 5b). Under the latter conditions, the net effect of insulin is a 6.4-fold increase in apparent receptor number in the absence of a change in receptor apparent K_a. In the absence of KCN, only ~ 20% of the cell-associated IGF-II is susceptible to removal by trypsin treatment in both the basal and insulin-stimulated cells; in the presence of KCN, this value is increased to more than 60% (not illustrated). Thus, when IGF-II binding to rat adipose cells is limited primarily to receptors on the surface of the cell by using KCN to prevent ligand internalization, insulin can be shown to stimulate IGF-II binding through the appearance of plasma membrane receptors. A similar conclusion is drawn by Oka et al. (1984), based on the binding of an Ig fraction of a rabbit anti-IGF-II receptor antiserum to basal and insulin-stimulated cells. However, these investigators do not clearly demonstrate what proportion of the total Ig bound actually remains on the cell surface.

The time courses of insulin's stimulatory action on the number of cell surface IGF-II receptors in the rat adipose cell and its reversal at 37 °C are illustrated in Fig. 6. Qualitatively, both processes parallel the corresponding time courses observed with glucose transport activity (Fig. 1); quantitatively, however, both are somewhat more rapid. The IGF-II receptor response to insulin has a $t_{\frac{1}{2}}$ of ~ 1.5 min and reaches steady state in ~ 10 min, values which are more comparable with those for the appearance of glucose transporters in the plasma membranes (Fig. 2) than for those for the slower stimulation of glucose transport activity in the intact cells (Fig. 1). The reversal of insulin's stimulatory effect on IGF-II receptor number with the collagenase technique for removing insulin has a $t_{\frac{1}{2}}$ of ~ 3 min, compared with ~ 5 min for glucose transport activity with the same collagenase technique (Fig. 1), and is complete by ~ 30 min.

Finally, Fig. 7 illustrates representative Scatchard plots of ^{125}I-IGF-II binding

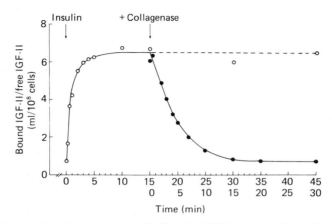

Fig. 6. *Stimulation of the appearance of cell surface IGF-II receptors by insulin at 37 °C in the isolated rat adipose cell*

Time course (○) in response to 0.67 nM-(100 μunits/ml) insulin and reversal (●) by removing insulin with crude collagenase (3 mg/ml).

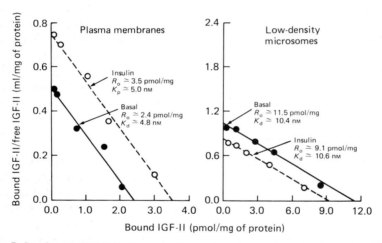

Fig. 7. *Steady-state IGF-II binding at 24 °C to subcellular membrane fractions from the isolated rat adipose cell*

Scatchard plots are shown for membrane fractions prepared from basal and maximally insulin-stimulated cells (from Wardzala et al., 1984).

to plasma membranes (Fig. 7*a*) and low-density microsomes (Fig. 7*b*) prepared from basal and maximally insulin-stimulated cells at steady state (Wardzala *et al.*, 1984). Virtually identical results are obtained whether or not KCN is added to the cells following their preincubation with or without insulin and 5 min prior to their homogenization. In essence, these observations qualitatively confirm those reported by Oppenheimer *et al.* (1983), and demonstrate (1) an almost sixfold enrichment in the concentration of IGF-II receptors per mg of membrane protein in the low-density microsomes compared with the plasma membranes from basal cells, and (2) an $\sim 55\%$ increase in the concentration of IGF-II receptors in the plasma membranes and concomitant $\sim 22\%$ decrease in their concentration in the low-density microsomes in response to insulin. In addition, while K_a for IGF-II binding to the plasma membranes are consistently about twofold greater than those in the low-density microsomes in both the basal and insulin-stimulated states, they are very similar to those observed in the intact cells in the presence of KCN (Fig. 5).

These results are interpreted as follows. (1) The effects of KCN on IGF-II binding to rat adipose cells represent a selective inhibition of endocytosis and exocytosis similar to its effects on the reversible translocation of glucose transporters, as reported by Kono *et al.* (1981). (2) In the absence of KCN, IGF-II receptors of constant affinity cycle between the cell's plasma membrane and a large intracellular pool, binding ligand at the cell surface and internalizing the ligand–receptor complex such that intracellular IGF-II accumulates, presumably in lysosomes. The latter is probably the consequence of a greater inhibition of IGF-II degradation than internalization at 24 °C compared with 37 °C. (3) Insulin stimulates IGF-II binding by inducing a steady-state redistribution of receptors from the intracellular pool to the plasma membrane either by stimulating the exocytic leg or inhibiting the endocytic leg of the cycling process. (4) The subcellular redistribution of IGF-II receptors in response to insulin observed by preparing membrane fractions is only detectable by Scatchard analysis of IGF-II binding to the intact cells when receptor cycling is prevented by KCN. The mechanisms of IGF-II receptor endocytosis/exocytosis and its modulation by insulin, and the relationship between this process and glucose transporter translocation are presently under active study.

Insulin Receptor Internalization

The initial step in the mechanism of insulin action is the binding of hormone to its specific cell surface receptor. Subsequent to binding, however, insulin is rapidly internalized by receptor-mediated endocytosis, much as IGF-II is as described above (Bergeron *et al.*, 1979; Carpentier *et al.*, 1979; Suzuki & Kono, 1979). The internalized insulin is then ultimately either degraded in the cell's lysosomes (Bergeron *et al.*, 1979; Carpentier *et al.*, 1979) or released undegraded back into the cell's environment (Suzuki & Kono, 1979). Internalized insulin in the rat adipose cell can be identified in specific non-lysosomal membrane vesicles by both sucrose density ultracentrifugation using ^{125}I-insulin (Suzuki & Kono, 1979) and electron microscopic techniques using insulin labelled with ferritin (Hammons & Jarett, 1980). In this cell type, internalized insulin can be identified in association with lysosomes only in the presence of lysosomotropic agents such as chloroquine (Suzuki & Kono, 1979). A representative time course of ^{125}I-insulin association with the isolated rat adipose cell, the net result of all these processes, is illustrated in Fig. 8; association is rapid but biphasic, with an apparent $t_{\frac{1}{2}}$ of ~ 2 min, and achieves steady state in ~ 30 min.

Until recently, however, little has been known regarding the internalization of the insulin receptor itself in rat adipose cells. Green & Olefsky (1982) have

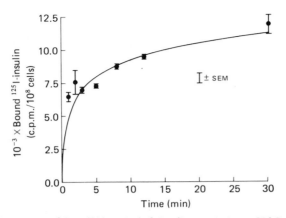

Fig. 8. *Time course of 6 nM-(900 μunits/ml) insulin association at 37 °C with the isolated rat adipose cell*

indirectly demonstrated an insulin-induced internalization of the insulin receptor by measuring the number of receptors in solubilized intact cells after trypsin digestion of the cell surface receptors. A partial processing of the internalized receptor under similar conditions has further been demonstrated by Berhanu *et al.* (1982), using a photoaffinity ^{125}I-insulin analogue. Nevertheless, these studies were carried out in the presence of Tris, a buffering agent which markedly promotes both the rate and degree of insulin receptor 'down regulation' (Marshall & Olefsky, 1981; Rennie & Gliemann, 1981). A detailed analysis of insulin receptor internalization in the rat adipose cell has now been carried out in our own laboratory (Wang *et al.*, 1983; Simpson *et al.*, 1984; Sonne & Simpson, 1984; Hedo & Simpson, 1984) and clearly demonstrates that insulin induces an internalization process with characteristics closely resembling those of the translocations of both glucose transporters and IGF-II receptors in response to insulin, but in the opposite direction. Many of these characteristics have recently been confirmed by Marshall *et al.* (1985) and Marshall (1985), using the trypsin digestion technique described above, but in cells incubated in the absence of Tris.

Fig. 9 illustrates the time course of the effect of 6 nM-insulin on the subcellular distribution of insulin receptors in rat adipose cells at 37 °C (Sonne & Simpson, 1984). In these experiments, membrane fractions are prepared by the same differential ultracentrifugation technique used in studying the subcellular distributions of glucose transporters and IGF-II receptors, and insulin receptors are assayed by direct ^{125}I-insulin binding for 18 h at 4 °C. Detailed binding isotherms obtained in each membrane fraction demonstrate that the tracer ^{125}I-insulin binding reported here is proportional to receptor number and not significantly influenced by the presence of any insulin carried over to the membrane fractions from the original incubation of the intact cells with hormone. In basal cells, indicated by the data obtained with a 0 min incubation, the subcellular distribution of insulin receptors directly parallels those of the two plasma membrane marker enzyme activities used in this laboratory, 5′-nucleotidase and

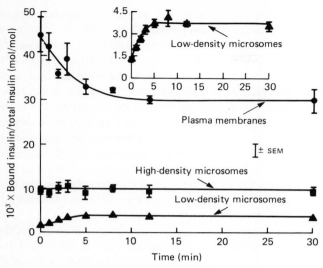

Fig. 9. *Time course of insulin-induced subcellular redistribution of insulin receptors at 37 °C in the isolated rat adipose cell*

Redistribution of insulin receptors in response to 6 nM-(900 μunits/ml) insulin (from Sonne & Simpson, 1984).

isoprenaline-stimulated adenylate cyclase (Simpson *et al.*, 1983). Thus, in the absence of insulin, most if not all of the cells' insulin receptors are associated with the plasma membrane.

In response to insulin, however, the concentration of insulin receptors per milligram of membrane protein in the plasma membranes rapidly decreases and that in the low-density microsomes concomitantly increases. Both changes occur with a $t_{\frac{1}{2}}$ of 2–3 min and achieve steady states in 5–10 min. At steady state after incubation of the intact cells with a saturating concentration of insulin, the decrease in receptor concentration in the plasma membranes amounts to $\sim 30\%$ of that observed in the basal state; the increase in receptor concentration in the low-density microsomes is approximately threefold. The concentration of insulin receptors in the high-density microsomes, on the other hand, does not appear to change. Both the increase in receptor concentration in the plasma membranes and decrease in the low-density microsomes are directly dependent on the concentration of insulin to which the intact cells are exposed, with identical half-maximal effects observed at 3–4 nM and overall concentration dependency curves which directly parallel cell surface receptor occupancy (Sonne & Simpson, 1984). Similar time courses and insulin concentration dependency curves for the appearance of insulin receptors in the low-density microsomes have been obtained in our laboratory, with a photoactive insulin analogue (B_{29}-NAPA-^{125}I-insulin) used to quantify the concentration of 135 kDa α-receptor subunits, the insulin binding subunit (Wang *et al.*, 1983).

Two features of the results illustrated in Fig. 9 are particularly worthy of note. First, the insulin receptors appearing in the low-density microsomes in response to insulin are detectable only in the presence of a small amount of digitonin.

Digitonin binds to the cholesterol in the membrane vesicles present in this subcellular fraction and renders them permeable to either the insulin used in the binding assay (Sonne & Simpson, 1984) or the photoactive insulin analogue used in labelling the α-receptor subunit (Wang *et al.*, 1983). No such effect of digitonin is observed on insulin binding to the plasma membranes and high-density microsomes. Thus, the orientation of the receptors in the low-density microsomes from insulin-stimulated cells appears to be reversed relative to that in the plasma membranes, a feature characteristic of endocytic internalization. Second, when the recoveries of membrane protein and marker enzyme activities from the original homogenates are used to estimate the recoveries of insulin receptors, only $\sim 50\%$ of the receptors lost from the plasma membranes in response to insulin can be accounted for by the appearance of receptors in the low-density microsomes (Sonne & Simpson, 1984). Examination of the mitochondrial/nuclear fraction further fails to reveal the missing receptors.

One of two alternative approaches to studying the stoichiometry of this insulin receptor internalization process in rat adipose cells is illustrated in Fig. 10 (Simpson *et al.*, 1984). Here, cells are maintained in primary tissue culture at 37 °C for 24 h in the presence of [^{3}H]glucosamine to label the sugar moieties of both the α- and 95 kDa β-glycoprotein receptor subunits, washed, and acutely incubated in the absence or presence of a saturating concentration of insulin. Membrane fractions are then prepared and solubilized in Triton X-100, the receptors immunoprecipitated with a specific anti-insulin receptor antiserum, and the labelled receptor subunits analysed by sodium dodecylsulphate/poly-acrylamide gel electrophoresis under reducing conditions and autoradiography and/or gel slicing and counting. Similar experiments in which both receptor subunits are biosynthetically labelled by using [^{35}S]methionine produce identical results (Simpson *et al.*, 1984).

The following points are demonstrated in Fig. 10. (1) Whereas the two receptor subunits are labelled somewhat differently, presumably owing to their distinct carbohydrate and amino acid compositions, their distributions among the three membrane fractions prepared from basal cells are the same, and directly correlate with the distribution of receptors as assessed by insulin binding (Fig. 9) and the distributions of plasma membrane marker enzyme activities (Simpson *et al.*, 1983). (2) In response to insulin, the concentrations of both receptor subunits decrease in the plasma membranes and concomitantly increase in the low-density microsomes in a fashion which directly parallels that observed with the insulin binding assay (Fig. 9). However, in contrast to the lack of change in receptor concentration in the high-density microsomes seen with hormone binding (Fig. 9), biosynthetic labelling reveals an increase (approximately two-fold) in receptor subunit concentration in this membrane fraction in response to insulin. The latter increase, together with that observed in the low-density microsomes, fully accounts for all of the receptors lost from the plasma membranes with insulin treatment of the intact cells. (3) Although the distribution of receptor subunits among the three membrane fractions substantially changes in response to insulin, the pattern of labelling between, and the relative molecular weights of, the α- and β-receptor subunits do not. Thus, receptors appear to be internalized intact, and if processing of either receptor subunit occurs during

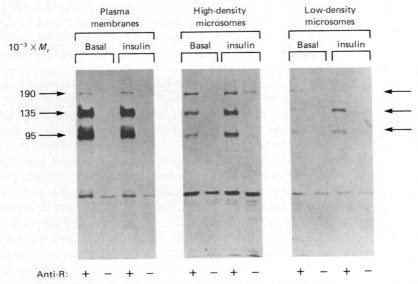

Fig. 10. *Steady-state subcellular distribution of α-(135 kDa) and β-(95 kDa) insulin subunits, biosynthetically labelled with [³H]glucosamine, at 37 °C in the isolated rat adipose cell*

Distribution of insulin receptor subunits in the basal state and in response to 30 nM-(4500 μunits/ml) insulin. Anti-R: +, immunoprecipitation with a human anti-insulin receptor antiserum; −, immunoprecipitation with control human serum (in part from Simpson *et al.*, 1984).

internalization, it does not involve detectable alterations in receptor subunit mobility on sodium dodecylsulphate gels.

Therefore, concomitant with the induction of glucose transporter translocation and a shift in the steady-state subcellular distribution of IGF-II receptors from large intracellular pools to the plasma membrane, insulin induces the internalization of its own receptor from the plasma membrane into at least two intracellular vesicular compartments. The latter are clearly distinguishable by their distinct sedimentation characteristics during the fractionation procedure, and their differential abilities to bind insulin. Indeed, Hedo & Simpson (1984) have proposed that the inability of the receptors appearing in the high-density microsomes to bind fresh insulin might be due to the presence of covalently bound insulin from the original exposure of the intact cells to hormone, as reported by Clark & Harrison (1982, 1983). The internalization process is rapid and a new steady-state subcellular distribution of receptors is achieved within 10 min. The likelihood that this new steady state represents the endocytic/ exocytic recycling of receptors is indicated by the rapid intracellular accumulation of insulin in the presence of chloroquine; under these conditions, degradation of the internalized hormone is inhibited in the absence of a significant effect on hormone-induced receptor subcellular redistribution (Sonne & Simpson, 1984). In addition, this internalization process must be clearly distinguished from the rapid 'down regulation' and partial processing of receptors induced by insulin in the presence of Tris (Green & Olefsky, 1982; Berhanu *et al.*, 1982) since no net loss of receptors is observed under the incubation conditions used here for at least several hours (Marshall & Olefsky, 1981; Rennie & Gliemann, 1981). Nevertheless, internalization may represent an early step in the down regulation process ultimately contributing receptors to the degradation pathway as they rapidly cycle between the plasma membrane and one or more intracellular compartments.

The stoichiometry of insulin receptor internalization in the isolated rat adipose cell can also be shown by using the $Na^{125}I$/lactoperoxidase technique to iodinate proteins, including insulin receptors, exposed on the surface of the intact cells (Hedo & Simpson, 1984). In these experiments, cells are surface labelled in the basal state, washed free of unbound label, and incubated in the absence or presence of a saturating insulin concentration to induce maximal receptor internalization. Subcellular membrane fractions are then prepared and their receptor content assessed by immunoprecipitation, sodium dodecyl-sulphate/polyacrylamide gel electrophoresis, and autoradiography as described above for biosynthetic labelling. In essence, although the pattern of labelling of the α- and β-receptor subunits with this technique is distinct from those observed when the receptor subunits are labelled biosynthetically with glucosamine or methionine, this pattern is unchanged among the membrane fractions prepared from either basal or insulin-stimulated cells, and the distribution of both subunits over the three membrane fractions shifts from one typical of a plasma membrane protein in the absence of insulin to one typical of an internalized membrane protein in response to insulin.

Because the lactoperoxidase iodination technique is side specific for membrane proteins on the surface of intact cells or sealed membrane vesicles and because the internalized insulin receptors appearing in the low-density microsomes of insulin-stimulated rat adipose cells appear to be inverted relative to those in the plasma membranes, this labelling procedure can be used in combination with subcellular fractionation to examine the disposition of the α- and β-receptor subunits within the plasma membrane (Hedo & Simpson, 1984). When plasma membranes from either basal or insulin-stimulated cells, both of which contain substantial concentrations of receptors, are labelled in this fashion, both receptor subunits are readily detectable after immunoprecipitation and gel analysis. The ratio of label incorporated into the α- and β-receptor subunits is $\sim 1:1$. When low-density microsomes from insulin-stimulated cells, containing internalized receptors, are labelled, on the other hand, only the β-receptor subunit is readily detectable. Lactoperoxidase iodination of these same membrane fractions after their solubilization with Triton X-100, however, labels both receptor subunits in both membrane fractions regardless of prior insulin treatment of the intact cells, although the ratio of label incorporated into the α- and β-receptor subunits is now $\sim 1:2$. These results are interpreted by Hedo & Simpson (1984) as an indication that the α-receptor subunit is primarily, if not exclusively, exposed on the extracellular surface of the plasma membrane and the β-receptor subunit spans the membrane. This interpretation is fully consistent with the prevailing thought that the α subunit, containing the insulin binding site, functions in hormone recognition, and the β-receptor subunit, containing an insulin-responsive tyrosine-specific protein kinase, in signal transduction (Hedo & Simpson, 1984).

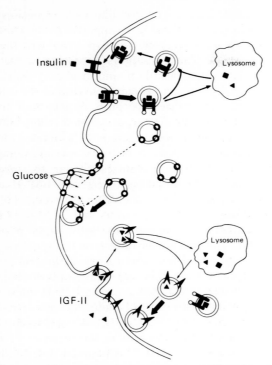

Fig. 11. *Schematic representation of hypothetical integral membrane protein translocations in the mechanism of insulin action in the isolated rat adipose cell*

Conclusions

A schematic representation of our working model of these integral membrane protein translocations in the mechanism of insulin action in the rat adipose cell is illustrated in Fig. 11. In the basal state, most if not all of the cell's insulin receptors are associated with the plasma membrane and exposed to the extracellular environment. In contrast, most of the cell's glucose transporters and a major portion of its IGF-II receptors are associated with intracellular membrane vesicles which fractionate in a low-density microsomal membrane fraction. Although the latter is enriched in membranes of the Golgi apparatus, the specific membrane species containing the intracellular glucose transporters and IGF-II receptors are probably distinct from this classic subcellular organelle, and perhaps from each other, but their exact identities and localizations in the intact cell remain major unresolved questions. The nature of any cycling of these three integral membrane proteins between the plasma membrane and their intracellular loci in the absence of ligands is totally unknown. In the presence of IGF-II and absence of insulin, IGF-II receptors do cycle at a basal rate between the plasma membrane and their intracellular pool in an apparently endocytic-/exocytic-like process, delivering an as yet unquantified portion of the internalized IGF-II to the cell's lysosomes, where it is degraded.

In response to insulin's interaction with its receptor at the cell surface, however, and through as yet unidentified signalling and propulsion mechanisms, insulin receptors are rapidly internalized by an apparently endocytic-like process into two intracellular compartments associated respectively with the high-density and low-density microsomes. Those receptors appearing in the high-density microsomes appear to contain covalently bound insulin. At least a portion of the simultaneously internalized insulin is then delivered to the cell's lysosomes for degradation while the internalized receptors recycle back to the plasma membrane by an apparently exocytic-like process. A new steady-state subcellular distribution of receptors is thus achieved, where $\sim 30\%$ of those initially present in the plasma membrane are now present in the cytoplasmic compartment.

Concomitant with, and with time courses similar to, insulin receptor internalization, intracellular glucose transporters and IGF-II receptors are rapidly translocated to the plasma membrane. After achievement of the maximal response, $\sim 60\%$ of those glucose transporters and $\sim 20\%$ of those IGF-II receptors initially associated with their respective intracellular pools are present in the plasma membrane. The appearance of glucose transporters in the plasma membrane is reflected in the classic stimulatory action of insulin on the maximum glucose transport velocity. The appearance of IGF-II receptors in the plasma membrane is reflected in (1) an increase in IGF-II internalization and degradation through an increased rate of endocytic-/exocytic-like receptor cycling, or (2) an increase in cell surface IGF-II binding if receptor recycling is inhibited during ligand association, such as in the presence of KCN. In the presence of IGF-II and insulin, the altered subcellular distribution of IGF-II receptors represents attainment of a new steady state; in the absence of IGF-II, insulin still shifts the subcellular distribution of receptors but the presence or absence of cycling is unknown. The presence or absence of glucose transporter cycling in the presence of insulin also remains to be established, although the shifts in the subcellular distributions of both IGF-II receptors and glucose transporters in response to insulin are rapidly reversible with insulin withdrawal.

Although these effects of insulin on its own binding and internalization, on glucose transport activity, and on IGF-II binding and internalization in the rat adipose cell are mediated primarily through modulating the subcellular distribution of insulin receptors, glucose transporters and IGF-II receptors, respectively, an additional level of regulation of these three processes by catecholamines, ACTH, and glucagon and their counter-regulatory hormones adenosine, nicotinic acid, and PGE_1 is now recognized. These agents are classically thought to elicit their responses through interaction with their own specific receptors, modulation of either a stimulatory or inhibitory GTP-binding regulatory subunit and, ultimately, stimulation or inhibition of adenylate cyclase activity and alteration of the cellular cyclic AMP concentration. In the case of insulin and IGF-II binding, adenylate cyclase stimulators appear to decrease and adenylate cyclase inhibitors appear to increase the cell surface concentrations of insulin (Pessin et al., 1983; Lönnroth & Smith, 1983) and IGF-II (unpublished observations) receptors; however, the relationship between the cell surface concentrations of these receptors and the translocations of these receptors between the plasma membrane and their corresponding intracellular pools is presently unknown. In contrast, in the case of glucose transport activity, these agents appear to decrease or increase glucose transporter intrinsic activity, respectively, without influencing glucose transporter translocation. Furthermore, the effects of these agents on glucose transporter intrinsic activity appear to be mediated by a cyclic AMP-independent mechanism, perhaps directly at the level of the glucose transporter itself. These agents may also influence the glucose transport response to insulin at the level of the signal generation process, as reflected in their marked effects on insulin sensitivity.

A mechanistic relationship among the three integral membrane protein translocations described here in the rat adipose cell remains to be established. Although the similar time courses and common endocytic-/exocytic-like character of these three processes are highly suggestive that insulin's stimulatory action on glucose transporter translocation and IGF-II receptor cycling may be mediated by the internalization and cycling of the insulin receptor itself, subtle qualitative and not so subtle quantitative differences, including those in the insulin concentration dependencies, are not yet explained. In addition, the mechanisms through which membrane translocation events might be linked are completely unknown, even with respect to the linkage between the endocytic and exocytic legs in the cycling of a single membrane protein species. Perhaps the tyrosine-specific protein kinase activity of the β-insulin receptor subunit will ultimately prove to play this very role: (1) this activity is specifically stimulated by insulin in the cell's plasma membrane (Kasuga et al., 1982; Häring et al., 1982; Petruzzeli et al., 1982), and (2) the altered phosphorylation state of the β-receptor subunit brought about by insulin is maintained with internalization. Thus, the activated kinase activity may be exposed to new substrates in previously inaccessible intracellular loci (Simpson & Hedo, 1984). On the other hand, Simpson & Hedo (1984) report the absence of detectable effects on the phosphorylation state of the β-receptor subunit by a human anti-insulin receptor antiserum with marked insulin-like effects on glucose transport, membrane protein phosphorylation, and insulin receptor internalization in the rat adipose cell. The mechanism of these integral membrane protein translocations in response to insulin is currently the subject of intensive investigation.

We thank our many colleagues, both former and current, for their indispensable contributions to the concepts and experimental results described here. These investigators include: Kenneth C. Appell, José A. Hedo, Paul J. Hissin, Rupert C. Honnor, Richard Horuk, Barbara B. Kahn, Eddy Karnieli, Masao Kuroda, Constantine Landos, Matthew M. Rechler, Lester B. Salans, Ole Sonne, Ulf Smith, Chih-Chen Wang, Lawrence J. Wardzala, Thomas J. Wheeler, Dena R. Yver and Mary Jane Zarnowski. We also thank Deborah L. Baly and Teresa M. Weber for their critical comments regarding this report, and Louie Zalc for her patience and expertise in typing the manuscript.

References

Bergeron, J. J. M., Sikstrom, R., Hand, A. R. & Posner, B. I. (1979) J. Cell Biol. 80, 427–443

Berhanu, P., Olefsky, J. M., Tsai, P., Saunders, D. T., Thamm, P. & Brandenburg, D. (1982) Proc. Natl. Acad. Sci. U.S.A. 79, 4069–4073

Carpentier, J.-L., Gorden, P., Freychet, P., Le Cam, A. & Orci, L. (1979) J. Clin. Invest. 63, 1249–1261

Clark, S. & Harrison, L. C. (1982) J. Biol. Chem. 257, 12239–12244

Clark, S. & Harrison, L. C. (1983) J. Biol. Chem. 258, 14–21

Crofford, O. B. & Renold, A. E. (1965a) J. Biol. Chem. 240, 14–21

Crofford, O. B. & Renold, A. E. (1965 *b*) *J. Biol. Chem.* **240**, 3237–3244
Cushman, S. W. & Wardzala, L. J. (1980) *J. Biol. Chem.* **255**, 4758–4762
Cushman, S. W., Wardzala, L. J., Simpson, I. A., Karnieli, E., Hissin, P. J., Wheeler, T. J., Hinkle, P. C. & Salans, L. B. (1984) *Fed. Proc.* **43**, 2251–2255
Diebert, D. C. & De Fronzo, R. A. (1980) *J. Clin. Invest.* **65**, 717–721
Ezaki, O. & Kono, T. (1982) *J. Biol. Chem.* **257**, 14306–14310
Foley, J. E., Cushman, S. W. & Salans, L. B. (1978) *Am. J. Physiol.* **234**, E112–E119
Gliemann, J. & Rees, W. D. (1983) *Curr. Top. Memb. Trans.* **18**, 339–379
Gliemann, J., Østerlind, K., Vinten, J. & Gammeltoft, S. (1972) *Biochim. Biophys. Acta* **286**, 1–9
Gorga, J. C. & Lienhard, G. E. (1984) *Fed. Proc.* **43**, 2237–2241
Green, A. (1983) *Biochem. J.* **212**, 189–195
Green, A. & Olefsky, J. M. (1982) *Proc. Natl. Acad. Sci. U.S.A.* **79**, 427–431
Guerre-Milo, M., Lavau, M., Horne, J. S. & Wardzala, L. J. (1985) *J. Biol. Chem.* **260**, 2197–2201
Hammons, G. T. & Jarett, L. (1980) *Diabetes* **29**, 475–486
Häring, J. U., Kasuga, M. & Kahn, C. R. (1982) *Biochem. Biophys. Res. Commun.* **108**, 1538–1545
Hedo, J. A. & Simpson, I. A. (1984) *J. Biol. Chem.* **259**, 11083–11089
Hissin, P. J., Karnieli, E., Simpson, I. A., Salans, L. B. & Cushman, S. W. (1982 *a*) *Diabetes* **31**, 589–592
Hissin, P. J., Foley, J. E., Wardzala, L. J., Karnieli, E., Simpson, I. A., Salans, L. B. & Cushman, S. W. (1982 *b*) *J. Clin. Invest.* **70**, 780–790
Honnor, R. C., Dhillon, G. & Londos, C. (1985 *a*) *J. Biol. Chem.* **260**, in the press
Honnor, R. C., Dhillon, G. & Londos, C. (1985 *b*) *J. Biol. Chem.* **260**, in the press
Horuk, R., Rodbell, M., Cushman, S. W. & Simpson, I. A. (1983) *FEBS Lett.* **164**, 261–266
Horuk, R., Rodbell, M., Cushman, S. W. & Wardzala, L. J. (1983) *J. Biol. Chem.* **258**, 7425–7429
Karnieli, E., Hissin, P. J., Simpson, I. A., Salans, L. B. & Cushman, S. W. (1981 *a*) *J. Clin. Invest.* **68**, 811–814
Karnieli, E., Zarnowski, M. J., Hissin, P. J., Simpson, I. A., Salans, L. B. & Cushman, S. W. (1981 *b*) *J. Biol. Chem.* **256**, 4772–4777
Kashiwagi, A., Heucksteadt, T. P. & Foley, J. E. (1983) *J. Biol. Chem.* **258**, 13685–13692
Kasuga, M., Karlsson, F. A. & Kahn, C. R. (1982) *Science* **215**, 185–187
King, G. L., Rechler, M. M. & Kahn, C. R. (1982) *J. Biol. Chem.* **257**, 10001–10006
Kono, T. (1984) *Fed. Proc.* **43**, 2256–2257
Kono, T., Suzuki, K., Dansey, L. E., Robinson, F. W. & Blevins, T. L. (1981) *J. Biol. Chem.* **256**, 6400–6407
Kono, T., Robinson, F. W., Blevins, T. L. & Ezaki, O. (1982) *J. Biol. Chem.* **257**, 10942–10947
Levine, R. & Goldstein, M. (1955) *Recent Prog. Horm. Res.* **11**, 343–380
Lienhard, G. E., Kim, H. K., Ransome, K. J. & Gorga, J. C. (1982) *Biochem. Biophys. Res. Commun.* **105**, 1150–1156
Londos, C., Honnor, R. C. & Dhillon, G. S. (1985) *J. Biol. Chem.* **260**, in the press
Lönnroth, P. & Smith, U. (1983) *Biochem. Biophys. Res. Commun.* **112**, 971–979
Marshall, S. (1985) *J. Biol. Chem.* **260**, 4136–4144
Marshall, S. & Olefsky, J. M. (1981) *Biochem. Biophys. Res. Commun.* **102**, 646–653
Marshall, S., Heidenreich, K. A. & Horikoshi, H. (1985) *J. Biol. Chem.* **260**, 4128–4135
Narahara, H. T. & Ozand, P. (1963) *J. Biol. Chem.* **238**, 40–49
Oka, Y. & Czech, M. P. (1984) *J. Biol. Chem.* **259**, 8125–8133
Oka, Y., Mottola, C., Oppenheimer, C. L. & Czech, M. P. (1984) *Proc. Natl. Acad. Sci. U.S.A.* **81**, 4028–4032
Oppenheimer, C. L., Pessin, J. E., Massague, J., Gitomer, W. & Czech, M. P. (1983) *J. Biol. Chem.* **258**, 4824–4830
Park, C. R., Reinwein, D., Henderson, M. J., Cadenas, E. & Morgan, H. E. (1959) *Am. J. Med.* **26**, 674–684
Pessin, J. E., Gitomer, W., Oka, Y., Oppenheimer, C. L. & Czech, M. P. (1983) *J. Biol. Chem.* **258**, 7386–7394
Pessin, J. E., Massague, J. & Czech, M. P. (1984) in *Investigation of Membrane Located Receptors* (Reid, E., Cook, G. N. W. and Moore, D. J., eds.), pp. 295–302, Plenum, London
Petruzzeli, L. M., Ganguly, S., Smith, C. S., Cobb, M. H., Rubin, C. S. & Rosen, O. M. (1982) *Proc. Natl. Acad. Sci. U.S.A.* **79**, 6792–6796
Rennie, P. & Gliemann, J. (1981) *Biochem. Biophys. Res. Commun.* **102**, 824–831
Robinson, F. W., Blevins, T. L., Suzuki, K. & Kono, T. (1982) *Anal. Biochem.* **122**, 10–19
Schoenle, E., Zapf, J. & Froesch, E. R. (1977) *Diabetologia* **13**, 243–249
Simpson, I. A. & Hedo, J. A. (1984) *Science* **223**, 1301–1304
Simpson, I. A., Yver, D. R., Hissin, P. J., Wardzala, L. J., Karnieli, E., Salans, L. B. & Cushman, S. W. (1983) *Biochim. Biophys. Acta* **763**, 393–407

Simpson, I. A., Hedo, J. A. & Cushman, S. W. (1984) *Diabetes* **33**, 13–18

Smith, U., Kuroda, M. & Simpson, I. A. (1984*a*) *J. Biol. Chem.* **259**, 8758–8763

Smith, M. M., Robinson, F. W., Watanobe, T. & Kono, T. (1984*b*) *Biochim. Biophys. Acta* **775**, 121–128

Sonne, O. & Simpson, I. A. (1984) *Biochim. Biophys. Acta* **804**, 404–413

Suzuki, K. & Kono, T. (1979) *J. Biol. Chem.* **254**, 9786–9794

Suzuki, K. & Kono, T. (1980) *Proc. Natl. Acad. Sci. U.S.A.* **77**, 2542–2545

Taylor, L. P. & Holman, G. D. (1981) *Biochim. Biophys. Acta* **642**, 325–335

Taylor, W. M. & Halperin, M. L. (1979) *Biochem. J.* **178**, 381–389

Vinten, J., Gliemann, J. & Østerlind, K. (1976) *J. Biol. Chem.* **251**, 794–800

Wang, C.-C., Sonne, O., Hedo, J. A., Cushman, S. W. & Simpson, I. A. (1983) *J. Biol. Chem.* **258**, 5129–5134

Wardzala, L. J. (1979) *Identification of the Glucose Transport System in Purified Rat Adipose Cell Plasma Membranes Using a Cytochalasin B Binding Assay: Effects of Insulin and Altered Physiological States* (Ph.D. dissertation), Dartmouth College, Hanover, NH

Wardzala, L. J. & Jeanrenaud, B. (1981) *J. Biol. Chem.* **256**, 7090–7093

Wardzala, L. J. & Jeanrenaud, B. (1983) *Biochim. Biophys. Acta* **730**, 49–56

Wardzala, L. J., Cushman, S. W. & Salans, L. B. (1978) *J. Biol. Chem.* **253**, 8002–8005

Wardzala, L. J., Simpson, I. A., Rechler, M. M. & Cushman, S. W. (1984) *J. Biol. Chem.* **259**, 8378–8383

Watanobe, T., Smith, M. M., Robinson, F. W. & Kono, T. (1984) *J. Biol. Chem.* **259**, 13117–13122

Wheeler, T. J., Simpson, I. A., Sogin, D. C., Hinkle, P. C. & Cushman, S. W. (1982) *Biochem. Biophys. Res. Commun.* **105**, 89–95

Biochem. Soc. Symp. **50**, 151–168
Printed in Great Britain

Secondary Active Nutrient Transport in Membrane Vesicles: Theoretical Basis for Use of Isotope Exchange at Equilibrium and Contributions to Transport Mechanisms

ULRICH HOPFER

Department of Developmental Genetics and Anatomy, Case Western Reserve University,
Cleveland, Ohio, 44106, U.S.A., and Max-Planck-Institute for Biophysics, Frankfurt,
Federal Republic of Germany

Synopsis

A detailed and quantitative analysis of secondary active transport mechanisms in membrane vesicles is complicated by heterogeneity of the vesicles. Functional heterogeneity can be demonstrated by the time-dependence of isotope exchange of any solute at equilibrium. The need for more than one rate constant in the fit proves functional heterogeneity. To treat the heterogeneity quantitatively, it is suggested to subject entire time curves of exchange to inverse Laplace transformations that yield the corresponding distribution of rate constants. The computer program CONTIN by Provencher (1982 *a*, *b*, *c*) can be used to carry out such a transformation. The distribution of rate constants under a particular set of conditions can be used to calculate a highly reliable initial rate. In addition, for spherical vesicles a mean, surface area-averaged permeability constant can be calculated if the size distribution of the vesicle population is known by other measurements and this size distribution is independent of the permeability distribution. Kinetic measurements under equilibrium conditions on the rabbit intestinal Na–glucose transporter indicate (using Cleland's nomenclature) an ordered iso-bi-bi mechanism with glide symmetry for substrate and co-substrate binding to the transporter at one interface and release at the other (first-in-first-out) (Hopfer & Groseclose, 1980). The kinetics are consistent with a gated pore mechanism of coupled Na–glucose cotransport. A similar mechanism seems to hold for renal Na–lactate cotransport (Mengual *et al.*, 1983).

Introduction

Isolated plasma membrane vesicles were introduced over a decade ago to investigate the properties of transporters mediating concentrative nutrient uptake (Hopfer *et al.*, 1973). The major advantages of the isolated plasma membrane approach are: 1, localization of transport processes to a particular region of the plasma membrane, e.g. apical or baso-lateral in epithelial cells; 2, wide range of experimental conditions under which transport can be investigated;

3, identification of driving forces and cofactors in a relatively simple system *in vitro*.

The existence of Na-dependent, secondary active transport as the mechanism for concentrative nutrient uptake by animal cells is now well established for many different substrates, ranging from the 'classical' substrates of glucose and amino acids, to tricarboxylic acid cycle intermediates and lactate. Several recent reviews are available on the topic of transport in vesicles (Kimmich, 1981; Nord *et al.*, 1982; Turner, 1983; Stevens *et al.*, 1984). Brush border membranes from the intestine and renal proximal tubule (both cortical and outer medullary segments) have provided a wealth of information on many different transport systems. The Na-dependent glucose transporters are probably the most extensively studied so far. With increasing information about properties of glucose transport in different membrane preparations, the number of different recognized transporters has become larger. Na-dependent amino acid transport is present not only in the luminal plasma membrane of epithelial cells from intestine and kidney, but also in the plasma membrane of most cells. Several amino acid transport systems with different specificity have been defined, although the exact relationship of Na-dependent transport systems from different cells, e.g. epithelial and non-epithelial, is not clear. One has to keep in mind that most transport systems remain only operationally defined in terms of activity (dependence on substrate concentration, activators, or inhibitors).

From a historical point of view, three different phases in the development of studies with isolated membrane vesicles can be distinguished: 1, qualitative assessment of the presence of transporters in isolated membranes and participation of putative co-substrates and transmembrane electrical gradients; 2, attempts to quantify transport reactions, both in terms of overall stoichiometries, as well as in terms of kinetic models that describe the dynamic behaviour of the transporter of interest; 3, attempts to chemically define the transporters, including the molecular weights, peptides involved and reactive groups that are essential for activity.

We may be approaching the end of phase 2, with the availability of theoretical and experimental tools to investigate quantitatively and describe solute transport in a population of membrane vesicles. However, concensus is not complete with respect to usefulness and limitations of different methods (Semenza *et al.*, 1984). Chemical information about transporters has been accumulated only slowly in spite of the early success by Crane's group (Crane *et al.*, 1976) with solubilization and reconstitution of Na-dependent glucose transporters. One of the reasons may be insufficient awareness of the difficulties associated with converting nutrient flux under 'energized' conditions to transporter activity *per se*. The apparent rate of Na-dependent nutrient transport is dependent on many factors extraneous to the membrane, such as the surface/volume ratio of vesicles or the driving force in the form of an electrochemical Na-gradient. These problems make it difficult to determine some kind of 'intrinsic' transporter activity that allows comparison of different preparations and evaluations of the success of protocols for purification. This point will be dealt with in greater detail below.

Most recently, Na–glucose cotransport has been reviewed by Semenza *et al.* (1984). The article is a good summary of much of the work that has been carried out in recent years in this field, particularly in the laboratory of the authors. Its availability allows me to concentrate on an area of special interest to me, namely experimental conditions that allow useful inferences from solute flux or uptake data with membrane vesicle populations. Clarity about the underlying theoretical framework relevant to this particular experimental system is important since it is used so much.

Our training in 'classical' biochemistry generally has equipped us well to deal with homogeneous systems in solution. This outlook has carried over to studies with membrane vesicles and produced a wealth of molecular models as well as critieria for experimental design and interpretation of flux data in terms of these models in homogeneous vesicle populations (one of the more elegant papers is that by Sanders *et al.*, 1984). My concern has been for some years whether the assumption of homogeneity is justified and, if not, how one can get interesting and valid information about the transporters of interest from heterogeneous vesicles (Hopfer, 1977, 1981).

Several sources can contribute to heterogeneity in isolated membrane preparations, which would influence interpretation of flux data in terms of underlying molecular models: 1, mixtures of different types of membranes (usually expressed as contamination of the membrane fraction of interest by other types); 2, degree of interaction of plasma membrane proteins with cytoskeletal elements (a common problem in renal brush border membranes, where tight association of the microvillus core with the membrane in some places and detachment in others can result in 'tennis racket' shapes); 3, surface/volume ratios (or shape and size) of membrane vesicles, which can vary considerably; 4, different types of transporters for the same substrate, but differing in either types or number of the co-substrates; 5, density of these transporters in different vesicles.

Why Isotope Exchange Measurements at Equilibrium?

Real membrane preparations constitute an ensemble of parallel compartments in contact with the medium. The different compartments usually do not behave homogeneously, which can be demonstrated in functional terms by the presence of more than one apparent rate constant for isotope flux under equilibrium exchange conditions (Hopfer, 1981). A major corollary of heterogeneity is that contributions of different vesicles to the macroscopic observations may vary with conditions. These considerations suggest the usefulness of methods to assess the functional heterogeneity of membrane preparations with respect to transport in a quantitative manner.

In a general way, permeability characteristics of biological and lipid bilayer membranes can be experimentally determined in two different physical configurations: (i) planar ('black' lipid membranes), whereby the membrane separates two macroscopic compartments and is supported at the edges by some kind of frame, and (ii) vesiculated, with the membrane separating a microscopic, intravesicular space from a macroscopic compartment (the incubation medium). In the latter system, measurements are generally not carried out on individual vesicles, but a population of vesicles. The planar configuration has the advantage that permeability of the membrane to a solute under a particular set of conditions can be normalized per unit area. In general, the same treatment is not possible for macroscopic flux data from a vesicle population as the normalization depends on information about the contributions of individual vesicles to the macroscopic flux data, the surface area of these vesicles, and, under some conditions, also the volume bounded by the respective surface.

A rigorous description of material properties of a heterogeneous ensemble of molecules or membrane vesicles requires careful selection of the investigative methods and information on the weighting of the properties of the individual members to the overall macroscopic average of the population. This point is well recognized in polymer science, and students of biochemistry are familiar with the differences in the number-average, the weight-average and the Z-average molecular weight of proteins. In the case of solute flux, the membrane permeability for a particular solute is generally of interest. Permeability is used here as a general term for the membrane property that determines solute flux and is defined as the proportionality constant derived from flux measurements after normalization for area, solute concentration and driving force for solute movement. It contains both transporter-mediated and non-mediated pathways. By choosing appropriate experimental conditions for flux measurements and methods of analysis, it is possible to determine average permeability properties of the membranes from heterogeneous populations.

Quantification of Heterogeneity of Transport in Membrane Vesicle Populations

The experimental conditions that allow quantitative inferences about hetero-geneity of transport in membrane vesicles become obvious when one considers the basic equations relating macroscopic flux measurements to events in individual vesicles. For example, solute flux is often measured by discrete sampling of solute uptake (or release) by (from) vesicles. In this case, the time-dependent macroscopic uptake is the sum of the events in the individual vesicles:

$$\text{Uptake } (t) = \sum_i \int_0^t v_i \cdot dt + \text{uptake } (0) \tag{1}$$

whereby v_i is the transport velocity into an individual vesicle and i is summed over all the vesicles in the measured sample. The membrane properties (P_m) can only be inferred from the velocities, v_i, after accounting for effects of surface area, substrate concentration and appropriate driving force:

$$v_i = P_m \cdot (\text{surface area}) \cdot f([\text{substrate}], \text{driving force}) \tag{2}$$

where f( ) means 'function of'.

In general, v_i can take on positive or negative values. For example, for solute–non-electrolyte cotransport the experimental conditions are often rigged in such a way that the non-electrolyte initially accumulates inside the vesicles, but is released later on when the driving force is dissipated. Obviously, if different vesicles have different dissipation times for the driving force, a time period must exist when both processes are going on simultaneously in the population. A consequence of eqn. (1) is that the v_i values cannot be evaluated without *a priori* knowledge of the time-dependence of v_i. In addition, if velocities (v) are taken as normalized (specific) values, the weights (contributions) of individual vesicles with a particular time-dependence of v to the macroscopic uptake need to be known or determinable.

An important consideration in the design of flux experiments in membrane vesicles is, then, the experimental condition for which the general time course of solute flux is known *a priori*. One such condition is isotope exchange between a small and large compartment in equilibrium with each other. For single-walled vesicles with a time-independent solute permeability (non-excitable membranes), the isotope flux is determined by a single exponential characterized by an apparent rate constant of exchange, whether solute flux is carrier-mediated or not. All parallel pathways for solute flux across a given membrane give rise to a combined rate constant. Under these experimental conditions it is possible to make inferences about the distribution (frequency) of rate constants in the population.

To demonstrate mathematically the relationship between the frequency of apparent rate constants of exchange in the population and the macroscopic uptake data it will be useful to express the experimental data in terms of a complementary exchange function $c \times f(t)$:

$$c \times f(t) = 1 - \text{fractional exchange} \quad \text{(dimensionless)} \tag{3}$$

$c \times f(t)$ is a time-dependent function, covering both the situations of isotope uptake and release. For an individual vesicle, the time-dependence of tracer exchange between an infinitely large and a small compartment is given by a single exponential:

$$c \times f(t) = \exp\{-k \cdot t\} \tag{4}$$

where k is a function of membrane-specific properties, such as surface, volume and permeability, as well as of experimental condition-dependent variables, such as substrate concentration, temperature, pH etc. k is assumed to be time-independent. For a vesicle population, the macroscopic complementary exchange function is then given by:

$$c \times f(t) = \sum f(k_i) \cdot \exp\{-k_i \cdot t\} \tag{5}$$

whereby the summation is carried out over all the vesicles in the sample. $f(k_i)$ measures the contribution that each set of vesicles with a particular k_i makes to the macroscopic flux. Eqn. (5) implies that $\sum f(k_i) = 1$.

For samples with large numbers of vesicles (and isolated membrane preparations typically contain more than 10^{10} vesicles/mg of protein), the time-dependence of $c \times f(\ )$ can be approximated as the integral:

$$c \times f(t) = \int_0^\infty f(k) \cdot \exp\{-k \cdot t\} \, dk \quad \text{with} \quad \int_0^\infty f(k) \, dk = 1 \tag{6}$$

Eqn. (6) illustrates that the exchange function $c \times f(t)$ is the Laplace transform of the frequency distribution of the apparent rate constants. The Laplace transform is a well understood operation, akin to the Fourier transform. The interesting information about the heterogeneity of the vesicle population is

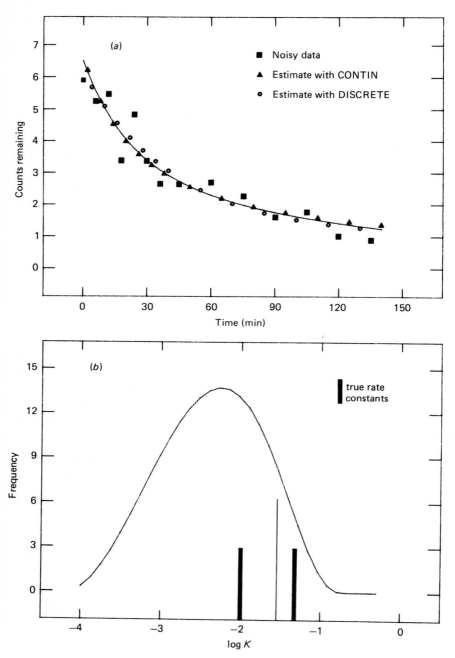

Fig. 1. *Simulation of isotope exchange*

Comparison of possible solutions with CONTIN (Provencher, 1982a, b, c) and DISCRETE (Provencher, 1976). Simulation of isotope efflux was carried out by using the built-in feature of CONTIN with added noise of about 20% of maximal counts remaining. The true rate constants are indicated in (b) (solid bars). (a) Time-dependent efflux of solute. Continuous line: data without noise. For clarity only every third grid point is plotted. (b) Comparison of estimated rate constants contained in f(k), which can be obtained from the experimentally measured $c \times f(t)$ by an inverse Laplace transformation. A practical inversion method is provided by the computer program CONTIN (Fig. 1). With sufficiently precise transport data, the frequency distribution f(k) may reveal distinct subpopulations.

Description of Transport in Terms of Mean Properties of the Population

Information on the frequency distribution of the apparent rate constant of a population allows calculation of a defined, mean apparent rate constant and initial rate which can both be used to characterize and compare vesicle populations, e.g. during purification of transporters. In addition, for spherical vesicles it is possible to calculate a mean permeability (i.e. essentially a rate constant normalized to surface area), provided the size distribution is known and is independent of the permeability distribution.

The mean rate constant for a population is defined in general (i.e. when f(k_i) is not normalized) as:

$$\langle k \rangle = \Sigma \, \mathrm{f}(k_i) \cdot k_i / \Sigma \, \mathrm{f}(k_i) \quad \text{(units: time}^{-1}) \tag{7}$$

Thus, when f(k_i) is obtained by a numerical inverse Laplace transformation of the transport data, $\langle k \rangle$ can be calculated in a straightforward manner. Interestingly, the mean rate constant gives a reliable estimate of the initial steady-state rate of the isotope exchange as is seen when eqn. (5) is differentiated with respect to time. The initial rate is of interest when comparing different membrane populations as it should be dependent only on the surface involved and thus can be used to define a specific activity. The initial velocity of isotope exchange $V_{(t=0)}$ is given by:

$$V_{(t=0)} = \langle k \rangle \cdot [\text{substrate}] \cdot (\text{volume}) \quad \text{(units: mass flux/time)} \tag{8}$$

[substrate] represents the substrate concentration at which the isotope exchange is carried out and 'volume' refers to the intravesicular volume involved (i.e. the small compartment where solute concentrations are changing). If the intravesicular volume is expressed per unit protein (typically, 1–5 μl/mg of protein) then the specific initial velocity is obtained. Because the initial velocity is dependent only on surface area, eqn. (8) is applicable to vesicles of any shape and can be used to compare the activity of different populations, to evaluate, for example, enrichment of transporters.

Although the determination of the specific transport activity on a protein basis is straightforward (eqn. 8), conversion to transport activity on a surface area basis is more difficult. Such information is desirable in many cases, particularly if activity in planar membrane systems is to be compared with that in vesicular

---

between CONTIN and DISCRETE. The high level of noise, that was added for demonstration purposes, makes retrieval of the true rate constants difficult with either program. The scales of the ordinate are arbitrary and are different for solutions with CONTIN and DISCRETE. The mean estimated rate constants were 22.9 and 35.1 ms^{-1} by CONTIN and DISCRETE respectively, as compared with the true one of 30 ms^{-1}.

Table 1. *Glucose permeability of brush border membranes*

Permeabilities were calculated from mean rate constants of isotope exchange at equilibrium and mean vesicle size according to eqn. (9). Mean rate constants were obtained from entire time curves of isotope efflux by inverse Laplace transformations with CONTIN (Provencher, 1982a, b, c). The size distribution was determined by photon correlation spectroscopy (PCS) also using CONTIN. Experimental conditions were similar to those reported by Hopfer & Groseclose (1980) for both transport and PCS, specifically, glucose (1 mM), Na (0.1–0.2 M) and phlorizin, when present (0.1 mM). The temperature was 25 °C. PCS was carried out at several angles between 60° and 120° by using a digital correlator. PCS experiments were carried out at the Max-Planck-Institute for Biophysical Chemistry (Goettingen, FRG) or at the Department of Macromolecular Science, Case Western Reserve University.

Tissue	Addition	Permeability (nm/s)
Intestine	—	3.3 ± 0.5
	+phlorizin	1.0 ± 0.1
Kidney	—	5.9 ± 3.4

ones. The conversion is possible in the case of spherical vesicle populations if the size distribution is determined by independent measurements, such as photon correlation spectroscopy, and the size distribution is independent of the permeability distribution. This situation is probably encountered frequently, although the alternative is conceivable. For example, when cells are disrupted and fractionated, the size of the membrane fragments and of the resulting resealed vesicles could depend on the carrier density so that vesicle size and permeability are no longer independent. Therefore, the assumption of independence of the two parameters needs to be stressed and may need to be evaluated experimentally for particular membranes.

With the above precautions in mind, it can be shown that eqn. (9) holds (see the Appendix for derivation):

$$\langle P \rangle = \langle k \rangle / (3 \cdot \langle 1/R \rangle) \quad \text{(units: length/time)} \qquad (9)$$

$\langle P \rangle$ is the surface area fraction averaged permeability and $\langle 1/R \rangle$ is the weight fraction averaged inverse radius. The determination of the latter parameter requires knowledge of the size distribution of the vesicle population. Estimates for glucose permeability of intestinal and renal brush border membranes are given in Table 1.

Usefulness and Limitations of Isotope Exchange in Vesicles

One of the major reasons for flux measurements is an interest in information about molecular properties of transport systems. Quantitative measurements are important in two areas: 1, reproducible determination of specific transport activity that is independent of the physical shape of the membrane so that different populations can be compared (e.g. during purification of transporters), and 2, for evaluation of kinetic mechanisms. For heterogeneous vesicle populations the quantification is not trivial since different vesicles will contribute differently to the macroscopic signal and information about the weighting factors (e.g. f(k), f(P), f(R)) need to be known. Functional non-homogeneity of membrane vesicle populations with respect to transport are immediately recognized in isotope exchange conditions at equilibrium if more than one rate constant is needed to achieve a fit to the time-dependent exchange. In this case, a major question arises with respect to the model to be employed for analysis. Are the data to be analysed in terms of a discrete number of exponentials (distinctly different, but in themselves homogeneous subpopulations) or in terms of a continuous distribution of exponentials [one or more (sub)populations with a spread in the transport properties and size]? With noisy data it is not possible by analytical means to arrive at an unequivocal solution as noise and a spread in the rate constants can have a similar effect on the observed exchange function (Fig. 1). Therefore, the answer has to depend on a consideration of the physical system. For vesicles of isolated membranes it is likely that, over a certain range, a continuous size distribution exists. Similarly, since the membranes are derived from many different cells, a distribution of solute permeability is also conceivable. Both of these factors would contribute to a broadening of the frequency peak in the rate constant domain.

For heterogeneous vesicle populations inferences about molecular mechanisms are then a two-step procedure: 1, estimation of appropriate weighting factors so that, for example, the mean rate constant can be determined; 2, dependence of the mean rate constant (or some related parameter) on independent variables such as substrate concentration, temperature, inhibitors etc. The measured dependence of the mean rate constant (or permeability) on independent (experimental) variables constitutes a meaningful and legitimate subject for kinetic modelling. For this modelling, one has to keep in mind that all parallel pathways that exist in a membrane for the solute of interest will contribute to the total permeability that is measured. The characteristics of the transporter-mediated pathway can only be inferred from the dependence of the rate constant (or related parameters) on independent variables, e.g. saturation behaviour, inhibition, activation by co-substrates etc.

The major advantages of the proposed framework can be summarized as follows: 1, information about homogeneity or heterogeneity of the vesicle population is readily available; 2, initial rates can be reliably estimated taking advantage of an entire time course of influx or efflux; 3, the equilibrium exchange conditions automatically ensure that the driving force for solute flux is constant and reproducible among laboratories (hence, fundamental kinetic constants, such as K_m or V_{max}, should become comparable); 4, for spherical vesicles, it is possible to estimate mean permeabilities, i.e. normalize flux data on the basis of surface area.

The major limitations are those resulting from the restriction to equilibrium conditions with an identical composition of the intra- and extra-vesicular medium. For kinetic experiments, these conditions can provide some, but certainly not exhaustive, information about the kinetic mechanism. An alternative to using an entire time course of flux measurements is the determination of so-called 'initial' rates of solute flux. This method dominates the field because it is applicable also to non-equilibrium conditions. For this condition, the major criterion available to judge the suitability of flux measurements for kinetic

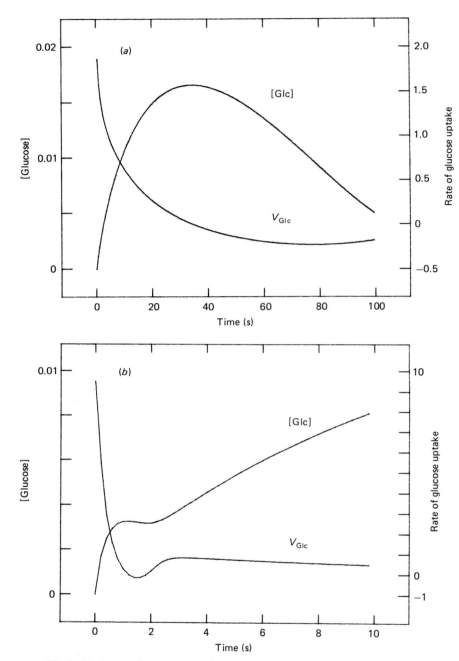

Fig. 2. *Simulation of concentrative glucose uptake by brush border membrane vesicles*

The simulation was carried out on an IBM-XT microcomputer with the TUTSIM (1982) program (for methods of modelling see Karnopp & Rosenberg, 1975). V_{Glc}, Velocity of glucose uptake; [Glc], glucose concentration. The assumptions in the model are existence of: 1, Na–glucose cotransport that follows Michaelis–Menten kinetics with respect to glucose as well as Na$^+$ concentration, and with voltage-dependent $V_{max.}$; 2, uncoupled glucose transport; 3, electroneutral NaCl transport;

analyses is linearity of the macroscopic flux data. It is obvious, however, from eqn. (1) that, in general (i.e. without additional information) the estimation of initial rates from discrete uptake data constitutes an ill-posed problem with many widely different solutions, all fitting the data without resembling the true initial velocity. Any error would carry over to estimates of membrane permeability and derived parameters, such as K_m for a transported solute. Therefore, inherent in 'initial' rate kinetics is an assumption of sufficiently homogeneous behaviour in the vesicle population, at least for the time period under study (i.e. a time window in which the velocities in all vesicles are constant and positive). Whether this assumption is justified in all cases is difficult to verify. One of the problems with vesicles is their high surface/volume ratio so that steady-state conditions are difficult to achieve (Hopfer, 1984). For vesicles with a diameter of 0.3 μm and an intravesicular volume of about 1 μl/mg of protein, the surface area is about 200 cm^2/mg and the depth of the fluid layer on the inside only 50–100 nm. Time constants for the decay of gradients could be very fast, and under these conditions it is easy to show with simulations the possible existence of rapid changes in uptake velocities (even changing sign) in heterogeneous populations if positive (influx) as well as negative (efflux) rates are found macroscopically, and hence must exist also in individual vesicles (Fig. 2).

The considerations on quantification of heterogeneity suggest that isotope exchange experiments at equilibrium may be useful in general to define membrane preparations and provide baseline information on transporters. Michaelis constants determined under these conditions reflect actual, although averaged for different transporter orientations, affinities for substrate binding and therefore may be able to bridge the gap between functional assays of transport and biochemical probes of 'active' sites. The conditions and extracted parameters of the transport systems should be reproducible among laboratories and therefore may help establish identity or differences of transport systems that are worked on by different investigators.

Application of Isotope Exchange at Equilibrium to Kinetic Models

Isotope exchange is an established method to probe mechanisms of enzymatic catalysis. The method turns out to be useful also for investigating coupled transport processes. Isotope exchange at equilibrium constitutes a very well-defined condition and therefore results can be interpreted usually with few ambiguities. Molecular models for different classes of coupled cotransport with 1:1 stoichiometries were considered by Hopfer & Liedtke (1981), and for an ordered iso-ter-ter reaction by Mengual et al. (1983). From a kinetic point of view, it is useful to distinguish between rapid-equilibrium and steady-state

---

4, Cl$^-$ conductance; 5, Na$^+$ conductance. The model solves for the quasi-steady-state potential, the velocity of glucose influx, Na$^+$ and Cl$^-$ influx, and intravesicular concentrations of these solutes. (a) A typical NaCl gradient-driven overshoot of glucose uptake assuming a homogeneous vesicle population. (b) The macroscopic appearance of the transport velocity when the membrane preparation contains two subpopulations for which the time constants of the overshoot differ by a factor of ten. The fast component is derived from only a small fraction of the vesicles. This Figure illustrates the possibility of large fluctuations in velocity that would be difficult to pick up with noisy uptake data.

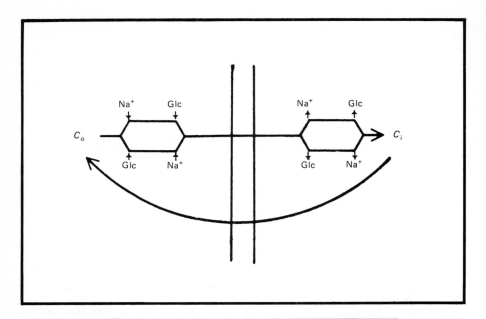

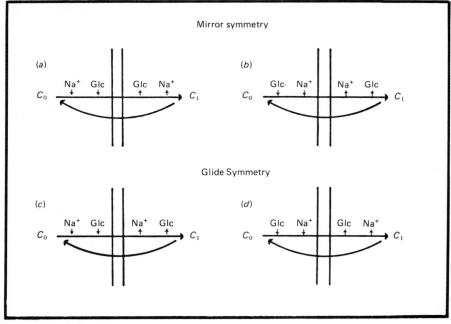

Fig. 3. *Symmetrical models for cotransport with a 1:1 stoichiometry*

Transport reactions are written as net flux from the 'o'-side to the 'i'-side. Top: random iso-bi-bi transport reaction. Bottom: ordered iso-bi-bi transport reaction. Reproduced with permission from Hopfer & Groseclose (1980).

mechanisms, depending on whether the translocation is slow (rapid-equilibrium of substrate binding) or fast (steady state) relative to the surface reactions. The models can again be subdivided on the basis of the order of interactions of substrate and co-substrate with the transporter as either random or ordered, depending on whether binding of a substrate is necessary before the co-substrate can be bound under net flux conditions (Fig. 3). A reaction could be ordered on one side of the membrane and random on the other. In Cleland's nomenclature, the symmetrical models (reactions at both interfaces are of the same type) are classified as iso-random-bi-bi and iso-order-bi-bi, respectively [see Fig. 3 and Semenza *et al.* (1984) for more detailed explanations]. For ordered transport, another distinction can be made with respect to the sequence of substrate and co-substrate interactions with the transporter at the two interfaces. The reaction sequence of substrate and co-substrate addition at one interface and release at the other could have a glide symmetry (first-on-first-off) or a mirror symmetry (last-on-first-off). The models have been described in detail (Hopfer & Groseclose, 1980; Hopfer & Liedtke, 1981).

The most interesting parameter for distinguishing between different models is the quantitative dependence of the activation of a substrate isotope flux on the co-substrate concentration (Table 2). For the two systems that have been investigated in depth by this method, i.e. rabbit intestinal Na–glucose cotransporter and horse renal Na–lactate cotransporter, pronounced biphasic activation curves were found for both rate of nutrient exchange as a function of Na^+ concentration and for the rate of Na^+ exchange as a function of nutrient concentration. High concentrations of the co-substrate suppressed the exchange rate of the substrate to very low levels or even back to control. This type of behaviour is predicted for steady-state models with predominantly ordered behaviour and glide symmetry of substrate and co-substrate binding and de-binding. The kinetic model has led to the interesting suggestion of a gated pore type mechanism with a fast translocation rate. One has to keep in mind that the model refers to a 'ground' state of the membrane without any electrical potential. It will be interesting to see how the model will hold up once the transporters become purified, and more refined kinetic as well as structural data are available. Of particular interest are the electrical characteristics of the transporter since the behaviour of the transport system strongly depends on which binding or transport steps are voltage sensitive (Sanders *et al.*, 1984).

Semenza *et al.* (1984) have criticized the approach of isotope exchange at equilibrium. They raised the following points: 1, the large experimental error of flux data and correction for 'leak' rates explaining differences in the K_m value for glucose which we reported (Hopfer, 1977; Hopfer & Groseclose, 1980); 2, identical model predictions for random and ordered models in terms of biphasic activation curves by the co-substrate; 3, a stoichiometry of more than 1 for Na^+ to glucose for the rabbit intestinal transporter, which was underlying the model for interpretation of the flux data.

With respect to the first point, the 'leak' rates are present whether glucose flux is measured under equilibrium (as isotope flux) or concentrative uptake conditions. The difference between the two experimental conditions is that appropriate corrections for the contribution of the 'leak' rate are straightforward

Table 2. *Velocity equations for isotope exchange of iso-bi-bi transport mechanisms*

Labelled form of a substrate is denoted by an asterisk (*). $*v$ and $*V_{max.}$ indicate the tracer velocities. $*V^1_{max.}$ indicates the maximal velocity for particular specified conditions of $[S_1]$, $[*S_1]$ and so on. $K^{s_1}_m$ is the apparent dissociation constant for S_1. It is a weighted average dissociation constant for both membrane interfaces. The concentration of S_2 has to be specified because $K^{s_1}_m$ may vary with $[S_2]$. The same applies to $K^{s_2}_m$ as related to $[S_1]$. K_1, K_2, K_A, and K_B are constants that are composites of rate and equilibrium constants. Reproduced with permission from Hopfer & Liedtke (1981).

Mechanism	Symmetry	Labelled substrate (S_2)	Condition	Equation
Activation of substrate (S_1) by its co-substrate (S_2)				
Random	—	—	$*S_1, S_1$ = constant; S_2 = variable	$*v_{s_1} = \dfrac{*V^1_{max.}[S_2]\{1 + K_A[S_2] + K_B[S_2]^2\}}{\{1 + [S_2]/K^{s_2}_m\}\{1 + K_1[S_2] + K_2[S_2]^2\}}$
Ordered	Mirror	First (outer) ligand	$*S_1, S_1$ = constant; S_2 = variable	$*v_{s_1} = \dfrac{*V^2_{max.}[S_2]}{\{1 + [S_2]/K^{s_2}_m\}\{1 + K_1[S_2]\}}$
	Mirror	Second (inner) ligand	$*S_1, S_1$ = constant; S_2 = variable	$*v_{s_1} = \dfrac{*V^3_{max.}[S_2]}{1 + [S_2]/K^{s_2}_m}$
Ordered	Glide	—	$*S_1, S_1$ = constant; S_2 = variable	$*v_{s_1} = \dfrac{*V^4_{max.}[S_2]}{\{1 + [S_2]/K^{s_2}_m\}\{1 + K_1[S_2]\}}$
K_m determination				
Random or ordered			$*S_1, S_2$ = constant; S_1 = variable	$*v_{s_1} = \dfrac{*V_{max.}}{K^{s_1}_m + [S_1]}$
			$*S_2, S_1$ = constant; S_2 = variable	$*v_{s_2} = \dfrac{*V_{max.}}{K^{s_2}_m + [S_2]}$

under equilibrium conditions, but much more complicated when concentrative uptake is measured. Under equilibrium conditions, solute flux through the 'leak' pathway is parallel to the Na-dependent one and solute flux through both pathways is in the same direction and experiences the same driving force. In contrast, under concentrative uptake conditions, the solute flux through the 'leak' experiences a driving force different from the one through the Na-dependent transporter, so that flux may be either in the same direction or in the opposite direction, depending on the ratio of solute concentration on the two sides of the membrane. With respect to the K_m values for the Na-dependent glucose transporter, I reported a value of 14.2 mM for rat intestine (Hopfer, 1977) and 2.4 mM for rabbit intestine (Hopfer & Groseclose, 1980). It is not surprising that the two species give different quantitative answers. Interestingly, the value for rabbit intestine is similar to 3.7 mM, the value reported by Kessler et al. (1978) for isotope exchange conditions, suggesting that they are useful for standardization and comparison of the properties of membrane preparations in different laboratories.

The predictions of random and ordered models of transporters have been discussed by Hopfer & Liedtke (1981). One has to keep in mind that the ordered model is an extreme case of the random model with zero probability for one possible mode of the random binding situation. Under isotope exchange at equilibrium both models predict first activation and then inhibition of the exchange rate with increasing concentration of the co-substrate; the difference is complete inhibition in an ordered mechanism by very high co-substrate concentration, because only one mode is operative, versus incomplete inhibition in a random model. In a way, the degree of inhibition by the co-substrate relative to the maximum transport rate indicates the extent of preference for this particular mode. In the case of the intestinal Na–glucose cotransport (Hopfer & Groseclose, 1980), the conclusion of an ordered model was based on the complete inhibition of the glucose-dependent Na-exchange rate by high glucose concentrations.

An important aspect of modelling, which cannot be determined from isotope exchange experiments, is the overall stoichiometry of cotransport. For rabbit intestinal Na–glucose transport, this question has become particularly interesting because of the availability of kinetic data. The question is controversial since precedence for higher stoichiometries exists in the kidney (Turner, 1985). Furthermore, chick enterocytes seem to contain a transporter with a ratio of 2:1 for Na:glucose (Kimmich, 1981). Earlier measurements on intact intestinal sheets from the rabbit had indicated a ratio (Na:glucose) of 1:1 (Goldner et al., 1969). In contrast, Kaunitz & Wright (1984) have presented vesicle data for a high capacitance system with a ratio of 1:1 and another system with a ratio of 3:1. These data were obtained under Na-gradient conditions that provide a very high driving force for the system, with a ratio of 3:1. It is likely that under equilibrium conditions the high capacity Na–glucose cotransporter with a 1:1 stoichiometry dominates, since the isotope gradient is the only driving force for any measurable fluxes. Because of the controversy, we re-examined the stoichiometry in rabbit jejunal brush border membranes by the static-head method introduced by Turner (1985). In this method, gradients of glucose and

Na^+ are set up across short-circuited membranes and the Na^+ gradient determined that prevents a net flux of glucose (at least for short times). The stoichiometry can be calculated from equilibrium thermodynamics (Kimmich, 1981; Turner, 1985). Our data for rabbit intestinal brush border membranes are consistent with a ratio of 1:1. In addition, the results with the Na–lactate transporter, for which a 2:1 stoichiometry has been established (Mengual & Sudaka, 1983; Mengual *et al.*, 1983), indicate similar kinetic behaviour in isotope flux experiments so that the higher stoichiometries would not appear to change the biphasic activation behaviour, useful for distinguishing between different models.

The above detailed discussion indicates that in my experience the isotope exchange method yields flux data that are interpretable with little ambiguity, and therefore contain valuable information. The points raised by Semenza *et al.* (1984) do not seem to have a real basis.

Conclusions

On theoretical grounds, isotope exchange experiments offer substantial advantages for investigating transporters in membrane vesicle populations that are functionally not homogeneous. Because of the simple time dependence of isotope fluxes in single vesicles, the quality of the inferences that are made from noisy data can be estimated relatively well. These types of experiments allow some quantitative estimates of heterogeneity in the transport of interest. Furthermore, they invite a two-tier process for making inferences from macroscopic data: first, about the contribution of different vesicles; second, about the contribution of transporters to the properties of vesicles.

I express my gratitude to Dr S. W. Provencher (Max-Planck-Institute for Biophysical Chemistry) for substantial contributions to the project, in particular modification of the CONTIN program for inverse Laplace transformations and analysis of photon correlation spectroscopy data for hollow spheres (vesicles). I am also indebted to Dr K. Neet, Dr G. Ehrenspeck, Dr A. Jamieson, Dr H. Ohno (Case Western Reserve University) and Professor L. DeMaeyer (Max-Planck-Institute for Biophysical Chemistry) for valuable discussions and/or use of their instruments. I also thank Ms Carola Koitz from the Max-Planck-Institute for Biophysical Chemistry, Professor K. J. Ullrich (Max-Planck-Institute for Biophysics) for his support and to Ms A. Zimmerschied from the Max-Planck-Institute for Biophysics and Mr T. Crowe (Case Western Reserve University) for technical help. Supported by grants from the PHS (AM 18265) and the National Science Foundation (PCM 78-07211).

References

Crane, R. K., Malathi, P. & Preiser, H. (1976) *FEBS Lett.* **67**, 214–216
Goldner, A. M., Schultz, S. G. & Curran, P. F. (1969) *J. Gen. Physiol.* **53**, 362–383
Hopfer, U. (1977) *J. Supramol. Struct.* **7**, 1–13
Hopfer, U. (1981) *Fed. Proc. Fed. Am. Soc. Exp. Biol.* **40**, 2480–2485
Hopfer, U. (1984) in *Mechanisms of Intestinal Electrolyte Transport and Regulation by Calcium* (Donowitz, M., ed.), pp. 27–33, Riss, New York.

Hopfer, U. & Groseclose, R. (1980) *J. Biol. Chem.* **255**, 4453–4462

Hopfer, U. & Liedtke, C. M. (1981) *Membr. Biochem.* **4**, 11–29

Hopfer, U., Nelson, K., Perrotto, J. & Isselbacher, K. J. (1973) *J. Biol. Chem.* **248**, 25–32

Karnopp, D. & Rosenberg, R. (1975) *System dynamics: a unified approach.* Wiley, London

Kaunitz, J. D. & Wright, E. M. (1984) *J. Membr. Biol.* **79**, 41–51

Kessler, M., Tannenbaum, V. & Tannenbaum, C. (1978) *Biochim. Biophys. Acta* **509**, 348–359

Kimmich, G. A. (1981) in *Physiology of the Gastrointestinal Tract* (Johnston, R. J., ed.), pp. 1035–1061, Raven Press, New York, NY

Mengual, R. & Sudaka, P. (1983) *J. Membr. Biol.* **71**, 163–171

Mengual, R., Leblanc, G. & Sudaka, P. (1983) *J Biol. Chem.* **258**, 15071–15078

Nord, E., Wright, S. H., Kippen, I. & Wright, E. M. (1982) *Am. J. Physiol.* **243**, F456–F462

Provencher, S. W. (1976) Technical Report. Max-Planck-Institut fuer biophysikalische Chemie, Goettingen, FRG

Provencher, S. W. (1982*a*) Technical Report EMBL-DAO5. European Molecular Biology Laboratory, Heidelberg, FRG

Provencher, S. W. (1982*b*) *Computer Physics Comm.* **27**, 213–227

Provencher, S. W. (1982*c*) *Computer Physics Comm.* **27**, 229–242

Sanders, D., Hansen, U. P., Gradmann, D. & Slayman, C. L. (1984) *J. Membr. Biol.* **77**, 123–152

Semenza, G., Kessler, M., Hosang, M., Weber, J. & Schmidt, U. (1984) *Biochim. Biophys. Acta* **779**, 343–379

Stevens, B. R., Kaunitz, J. D. & Wright, E. M. (1984) *Annu. Rev. Physiol.* **46**, 417–433

Turner, R. J. (1983) *J. Membr. Biol.* **76**, 1–15

Turner, R. J. (1985) *Ann N.Y. Acad Sci.*, in the press.

TUTSIM (1982) User's Manual, Twente University of Technology, Enschede, The Netherlands

APPENDIX

Derivation of the Relationship Between Rate Constant, Membrane Permeability and Size in a Vesicle Population

If all vesicles in a population are spherical in shape, then the surface/volume ratio is a function solely of the radius. Several different methods exist for the determination of size distributions of vesicles (e.g. gel permeation, photon correlation spectroscopy). Knowledge of the size distribution allows conversion of the mean apparent rate constant $\langle k \rangle$ for isotope exchange in a vesicle population to a mean permeability $\langle P \rangle$, i.e. normalization on a surface area basis, provided that the size distribution is independent from the permeability distribution. In practical terms, this condition means for isolated biological membranes that, for instance, the fragmentation pattern that occurs during cell fractionation is random and not influenced by the carrier density in a particular piece of membrane. This situation may not always be the case. However, if the above condition is correct, the following relationship can be shown to hold for a population of spherical vesicles:

$$\langle k \rangle = 3 \langle P \rangle \langle 1/R \rangle \qquad (A1)$$

where $\langle P \rangle$ is a mean permeability (units: length/time) and $1/\langle 1/R \rangle$ a mean radius (units: length). The relationship in a single vesicle between k, P and R is:

$$k = (3P)/R \qquad (A2)$$

as the apparent rate constant of exchange is directly proportional to the surface area and the permeability and inversely proportional to the volume of the vesicle.

The complementary exchange function of a population of vesicles can therefore be expressed as a function of k or of P and R:

$$c \times f(t) = \sum f(k_i) \cdot \exp\{-k_i t\} \quad \text{(see eqn. 4)} \tag{A3}$$

$$c \times f(t) = \sum_m f(P_m) \sum_j f(R_j) \cdot \exp\{-3P_m t/R_j\} \tag{A4}$$

where $f(P_m)$ is the surface area fraction of the permeability P_m and $f(R_j)$ is the weight ($=$ volume) fraction with radius R_j. In this formulation, $f(k_i)$, $f(P_m)$ and $f(R_j)$ are the normalized frequencies, i.e. their respective sums equal 1. The involvement of the weight fraction is easily seen when one considers vesicles with the same membrane permeability, but different radii. The contribution that each vesicle makes to the macroscopic signal is proportional to its volume (weight) fraction. Rearrangement of eqn. (A4) to:

$$c \times f(t) = \sum_m \sum_j f(P_m) \cdot f(R_j) \cdot \exp\{-3P_m t/R_j\} \tag{A5}$$

and comparison with eqn. (A3) indicates that for each vesicle there is

$$f(k_i) = f(P_m) \cdot f(R_j) \tag{A6}$$

Using this latter relationship, eqn. (A1) can be derived from the definition of $\langle k \rangle$:

$$\langle k \rangle = \sum f(k_i) \cdot k_i = \sum_m \sum_j f(P_m) \cdot f(R_j) \cdot 3 \cdot P_m/R_j \tag{A7}$$

or:

$$\langle k \rangle = 3 \sum_m f(P_m) \cdot P_m \sum_j f(R_j) \cdot 1/R_j = 3 \cdot \langle P \rangle \cdot \langle 1/R \rangle \tag{A8}$$

Eqn. (A8) defines the weighting factors for averaging, namely the surface area fraction for the permeability and the weight fraction for the inverse radius.

Biochem. Soc. Symp. **50**, 169
Printed in Great Britain

Translocation of Proteins Across Endoplasmic Reticulum Membranes Probed with Short Polypeptides (Abstract)

BERNHARD DOBBERSTEIN

European Molecular Biology Laboratory, Postfach 10.2209, 6900 Heidelberg, Federal Republic of Germany

Translocation of proteins across endoplasmic reticulum (ER) membranes is mediated by specific receptors, Signal Recognition Particle (SRP) and Docking Protein (DP) (Meyer *et al.*, 1982; Walter *et al.*, 1984). SRP interacts with the signal sequence in nascent secretory and membrane proteins and, in the absence of membranes, can arrest chain elongation after about 70 amino acid residues have been polymerized.

For the small coat protein of phage M13 a receptor-independent mechanism for membrane insertion has been found. In order to investigate requirements for membrane insertion of small proteins we used a coupled transcription–translation system in which different sized lysozyme fragments can be generated (Stueber *et al.*, 1984). It was found that peptides of a size larger than 70 amino acid residues are translocated across microsomal membranes by an SRP and DP dependent mechanism. A peptide comprising 51 amino acid residues could not any more be translocated. It is concluded that there is a critical size limit for proteins to be translocated across the ER membrane.

References
Meyer, D., Krause, E. & Dobberstein, B. (1982) *Nature (London)* **297**, 647–652
Stueber, D. and others (1984) *EMBO J.* **3**, 3143–3148
Walter, P., Gilmore, R. & Blobel, G. (1984) *Cell* **38**, 5–8

Biochem. Soc. Symp. **50**, 171–191
Printed in Great Britain

Entry Mechanisms of Protein Toxins and Picornaviruses

SJUR OLSNES, KIRSTEN SANDVIG, INGER HELENE MADSHUS and
ANDERS SUNDAN

*Norsk Hydro's Institute for Cancer Research, and The Norwegian Cancer Society, Montebello,
Oslo 3, Norway*

Synopsis

The mode of entry into cells of a number of protein toxins with intracellular
sites of action and of three picornaviruses is discussed. Of the different toxins
in this group, diphtheria toxin has been most thoroughly studied with respect
to its uptake mechanism. This toxin binds to cell surface receptors which are
possibly part of the major anion-transport system in the cells. The bound toxin
is then endocytosed and, when the pH drops below pH 5, a normally hidden
hydrophobic domain is exposed and inserted into the membrane. By a process
which, in addition to low pH, requires chloride transport and a proton gradient
across the membrane, the toxin A fragment is translocated to the cytosol. When
diphtheria toxin is bound at the cell surface, rapid entry through the surface
membrane can be induced by treatment with low pH. Modeccin and *Pseudomonas*
exotoxin A also require low pH for entry, but low pH is not able to induce rapid
entry of these toxins from the cell surface. Another group of toxins, abrin, ricin
and viscumin, is characterized by the fact that low pH in the medium prevents
the toxins from entering the cytosol, but not from entering endocytic vesicles.
However, when the pH is subsequently returned to neutrality the endocytosed
toxins are able to enter the cytosol. In the picornaviruses the entry of a single
hydrophilic macromolecule per cell is also sufficient to induce maximal biological
effect. Poliovirus, like diphtheria toxin, appears to enter the cytosol from an
acidic intracellular compartment which may be the endosome. Also human
rhinovirus 2 requires low pH for entry, whereas encephalomyocarditis virus
does not enter at low pH. The similarities and differences between the uptake
mechanisms of toxins and viruses are discussed.

Introduction

The biological effects of a number of protein toxins are due to their ability
to enzymatically modify intracellular components (for reviews, see Olsnes & Pihl,
1976, 1982; Olsnes & Sandvig, 1983; Pappenheimer, 1977; Uchida, 1983). Thus
the plant toxins abrin, modeccin, ricin and viscumin and the bacterial *Shigella*
toxin all inactivate enzymatically the 60 S ribosomal subunits and thereby block
protein synthesis. Diphtheria toxin and exotoxin A from *Pseudomonas aeruginosa*
(PEA) inhibit protein synthesis by inactivating elongation factor 2.

With the possible exception of PEA, the toxins here discussed all consist of

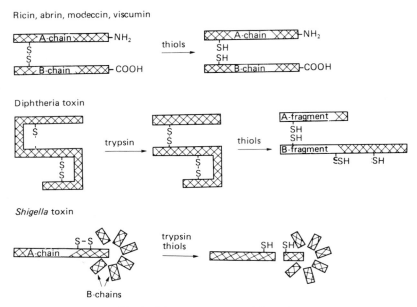

Fig. 1. *Schematic structure of the toxins*

two disulphide-linked, functionally different parts, namely a B-moiety, which consists of one or several polypeptides and which binds the toxins to receptors at the cell surface, and an A-moiety, which has enzymatic properties and which therefore is able to inactivate a large number of target molecules (Fig. 1). In fact, it has been shown for some of these toxins that the entry of a single molecule into the cytosol may be sufficient to kill a cell (Yamaizumi *et al.*, 1978; Eiklid *et al.*, 1980).

The enzymatically active A-moiety of the toxins consists in each case of a single polypeptide chain or a polypeptide fragment with molecular weight in the range 20000–30000. The main problem considered in this paper is how a hydrophilic protein of this size is able to penetrate the lipid bilayer of cellular membranes and enter the cytosol.

Picornaviruses are positive-strand RNA viruses which are approximately 100 times larger than the toxins here discussed. In spite of the difference in size, there are a number of similarities in the mode of entry of toxins and viruses. The picornaviruses bind by their protein capsid to receptors at the cell surface (Fig. 2). The only requirement for infection is that a copy of the RNA genome is translocated to the cytosol. Once present in the cytosol, the genome serves as messenger RNA for the production of proteins that are required for replication of the virus.

It has recently been shown that the nucleocapsid of Semliki forest virus, influenza virus, vesicular stomatitis virus and a large number of other coated viruses enters the cytosol after fusion of the lipid envelope with membranes in the cells (for review, see Lenard & Miller, 1983). Since the picornaviruses do not have a lipid coat that can fuse with the cellular membranes to allow entry of

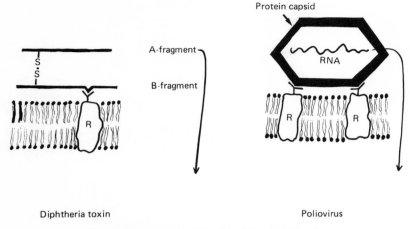

Fig. 2. *Similarities in entry of toxins and picornaviruses*

The toxins bind by their B-moiety to cell surface receptors (R), and picornaviruses bind to their receptors by the capsid proteins. The requirement for intoxication is that one copy of the A-moiety enters the cytosol; the requirement for infection is the entry of one copy of the RNA genome.

the genome to occur, we have considered the possibility that the entry mechanism of the picornaviruses may resemble that of the protein toxins.

Mode of Entry of Diphtheria Toxin

Among the different protein toxins with intracellular sites of action, diphtheria toxin has been studied in most detail (for review, see Pappenheimer, 1977; Uchida, 1983). This toxin is synthesized as a single polypeptide chain which is easily cleaved by trypsin-like enzymes to yield two polypeptide fragments, the enzymatically active A-fragment, and the B-fragment which is involved in binding of the toxin to cell surface receptors (Fig. 1). Some of the toxin molecules contain one molecule of the dinucleotide ApUp, bound non-covalently to a site on the toxin designated the P-site (Lory & Collier, 1980).

Role of low pH

The first indication as to how diphtheria toxin may enter cells was the observation by Boquet & Pappenheimer (1976) that the B-fragment contains a hydrophobic region which is normally not exposed to the solvent. They speculated that this region may become inserted into cellular membranes and allow the A-fragment to penetrate into the cytosol.

Some years later it was found that, for entry into the cytosol to occur, diphtheria toxin must pass through a low-pH compartment (Draper & Simon, 1980; Sandvig & Olsnes, 1980). Fifteen years earlier Kim & Groman (1965) observed that NH_4Cl strongly protects cells against diphtheria toxin. By studying how variation of the pH in the medium influenced the process, they concluded that the protection is provided by free NH_3 and not by the salt. In

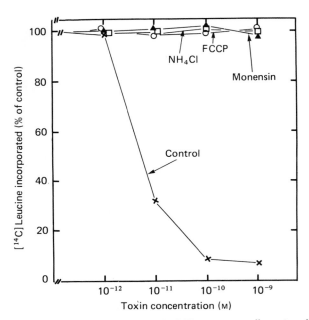

Fig. 3. *Ability of NH₄Cl, monensin and FCCP to protect cells against diphtheria toxin*

To cells in Hepes medium were added the indicated compounds and 15 min later increasing concentrations of toxins. After incubation at 37 °C for 1 h the rate of [³H]leucine incorporation was measured. The additions were: □, 10 mM-NH₄Cl; ○, 10 μM-FCCP; ▲, 10 μM-monensin; ×, none (Sandvig & Olsnes, 1980).

spite of this observation the mechanism behind the protection remained obscure until 1980, when it was found that not only NH_4Cl, but also monensin and other compounds which increase the pH of intracellular vesicles, protect against the toxin (Fig. 3). Furthermore, it was found that the protection is circumvented if cells with surface-bound toxin are briefly exposed to low pH. These results indicated that the toxin normally enters from acidic intracellular vesicles.

Treatment with low pH was also found to strongly reduce the lag period from addition of toxin to the cells until protein synthesis inhibition becomes evident, indicating that under these conditions the toxin enters directly from the cell surface (Fig. 4). On the basis of these experiments we proposed that the toxin enters from endocytic vesicles as soon as the pH of the vesicles has reached a sufficiently low value and that transfer to the lysosomes is not necessary (Sandvig & Olsnes, 1981 *a*). It has later been found that fusion of the endosomes with other vesicular compartments does not appear to be required for the action of diphtheria toxin (Sandvig *et al.*, 1984; Marnell *et al.*, 1984).

One reason that low pH is required for entry of diphtheria toxin is that the hidden hydrophobic domain in the B-fragment of the toxin becomes exposed when the pH is reduced below pH 5 (Sandvig & Olsnes, 1981 *a*; Blewitt *et al.*, 1984). Electrophysiological studies with lipid bilayers showed that under these conditions the hydrophobic domain inserts itself into the artificial membrane

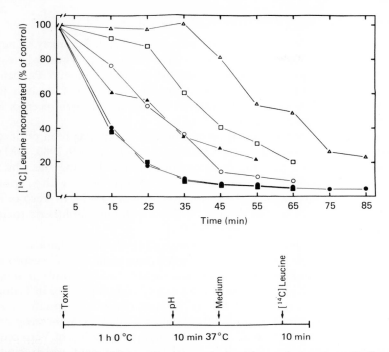

Fig. 4. *Effect of pH on the rate of protein synthesis inhibition after exposure of cells to diphtheria toxin*

Diphtheria toxin was added to cells in 24-well disposable trays. After incubation for 1 h at 0 °C to allow maximal toxin binding, the medium (and unbound toxin) was removed and phosphate buffer with the indicated pH was added (time zero). After 10 min the buffer was removed, and medium was added. After various periods of time, the ability of the cells to incorporate [14C]leucine into acid-precipitable material during a 10 min period was measured. △, 1 μg of toxin/ml, buffer pH 7.2; □, 10 μg of toxin/ml, buffer pH 7.2; ○, 100 μg of toxin/ml, buffer pH 7.2; ▲, 0.1 μg of toxin/ml, buffer pH 4.5; ●, 1 μg of toxin/ml, buffer pH 4.5; ■, 10 μg of toxin/ml, buffer pH 4.5 (Sandvig & Olsnes, 1980).

and forms an ion-permeable channel across the bilayer (Donovan *et al.*, 1981; Kagan *et al.*, 1981; Kayser *et al.*, 1981; Misler, 1983; Deleers *et al.*, 1983).

The hydrophobic region can also be inserted into liposomes, and the protein then forms a channel with an estimated diameter of 1.8–2.4 nm (Kagan *et al.*, 1981; Zalman *et al.*, 1984). This may be sufficient to allow the extended form of the A-fragment to pass the membrane. In fact, it has been claimed that when liposomes are exposed to diphtheria toxin at low pH, the A-fragment is translocated to the interior (Donovan *et al.*, 1982).

A number of findings indicate that the entry of the A-fragment into the cytosol of living cells is more complicated than suggested by the model studies with artificial membranes and liposomes. Thus, toxin entry does not occur at low pH if the pH gradient across the membrane is reduced by acidification of the cytosol, or if chloride entry into the cells is prevented (Sandvig & Olsnes, 1984). We have therefore studied in more detail the uptake of diphtheria toxin by living cells.

Properties of the diphtheria toxin receptor in Vero cells

Middlebrook *et al.* (1978) observed that Vero cells and related monkey kidney cells are particularly rich in binding sites for diphtheria toxin. Thus, whereas HeLa cells have only approximately 5000 receptors per cell (Boquet & Pappenheimer, 1976), Vero cells have 100000, and these cells are also very sensitive to the toxin (Middlebrook *et al.*, 1978). Studies of the diphtheria toxin receptor is therefore most conveniently done on Vero cells.

A glycoprotein (M_r 150000) which binds diphtheria toxin has been purified from guinea-pig thymocytes and characterized by Proia *et al.* (1979, 1980, 1981). They proposed a model where the receptor contains one binding site which interacts with the P-site of the toxin and another site which interacts with a region of the toxin which is distant from the P-site (Eidels *et al.*, 1982). In Vero cells they found two proteins (M_r 140000 and 70000) which bind diphtheria toxin (Eidels *et al.*, 1983).

So far there are no data available demonstrating that the characterized proteins which bind diphtheria toxin are indeed functional toxin receptors. Experiments in our laboratory have shown that there is a considerable amount of unspecific binding of diphtheria toxin to cells. Also certain proteins in Triton X-100 lysates of cells bind the toxin in an apparently non-specific way (A. Sundan, unpublished results). To be able to identify the functional receptor, we have therefore attempted to find ways of modifying the ability of Vero cells to bind the toxin and then study if a membrane component which binds diphtheria toxin is concomitantly altered.

Several years ago Moehring & Crispell (1974) found that when cells were treated with phospholipase C, or with trypsin, they became less sensitive to diphtheria toxin. We have recently repeated their experiments with purified enzymes and found that the treatments strongly reduce the ability of Vero cells to bind diphtheria toxin (Fig. 5a). In these experiments the relative amount of toxin bound was estimated from the toxic effect induced by the bound toxin.

Treatment of cells with phospholipase C may activate protein kinase C (Nishizuka, 1984). To study if the protective effect observed was an indirect one due to phosphorylation by this enzyme, we repeated the experiment in cells treated with 2-deoxyglucose and NaN_3 to strongly reduce the level of ATP in the cells. Essentially the same reduction in the ability of the cells to bind the toxin was found also in this case (Fig. 5a). It is therefore possible that phospholipids as well as protein at the cell surface are involved in the binding of diphtheria toxin.

In the search for compounds and conditions which inhibit the entry of diphtheria toxin, we found that chloride deprivation and treatment with compounds like SITS (4-acetamido-4′-isothiocyanostilbene-2,2′-disulphonic acid), which inhibit chloride entry, reduced the ability of the cells to bind diphtheria toxin. Furthermore, such treatments prevented the entry of already bound toxin (Sandvig & Olsnes, 1984; S. Olsnes *et al.*, unpublished work). SITS contains an isothiocyano group which is able to form a covalent bond with the anion channel. Since treatment with SITS was highly efficient also when it was carried out in ATP-depleted cells (Fig. 5a), as well as at 0 °C, it is likely that

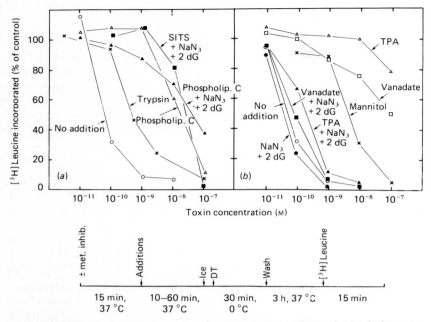

Fig. 5. *Ability of phospholipase C, trypsin, SITS, TPA, vanadate and chloride deprivation to reduce the sensitivity of Vero cells to diphtheria toxin in ATP-depleted and in metabolically active cells*

Vero cells in 24-well disposable trays were incubated for 15 min at 37 °C in the absence (open symbols) and presence (filled symbols) of 50 mM-2-deoxyglucose (2dG) and 10 mM-NaN₃. The cells were then treated with enzymes and compounds as indicated below and then chilled on ice. Increasing concentrations of diphtheria toxin (DT) were added and allowed to bind for 30 min at 0 °C. Subsequently the cells were washed to remove unbound toxin and then incubated at 37 °C for 3 h to allow the bound toxin sufficient time to express its toxic effect. The toxic effect was finally measured as reduction in the ability of the cells to incorporate [³H]leucine. (*a*) Additions: ○, none; △, ▲, 20 units of phospholipase C/ml; ×, 50 μg of trypsin/ml; ■, 0.1 mM-SITS. Incubation with these compound was 30 min at 37 °C. (*b*) Additions: ○, none; △, ▲, 0.1 μM-TPA; □, ■, 0.1 mM-sodium vanadate; ×, in this case the cells were incubated in Cl⁻-free buffer made isotonic with mannitol. The incubations at 37 °C under these conditions were 10 min with TPA, 60 min with sodium vanadate and 30 min in mannitol-balanced buffer.

the binding of SITS to the diphtheria toxin receptor is sufficient to prevent binding of the toxin.

Also, when cells were incubated in chloride-free buffer made isotonic with mannitol, their ability to bind diphtheria toxin was strongly reduced (Fig. 5*b*). However, in this case it was necessary to preincubate the cells in the mannitol buffer for 30 min at 37 °C to obtain maximal effect.

We observed several years ago that treatment of cells with the tumour-promoter TPA (12-*O*-tetradecanoyl-phorbol-13-acetate) protected against diphtheria toxin (Sandvig & Olsnes, 1981*a*). Recent results have shown that this protection can be strongly increased if the cells are preincubated with TPA (Guillemot *et al.*, 1985). Metabolic energy appears to be necessary, since TPA-treatment of ATP-depleted cells did not reduce their ability to bind the toxin (Fig. 5*b*). Similar results were obtained with sodium vanadate and NaF.

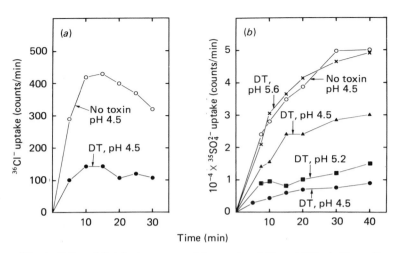

Fig. 6. *Ability of diphtheria toxin to inhibit uptake of Cl⁻ and SO₄²⁻ in Vero cells*

Vero cells were incubated in the absence and presence of 30 nM-diphtheria toxin (DT) for 20 min at 37 °C and then treated for 1 min at 37 °C with Hepes medium adjusted to the pH indicated. The cells were then washed twice with 260 nM-mannitol containing 1 mM-Ca(OH)$_2$ and 20 mM-2-[N-morpholino]ethanesulphonic acid adjusted to pH 6 with Tris, and incubated with the same buffer containing ^{36}Cl⁻ (a) or ^{35}SO$_4^{2-}$ (b). In ^{36}Cl⁻ uptake the buffer was adjusted to pH 7. After the indicated periods of time, cells were washed and treated with 5% (w/v) trichloroacetic acid and the acid-extracted radioactivity was measured.

Treatment of cells with TPA, vanadate and fluoride all increase the phosphorylated state of the cells, albeit by different mechanisms. Thus TPA stimulates protein kinase c (Nishizuka, 1984), whereas vanadate and fluoride inhibit dephosphorylation reactions (Swarup *et al.*, 1982; Mumby & Traugh, 1979). Possibly, therefore, the diphtheria toxin receptor is only able to bind the toxin when the receptor exists in a non-phosphorylated state.

Is the main anion transporter involved in the binding of diphtheria toxin to Vero cells?

Since chloride deprivation was found to reduce the ability of Vero cells to bind diphtheria toxin by a process requiring metabolic energy, the possibility existed that the receptor is an anion channel and that its functional state is regulated by phosphorylation and dephosphorylation reactions. If this is so, treatment of the cells with diphtheria toxin might alter their ability to transport anions. We therefore tested if diphtheria toxin bound at the cell's surface alters the ability of Vero cells to transport ^{35}SO$_4^{2-}$ and ^{36}Cl⁻. The results showed that, when the entire experiment was carried out at pH 6–7, very little inhibition took place. However, if cells with surface-bound toxin were briefly exposed to medium adjusted to pH 5.2 or lower, and the cells were then brought back to medium of pH 6–7, their ability to take up ^{35}SO$_4^{2-}$ and ^{36}Cl⁻ was strongly reduced (Fig. 6).

The reduction in uptake cannot be due to the fact that diphtheria toxin induces

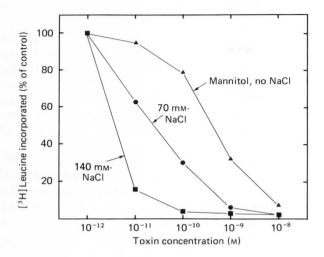

Fig. 7. *Effect of NaCl on the entry of diphtheria toxin induced by low pH*

Increasing amounts of diphtheria toxin were added to Vero cells growing in 24-well disposable trays immediately after addition of 10 μM-monensin. After 15 min incubation at 37 °C the medium was removed, the cells were washed rapidly with cold Cl⁻-free buffer (pH 7.2) containing 260 mM-mannitol and then exposed for 2 min to the buffers detailed below, adjusted to pH 4.5, and then medium (pH 7.2) with 10 μM-monensin was added. After 1 h further incubation protein synthesis during a 10 min interval was measured. The buffers contained 10 μM-monensin, 20 mM-2-[*N*-morpholino]-ethanesulphonic acid adjusted to pH 4.5 with tris(hydroxymethyl)aminomethane and were osmotically balanced with: ■, 140 mM-NaCl; ●, 70 mM-NaCl and 130 mM-mannitol; ▲, 260 mM-mannitol.

inhibition of protein synthesis, Thus, similar results were obtained in a diphtheria toxin-resistant cell line with altered elongation factor 2 and normal binding capability for the toxin previously described by Sundan *et al.* (1982). Furthermore, identical results were obtained when protein synthesis, both in control cells and toxin-treated cells, was completely blocked by treatment with cycloheximide. It therefore appears that the insertion of the hydrophobic region of diphtheria toxin into Vero cells occurs in close enough proximity to the main anion transporter to inhibit its function. Possibly, the anion transporter itself or part of it carries the binding site for diphtheria toxin in Vero cells.

In cells incubated in the absence of chloride, diphtheria toxin does not enter the cytosol (Sandvig & Olsnes, 1984). In spite of this low pH was found to induce insertion of the toxin into the membrane also under this condition, albeit somewhat more slowly than in cells kept in normal medium. Since anion transport is necessary for vesicle acidification, we investigated whether low pH is able to induce entry of surface-bound diphtheria toxin into chloride-deprived cells. As shown in Fig. 7 this was not the case. It therefore appears that chloride transport is required, not only for vesicle acidification, but also for the transfer of diphtheria toxin into the cytosol. Furthermore, the results showed that insertion of the toxin into the membrane is not sufficient to induce entry of the A-fragment into the cytosol.

Requirement for a pH gradient

In addition to chloride transport and low pH, we have identified a third factor required for entry of surface-bound diphtheria toxin into the cytosol. Thus, it is also necessary that the pH of the cytosol is close to neutrality, to ensure the existence of a sufficiently steep pH gradient across the membrane. In experiments where we acidified the cytosol by incubating the cells in the presence of acetate, the toxin was unable to enter even at low pH (Fig. 8). Similar results were obtained when the cytosol was acidified by other means (K. Sandvig & S. Olsnes, unpublished work).

We also tested the possibility that the membrane potential could provide a vectorial driving force for the entry of the toxin. For this purpose we depolarized the cells by incubating them in the presence of 0.2 M-KCl; the normal entry route for the toxin was then blocked with monensin and finally we tested the ability of low pH to induce toxin entry. The results showed that under these conditions the toxin entered as efficiently in the presence of KCl as in the presence of NaCl.

Hypothetical model for diphtheria toxin entry

On the basis of the available data, we propose the following model for diphtheria toxin entry (Fig. 9). The toxin binds to its receptor, which in Vero cells may be the main anion transporter, or it may be located in close proximity to it. The complex is then endocytosed by coated vesicles and reaches the endosomes. When the pH is reduced to 5.2, the hydrophobic region in the B-fragment becomes exposed and inserted into the membrane. The low pH may also induce conformational changes in the A-fragment resulting in unfolding of the polypeptide chain. The A-fragment could then in its extended form pass through the channel formed by the B-fragment or by the B-fragment in conjunction with one or more polypeptides of the anion transporter. Either influx of chloride, or protons, or both may serve as driving force for the entry of the A-fragment.

Entry of Modeccin

In cell-free systems the A-moieties of the toxins here discussed start to inhibit protein synthesis immediately, whereas when the toxins are given to whole cells, there is always a lag period before the rate of protein synthesis starts to decline (Moynihan & Pappenheimer, 1981; Refsnes *et al.*, 1974; Olsnes *et al.*, 1976; Sandvig *et al.*, 1984). At least part of this lag time must be due to the time required for the toxins to enter the cytosol. When diphtheria toxin and modeccin are incubated with Vero cells overnight, they are approximately equally toxic. In spite of this, the lag time is much longer with modeccin than with diphtheria toxin (Fig. 10). Thus, whereas 1 nM-diphtheria toxin reduced protein synthesis to half the control value already after 25 min, approximately 140 min was required after treatment with the same concentration of modeccin.

With increasing concentrations of toxins a limit was reached where the lag time could not be further reduced. This occurred after 15 min in the case of

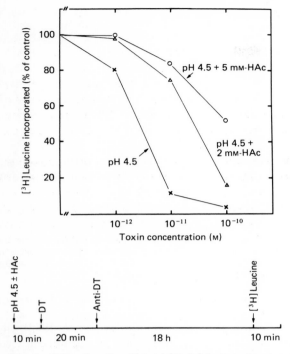

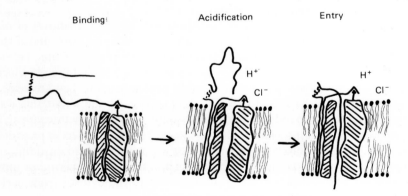

Fig. 8. *Ability of acetic acid to inhibit diphtheria toxin entry at low pH*

Vero cells were incubated for 10 min with or without the indicated concentrations of acetic acid (HAc) adjusted to pH 4.5, and then increasing concentrations of diphtheria toxin (DT) were added. After 20 min the low pH medium was removed, the cells were incubated overnight in growth medium containing anti-diphtheria toxin (anti-DT), and then the incorporation of [³H]leucine during 10 min was measured.

Fig. 9. *Hypothetical model of diphtheria toxin entry*

The toxin binds to the receptor, which is possibly the main anion transporter in the cells. When pH is reduced below 5.4 the hydrophobic domain of the B-fragment inserts itself into the membrane in such a way that it inhibits the normal anion exchange. The inserted part of the B-fragment could, together with the anion channel, form a pore which is sufficiently large to allow the A-fragment to enter. A proton gradient across the membrane is required for entry and so is chloride transport.

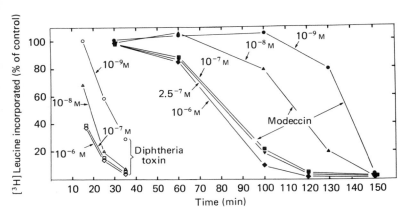

Fig. 10. *Rate of protein synthesis inhibition by diphtheria toxin and modeccin*

The indicated amounts of toxin were added to Vero cells growing in 24-well disposable trays. After different periods of incubation at 37 °C, the medium was removed and the cells were incubated for 10 min with [³H]leucine to measure the rate of protein synthesis. The toxin concentrations were: $\bigcirc$, $\bullet$, 10^{-9} M; $\triangle$, $\blacktriangle$, 10^{-8} M; $\square$, $\blacksquare$, 10^{-7} M; $\triangledown$, $\blacktriangledown$, 2.5×10^{-7} M; $\diamondsuit$, $\blacklozenge$, 10^{-6} M. Open symbols: nicked diphtheria toxin. Filled symbols: modeccin (Sandvig *et al.*, 1984).

diphtheria toxin and 75 min with modeccin (Sandvig *et al.*, 1984). One possible reason for this difference could be that modeccin is bound and endocytosed more slowly than diphtheria toxin. To test this, cells were incubated with the toxins for increasing periods of time and then washed and treated with antitoxins to neutralize any toxin remaining at the cell surface. The cells were subsequently incubated overnight to allow endocytosed toxin time to exert its maximal toxic effect. As shown in Fig. 11, modeccin was no less active than diphtheria toxin in this assay, indicating that the reason for the long lag-time with modeccin is not slow binding and endocytosis of the toxin.

Like diphtheria toxin, modeccin requires low pH for entry (Sandvig *et al.*, 1979). If the step requiring low pH occurs late in the entry process, and if the toxin remains in intracellular vesicles for a considerable period of time before it enters the cytosol, addition of NH_4Cl, monensin and other compounds which increase the pH of intracellular vesicles, might protect even if they are added comparatively late in the lag-period. The data in Fig. 11 show that this was indeed so. Thus monensin, NH_4Cl, carbonyl cyanide-*p*-trifluoromethoxyphenyl-hydrazone) (FCCP) and procaine added as late as 2 h after modeccin provided good protection against intoxication. In contrast, with diphtheria toxin, addition of the same components did not provide protection for a longer period of time than the antiserum, indicating that this toxin enters the cytosol from early endosomes as soon as the pH is sufficiently reduced.

Further evidence that the two toxins enter the cytosol from different compartments was obtained in experiments carried out at 20 °C. At this temperature, diphtheria toxin was able to enter the cytosol, whereas modeccin was not (Sandvig *et al.*, 1984; Draper *et al.*, 1984). It has been found that at 20 °C the fusion of endosomes with lysosomes is inhibited (Dunn *et al.*, 1980). Possibly,

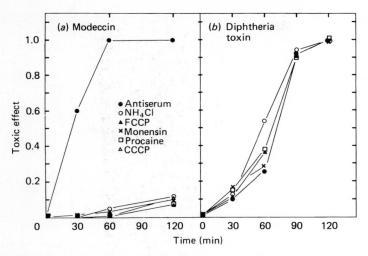

Fig. 11. *Protective effect of various compounds added to cells exposed to (a) modeccin and (b) diphtheria toxin for different periods of time*

Toxin was added at time 0 and the compounds to be tested were added to the cells at the times indicated on the abscissa. The extent of intoxication was measured after 18 h. The additions were: ●, 10 μl of antiserum/ml; ○, 10 mM-NH$_4$Cl; ▲, 10 μM-FCCP; ×, 10 μM-monensin; □, 10 mM-procaine; △, 15 μM-FCCP (Sandvig *et al.*, 1984).

other intracellular fusion events are also inhibited at this temperature. This could imply that at the low temperature, modeccin is unable to reach the particular compartment that is suitable for penetration of this toxin. This compartment is likely to be acidic and could be the lysosomes. However, since acidic vesicles have recently been found also in the Golgi apparatus (Glickman *et al.*, 1983), it is perhaps more likely that this is the site of entry. Possibly a modification of the protein is necessary for penetration. It is in accordance with this possibility that we have been unable to induce entry of modeccin from the cell surface by exposure of cells to low pH, in contrast to the situation with diphtheria toxin.

Entry of *Pseudomonas* Exotoxin A

Exotoxin A from *Pseudomonas aeruginosa* (PEA) also requires low pH for entry. At least this is the case for the efficient uptake mechanism in highly sensitive cells like L-cells and mouse 3T3 cells (FitzGerald *et al.*, 1980; Sundan *et al.*, 1984a). A number of other cells, e.g. BHK cells, are much less sensitive to this toxin and they are only slightly protected by compounds that increase the pH of intracellular vesicles. Trifluoperazine and a number of other compounds that inhibit calmodulin-dependent processes, and protein kinase c increased the sensitivity of BHK cells to PEA approximately 100-fold (Fig. 12). Similar results were obtained with a number of other comparatively insensitive cells, whereas the highly sensitive L-cells and 3T3 cells were not further sensitized by trifluoperazine.

It is also interesting that the increased sensitivity induced by trifluoperazine

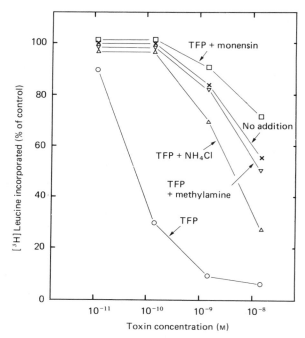

Fig. 12. *Effect of NH$_4$Cl, methylamine and monensin on the toxic effect of PEA in BHK cells treated with trifluoperazine*

The cells were preincubated for 15 min in Hepes-buffered medium with the compounds indicated and then increasing amounts of toxin were added. After 3 h incubation the rate of protein synthesis was measured. ×, No addition; 0, 10 M-trifluoperazine (TFP); △, 10 μM-trifluoperazine and 10 mM-NH$_4$Cl; □, 10 μM-trifluoperazine and 0.1 μM-monensin; ▽, 10 μM-trifluoperazine and 10 mM-methylamine (Sundan *et al.*, 1984*a*).

and related compounds was eliminated in the presence of NH$_4$Cl, monensin and other compounds that increase the pH of intracellular vesicles (Sundan *et al.*, 1984*a*). Apparently, therefore, in cells that are comparatively insensitive to the toxin, trifluoperazine is able to induce an uptake mechanism which resembles that normally occurring in highly sensitive cells. We could induce sensitization even when trifluoperazine was added after the toxin had been endocytosed. Possibly, trifluoperazine and compounds with related action alter the routing of the internalized toxin by changing an inefficient route into a highly efficient one.

Malignant transformation was found to alter the response of cells to PEA (Sundan *et al.*, 1984*b*). Thus a number of mouse and rat cells were found to be less sensitive to the toxin in the transformed than in the non-transformed state. The transformed cells could, however, be sensitized by treatment with trifluoperazine and related drugs. Possibly, therefore, transformation alters the routing of physiological ligands which bind to the structure that also functions as receptor for PEA.

Entry of *Shigella* Toxin

Shigella toxin has a more complicated structure than the other toxins here discussed (Fig. 1), but also this toxin consists of two functionally different moieties, namely an enzymatically active A-fragment which is linked by a disulphide bridge to a receptor-binding B-moiety (Olsnes & Eiklid, 1980; Olsnes *et al.*, 1981; O'Brien & La Veck, 1983; Donohue-Rolfe *et al.*, 1984). Only few cell lines were found to be sensitive to *Shigella* toxin (Eiklid & Olsnes, 1980). It is not clear if low pH is required for entry of this toxin. Thus FCCP (carbonyl cyanide-*p*-trifluoro-methoxyphenylhydrazone), carbonyl cyanide-*m*-chlorophenylhydrazone (CCCP), *N*,*N'*-dicyclohexylcarbodi-imide (DCCD), monensin and chloroquine protected against the toxin (Eiklid & Olsnes, 1984). It should be noted, however, that comparatively high concentrations of monensin were required. In our hands NH_4Cl did not inhibit, and Keusch & Jacewicz (1984) found protection only at high concentrations of NH_4Cl. Possibly only a slight acidification is required for entry of this toxin.

Shigella toxin may require an intracellular fusion for activity. Thus the toxin was approximately equally as active at 26 °C as at 37 °C, whereas at 20 °C the activity was strongly reduced (Eiklid & Olsnes, 1983). Low pH did not induce entry of *Shigella* toxin bound at the cell surface, and, in fact, when the pH was kept at pH 6.4 and lower, the toxin did not enter the cytosol.

Entry of Ricin, Abrin and Viscumin

In contrast to the toxins described above, there is no evidence that these three plant toxins require low pH for entry. In fact compounds like NH_4Cl, chloroquine and monensin, which increase the pH of intracellular vesicles, sensitize cells to these toxins (Sandvig *et al.*, 1979; Sandvig & Olsnes, 1982*b*; Mekada *et al.*, 1981; Ray & Wu, 1981). Furthermore, when the pH of the medium is slightly alkaline, these toxins enter most efficiently, whereas they do not intoxicate cells at pH 6.5 and below (Sandvig & Olsnes, 1982*b*; Stirpe *et al.*, 1982).

Toxins of this group do not enter the cytosol if the cells are depleted of Ca^{2+} or when the Ca^{2+} transport is inhibited (Sandvig & Olsnes, 1982*a*). This requirement for Ca^{2+} was shown to be linked to a function of the A-chain, possibly the transfer across the membrane as such.

The possibility existed that the toxins of this group enter directly from the cell surface rather than from intracellular vesicles. To test this, we took advantage of the observations that at low pH, as well as in the absence of Ca^{2+}, the toxins are endocytosed, but they do not enter the cytosol (Fig. 13). If the cells were then treated with antitoxin to inactivate any toxin remaining at the cell surface and subsequently incubated overnight in normal growth medium to allow the endocytosed toxin time to enter the cytosol and inhibit protein synthesis, we found that there was no difference whether or not Ca^{2+} was present and whether the pH was 6 or 7 during the time period when the cells were exposed to the toxins (Sandvig & Olsnes, 1982*b*). It is therefore clear that these

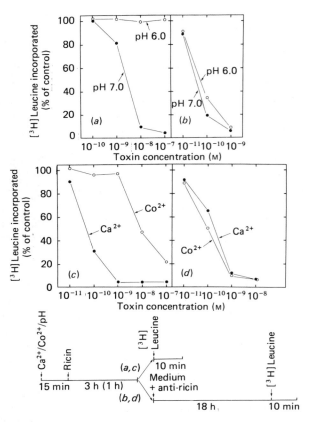

Fig. 13. *Ability of ricin endocytosed in the absence of Ca^{2+} or at pH 6 to intoxicate cells after transfer to normal medium*

(*a*) and (*b*), Vero cells were incubated for 15 min in 0.14 M-NaCl, 20 mM-Hepes, pH 6 or pH 7 as indicated, 5 mM-glucose, 2 mM-CaCl$_2$, and then increasing amounts of ricin were added. (*a*) The buffers were removed after 1 h, medium containing [^{3}H]leucine was added and the rate of protein synthesis during 10 min was measured. (*b*) The medium was removed after 1 h, growth medium containing neutralizing amounts of antiricin was added, the cells were incubated overnight and then the rate of protein synthesis was measured. (*c*) and (*d*), The cells were incubated for 15 min in 0.14 M-NaCl, 20 mM-Hepes, pH 7.2, 5 mM-glucose and either 2 mM-CaCl$_2$ or 2 mM-CoCl$_2$, and then increasing amounts of ricin were added. (*c*) The buffers were removed after 3 h and the rate of protein synthesis was measured. (*d*) The buffers were removed after incubation with ricin for 1 h and growth medium with 1.8 mM-CaCl$_2$ and neutralizing amounts of antiricin were added. The cells were incubated overnight and then the rate of protein synthesis was measured (Sandvig & Olsnes, 1982*b*).

toxins also are able to enter from intracellular vesicles and that entry from such vesicles represents a major entry route.

Possibly ricin, abrin and viscumin enter the cytosol from neutral rather than from acidic vesicles. If this is so, the sensitizing effect of compounds that increase the pH of acidic vesicles could be to increase the amounts of toxin present in non-acidified vesicles. Furthermore, the fact that the cells are more sensitive to these toxins when the medium has a slightly alkaline pH (Sandvig & Olsnes, 1982*b*), could be due to engulfment of the buffered alkaline medium which is then likely to increase the pH in the newly formed vesicles.

Role of pH in the Entry of Picornaviruses

Since low pH is required for the entry of several toxins and a large number of enveloped viruses, we decided to test if low pH is also required for the entry of picornaviruses which do not have a lipid envelope. For this purpose we took advantage of the fact that when such viruses are grown in the presence of neutral red or acridine orange, some dye is trapped within the virus capsid in close proximity to the RNA genome (Crowther & Melnick, 1961; Mandel, 1967). As a consequence the virus becomes sensitive to visible light (Fig. 14). However, when the light-sensitive virus penetrates into cells, the dye that is trapped within the virus capsid diffuses away from the RNA, which then becomes light resistant. The transition from light sensitivity to light resistance may therefore be taken as a measure of virus penetration (Mandel, 1967). This allowed us to limit to a short period of time the exposure of the cells to different compounds inhibiting cellular functions. Since many of these compounds are toxic and can only be tolerated by the cells for a short period of time, the use of light-sensitive virus offers a great advantage in these experiments.

The data in Fig. 15 show that when cells were exposed to poliovirus in the presence of monensin, the infection was strongly inhibited. This was also the case in the presence of CCCP, tributyltin and chloroquine. These compounds did not inhibit the production of virus in the cells. Some compounds, like NH_4Cl, increased the production of virus slightly (Madshus et al., 1984a).

To test if poliovirus bound at the cell surface is able to enter when the cells are exposed to low pH, cells with prebound virus were treated with pH 5.5 in the presence of monensin to stop the normal entry mechanism. As shown in Fig. 16 the low pH largely eliminated the protection afforded by monensin. The entry of poliovirus was half-maximal at pH 6.1 (Madshus et al., 1984b). This value is considerably higher than that required for half-maximal entry of diphtheria toxin (pH 5.4). In the absence of monensin the low pH treatment did not strongly increase the sensitivity of the cells. It therefore appears that, as with diphtheria toxin, low pH is able to induce entry of poliovirus from the cell surface.

With diphtheria toxin, low pH induces exposure of a hydrophobic region in the B-fragment (Sandvig & Olsnes, 1981a; Blewitt et al., 1984). To see if a similar situation applies to poliovirus, we measured the ability of the virus to enter the detergent Triton X-114. A number of hydrophobic proteins have been found to enter this detergent, whereas hydrophilic proteins do not (Bordier, 1981). We found that at neutral pH, [35S]methionine-labelled poliovirus remained in the water phase, but when the pH was reduced below pH 5.5, increasing amounts of virus entered the detergent phase. Binding of the virus to the cell surface receptors reduced the extent of acidity required for entry into the detergent (Madshus et al., 1984b). When the cells were preincubated with acetic acid, the extent of virus entry was strongly reduced (Madshus et al., 1984b). This indicates that, similarly to diphtheria toxin, also poliovirus requires the presence of a pH gradient across the membrane for entry to occur.

We also studied the pH requirement for entry into the cytosol of two other picornaviruses, human rhinovirus 2 and encephalomyocarditis (EMC) virus (Madshus et al., 1984c). The entry of human rhinovirus 2 was dependent on

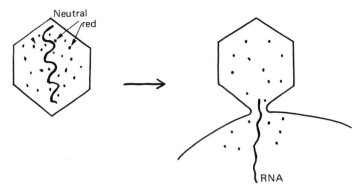

Fig. 14. *Scheme of infection by light-sensitive virus*

Neutral red trapped within the virus capsid renders the virus light sensitive. During infection the dye diffuses away from the RNA, which thereby becomes light resistant.

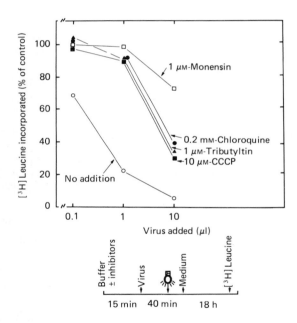

Fig. 15. *Effect of different compounds on poliovirus entry*

The indicated compounds were added to HeLa S_3 cells in 24-well disposable trays and, after 15 min, increasing amounts of virus containing neutral red were added. After incubation in the dark for 40 min at 37 °C, the cells were exposed to light and then transferred to normal medium and incubated overnight. The rate of [³H]leucine incorporation was measured the next day. The additions were: ●, 0.2 mM-chloroquine; ■, 10 μM-CCCP; ▲, 1 μM-tributyltin; □, 1 μM-monensin; ○, control, no addition (Madshus *et al.*, 1984*a*).

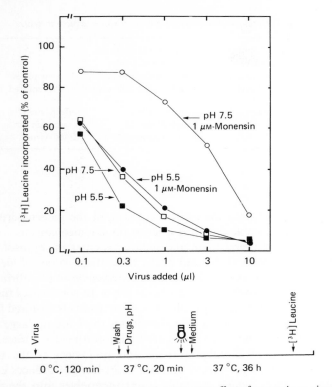

Fig. 16. *Ability of low pH to counteract the protective effect of monensin on virus entry*

Poliovirus was allowed to bind to cells at 0 °C for 2 h. The cells were then washed and Hepes medium containing 1 μM-monensin adjusted to pH 5.5 or 7.5, and prewarmed to 37 °C, was added. After 20 min the cells were exposed to light, transferred to normal medium and incubated for 36 h. Finally, the ability of the cells to incorporate [³H]leucine was measured (Madshus *et al.*, 1984*b*).

low pH, but, in contrast to the findings with poliovirus, entry of rhinovirus 2 could not be induced by treatment of the cells with acidic medium. In addition to this inhibitory effect on the entry of human rhinovirus 2 as such, compounds which increase the pH of intracellular vesicles also reduced the ability of the cells to bind the virus. This is probably due to inhibition of the recycling of endocytosed receptors back to the cell surface.

EMC virus does not appear to require low pH for entry. Thus, treatments which increase the pH of intracellular vesicles, did not protect against this virus. Furthermore, as with abrin, ricin and modeccin, EMC virus did not enter when the medium was acidic (Madshus *et al.*, 1984*c*).

It therefore appears that different picornaviruses enter the cytosol by different mechanisms, similarly to the protein toxins. Possibly, different toxins and viruses have adapted themselves to take advantage of different uptake mechanisms in the cells.

Table 1. *Uptake mechanisms for protein toxins and picornaviruses*

Low pH is required for normal entry and also induces entry from the cell surface	Low pH is required for normal entry, but does not induce entry from the cell surface	Low pH is not required
Diphtheria toxin	Modeccin, PEA, *Shigella* toxin (?)	Abrin, ricin, viscumin
Poliovirus	Human rhinovirus 2	EMC virus

Conclusions

From the presented data the protein toxins and the picornaviruses can be grouped as shown in Table 1 with respect to uptake mechanisms.

It is likely that diphtheria toxin and poliovirus enter the cytosol as soon as the pH of the endosomes is reduced below the critical value. Modeccin enters the cytosol particularly slowly, but also PEA, *Shigella* toxin, abrin, ricin and viscumin enter more slowly than diphtheria toxin. In some cases fusion of the endocytic vesicles with other intracellular compartments is required for transfer of the A-moiety into the cytosol. For the viruses there are no large differences in rate of penetration once the viruses are bound to the cell surface.

Only in the case of diphtheria toxin and poliovirus is the penetration mechanism known to some extent. In both cases low pH induces exposure of hydrophobic domains, which probably insert themselves into the membrane. For diphtheria toxin this insertion appears to inhibit the transport of anions into the cells, indicating that the insertion occurs in close proximity to the anion transporter.

For entry of the A-fragment into the cytosol low pH is not sufficient. Also chloride transport and a proton gradient across the membrane are required, possibly as energy sources for the entry. A proton gradient appears to be required also for the entry of poliovirus. It is therefore clear that the entry depends upon a complicated co-operation between the intruding agent and the attacked cell.

References

Blewitt, M. G., Zhao, J.-M., MsKeever, B., Sarma, R. & London, E. (1984) *Biochem. Biophys. Res. Commun.* **120**, 286–290

Boquet, P. & Pappenheimer, A. M., Jr. (1976) *J. Biol. Chem.* **251**, 5770–5778

Bordier, C. (1981) *J. Biol. Chem.* **256**, 1604–1607

Crowther, D. & Melnick, J. L. (1961) *Virology* **14**, 11–21

Deleers, M., Beugnier, N., Falmagne, P., Cabiaux, V. & Ruysschaert, J.-M (1983) *FEBS Lett.* **160**, 82–86

Donohue-Rolfe, A., Keusch, G. T., Edson, C., Thorley-Lawson, D. & Jacewicz, M. (1984) *J. Exp. Med.* **160**, 1767–1787

Donovan, J. J., Simon, M. I., Draper, R. K. & Montal, M. (1981) *Proc. Natl. Acad. Sci. U.S.A.* **78**, 172–176

Donovan, J., Simon, M. & Montal, M. (1982) *Biophys. J.* **37**, 256a

Draper, R. K. & Simon, M. I. (1980) *J. Cell Biol.* **87**, 849–854

Draper, R. K., O'Keefe, D. O., Stookey, M. & Graves, J. (1984) *J. Biol. Chem.* **259**, 4083–4088

Dunn, W. A., Hubbard, A. L. & Aronsen, N. N. (1980) *J. Biol. Chem.* **255**, 5971–5978

Eidels, L., Ross, L. L. & Hart, D. A. (1982) *Biochem. Biophys. Res. Commun.* **109**, 493–499
Eidels, L., Proia, R. L. & Hart, D. A. (1983) *Microbiol. Rev.* **47**, 596–620
Eiklid, K. & Olsnes, S. (1980) *J. Recept. Res.* **1**, 199–213
Eiklid, K. & Olsnes, S. (1983) *Infect. Immun.* **42**, 771–777
Eiklid, K., Olsnes, S. & Pihl, A. (1980) *Exp. Cell Res.* **126**, 321–326
FitzGerald, D., Morris, R. E. & Saelinger, C. B. (1980) *Cell* **21**, 867–873
Glickman, J., Croen, K., Kelly, S. & Al-awqati, Q. (1983) *J. Cell Biol.* **97**, 1303–1308
Guillemot, J. C., Sundan, A., Olsnes, S. & Sandvig, K. (1985) *J. Cell. Physiol.* **122**, 193–199
Kagan, B. L., Finkelstein, A. & Colombini, M. (1981) *Proc. Natl. Acad. Sci. U.S.A.* **78**, 4950–4954
Kayser, G., Lambotte, P., Falmagne, P., Capiau, C., Zanen, J. & Ruysschaert, J.-M. (1981) *Biochem. Biophys. Res. Commun.* **99**, 358–363
Keusch, G. T. & Jacewicz, M. (1984) *Biochem. Biophys. Res. Commun.* **121**, 69–76
Kim, K. & Groman, N. B. (1965) *J. Bacteriol.* **90**, 1557–1562
Lenard, J. & Miller, D. K. (1983) in *Receptors and Recognition Series B: Receptor-Mediated Endocytosis* (Cuatrecasas, P. & Roth, T., eds.), vol. 15, pp. 119–138, Chapman and Hall, London
Lory, S. & Collier, R. J. (1980) *Proc. Natl. Acad. Sci. U.S.A.* **77**, 267–271
Madshus, I. H., Olsnes, S. & Sandvig, K. (1984a) *J. Cell Biol.* **98**, 1194–1200
Madshus, I. H., Olsnes, S. & Sandvig, K. (1984b) *EMBO J.* **3**, 1945–1950
Madshus, I. H., Olsnes, S. & Sandvig, K. (1984c) *Virology* **139**, 346–357
Mandel, B. (1967) *Virology* **31**, 702–712
Marnell, M. H., Shia, S.-P., Stookey, M. & Draper, R. D. (1984) *Infect. Immun.* **44**, 145–150
Mekada, E., Uchida, T. & Okada, Y. (1981) *J. Biol. Chem.* **256**, 1225–1228
Middlebrook, J. L., Dorland, R. B. & Leppla, S. H. (1978) *J. Biol. Chem.* **253**, 7325–7330
Misler, S. (1983) *Proc. Natl. Acad. Sci. U.S.A.* **80**, 4320–4324
Moehring, T. J. & Crispell, J. P. (1974) *Biochem. Biophys. Res. Commun.* **60**, 1446–1452
Moynihan, M. R. & Pappenheimer, A. M., Jr. (1981) *Infect. Immun.* **32**, 575–582
Mumby, M. & Traugh, J. A. (1979) *Biochemistry* **18**, 4548–4556
Nishizuka, Y. (1984) *Nature (London)* **308**, 693–698
O'Brien, A. D. & La Veck, G. D. (1983) *Infect. Immun.* **40**, 675
Olsnes, S. & Eiklid, K. (1980). *J. Biol. Chem.* **255**, 284–289
Olsnes, S. & Pihl, A. (1976) in *Receptors and Recognition, Series B. The Specificity and Action of Animal, Bacterial and Plant Toxins* (Cuatrecasas, P., ed.), pp. 129–173, Chapman and Hall, London
Olsnes, S. & Pihl, A. (1982) in *The Molecular Action of Toxins and Viruses* (van Heyningen, S. & Cohen, P., eds.), pp. 51–105, Elsevier/North Holland, Amsterdam
Olsnes, S. & Sandvig, K. (1983) in *Receptor-Mediated Endocytosis, Receptors and Recognition* (Cuatrecasas, P. & Roth, T. F., eds.), series B, vol. 15, pp. 187–236, Chapman and Hall, London
Olsnes, S., Reisbig, R. & Eiklid, K. (1981) *J. Biol. Chem.* **256**, 8732–8738
Olsnes, S., Sandvig, K., Refsnes, K. & Pihl, A. (1976) *J. Biol. Chem.* **251**, 3985–3992
Pappenheimer, A. M., Jr. (1977) *Annu. Rev. Biochem.* **46**, 69–94
Proia, R. L., Hart, D. A. & Eidels, L. (1979) *Infect. Immun.* **26**, 942–948
Proia, R. L., Wray, S. K., Hart, D. A. & Eidels, L. (1980) *J. Biol. Chem.* **255**, 12025–12033
Proia, R. L., Eidels, L. & Hart, D. A. (1981) *J. Biol. Chem.* **256**, 4991–4997
Ray, B. & Wu, H. C. (1981) *Mol. Cell. Biol.* **1**, 544–551
Refsnes, K., Olsnes, S. & Pihl, A. (1974) *J. Biol. Chem.* **249**, 3557–3562
Sandvig, K. & Olsnes, S. (1980) *J. Cell Biol.* **87**, 828–832
Sandvig, K. & Olsnes, S. (1981a) *J. Biol. Chem.* **256**, 9068–9076
Sandvig, K. & Olsnes, S. (1981b) *Biochem. J.* **194**, 821–827
Sandvig, K. & Olsnes, S. (1982a) *J. Biol. Chem.* **257**, 7495–7503
Sandvig, K. & Olsnes, S. (1982b) *J. Biol. Chem.* **257**, 7504–7513
Sandvig, K. & Olsnes, S. (1984) *J. Cell. Physiol.* **119**, 7–14
Sandvig, K., Olsnes, S. & Pihl, A. (1979) *Biochem. Biophys. Res. Commun.* **90**, 648–655
Sandvig, K., Sundan, A. & Olsnes, S. (1984) *J. Cell Biol.* **98**, 963–970
Stirpe, F., Sandvig, K., Olsnes, S. & Pihl, A. (1982) *J. Biol. Chem.* **257**, 13271–13277
Sundan, A., Olsnes, S., Sandvig, K. & Pihl, A. (1982) *J. Biol. Chem.* **257**, 9733–9739
Sundan, A., Sandvig, K. & Olsnes, S. (1984a) *J. Cell. Physiol.* **119**, 15–22
Sundan, A., Sandvig, K. & Olsnes, S. (1984b) *Cancer Res.* **44**, 4919–4923
Swarup, G., Cohen, S. & Garbers, D. L. (1982) *Biochem. Biophys. Res. Commun.* **107**, 1104–1109
Uchida, T. (1983) *Pharmacol. Therap.* **19**, 107–122
Yamaizumi, M., Mekada, E., Uchida, T. & Okada, Y. (1978) *Cell* **15**, 245–250
Zalman, L. S. & Wisnieski, B. J. (1984) *Proc. Natl. Acad. Sci. U.S.A.* **81**, 3341–3345

Biochem. Soc. Symp. **50**, 193–204
Printed in Great Britain

Protein Accumulation in the Cell Nucleus

C. DINGWALL

CRC Molecular Embryology Research Unit, Department of Zoology, University of Cambridge, Downing Street, Cambridge CB2 3EJ, U.K.

Synopsis

Proteins which have been extracted from nuclei can re-enter and accumulate in the nucleus when deposited in the cytoplasm of the cell. This phenomenon has been investigated in two nuclear proteins having widely different properties. The same experimental strategy has been used in both cases, that is, microinjection of proteolytic fragments of these proteins into *Xenopus* oocytes and observing which of these fragments can accumulate in the nucleus. In the case of nucleoplasmin, a large, pentameric acidic protein, which is the most abundant protein of the *X. laevis* oocyte nucleus, a small fragment has been isolated which is both necessary and sufficient for accumulation in the oocyte nucleus. In the case of calf thymus histone H1, a small basic protein, a *C*-terminal fragment of 87 amino acids can accumulate in the oocyte nucleus. The amino acids lysine, proline and alanine constitute 73 of the 87 amino acids. Since the other 14 amino acids are scattered, not clustered, these three amino acids must presumably predominate in any sequence which specifies accumulation of the fragment in the nucleus. By using the expression vector λgt 11, cDNA clones of nucleoplasmin have recently been obtained and their properties are described.

Introduction

When compared with our knowledge of the mechanisms of protein secretion and the uptake of proteins by mitochondria and chloroplasts, our current understanding of how proteins accumulate in the cell nucleus is at a very rudimentary level. In the case of nuclear proteins some property of the mature protein causes them to become concentrated in the nucleus soon after they are synthesized or soon after they are introduced into the cytoplasm artificially, e.g. by microinjection. Studies of the accumulation of nucleoplasmin, a protein present in the nucleus of oocytes of *Xenopus laevis*, have demonstrated that a proteolytic fragment of the protein encompassing a discrete polypeptide domain contains all the molecular features responsible for entry into the nucleus. Nucleoplasmin molecules lacking this domain can no longer enter the nucleus.

In the case of histone H1, a proteolytic fragment with an extremely simple amino acid composition has been isolated and this fragment has the ability to accumulate in the nucleus. This suggests that the amino acids which predominate in this fragment, namely lysine, proline and alanine, are represented in the primary sequence that specifies nuclear entry.

More direct evidence that these amino acids are involved comes from studies of the transport of nuclear proteins in yeast and the transport of SV40 large T antigen into the nucleus of infected cells.

Purified Nuclear Proteins Can Accumulate in the Nucleus

The mechanism of accumulation of nuclear proteins within the cell nucleus can be investigated by exploiting the unique properties of the oocyte of the African clawed toad *Xenopus laevis*. The oocyte, arrested in the first meiotic prophase, is a giant cell of about 1 mm in diameter with a large nucleus (~ 0.2 mm in diameter) that can be manually dissected from the oocyte. This makes it a simple matter to microinject radiolabelled nuclear proteins into the cytoplasm of an oocyte and then, at some later time, separate the nucleus and cytoplasm in order to determine the intracellular location of the proteins.

Using this system, Gurdon (1970) demonstrated that purified histones have the ability to accumulate in the nucleus after microinjection into the cytoplasm. Later Bonner (1975b) radiolabelled the proteins of the oocyte by incubation in [^{35}S]methionine. He then isolated the nuclei from these oocytes and microinjected the labelled nuclear proteins into the cytoplasm of another (unlabelled) oocyte. After a suitable incubation period, he examined the distribution of the labelled nuclear proteins in the nucleus and cytoplasm of the recipient oocytes using SDS–polyacrylamide gel electrophoresis. He observed that the nuclear proteins were sequestered exclusively in the nucleus of the recipient oocytes. Therefore nuclear proteins must have within their mature structure a signal for accumulation in the nucleus. If so, this contrasts with the co-translational transport of secretory proteins and the transport of proteins into mitochondria or chloroplasts. These processes invariably involve a transient precursor form of the protein, which is processed at some stage to yield the mature protein, which then lacks the ability to be transported to its correct extracellular or subcellular location (reviewed by Kreil, 1981).

By examining the accumulation of one particular nuclear protein from the *Xenopus* oocyte nucleus we have produced evidence that the signal for nuclear entry is located within a discrete polypeptide domain and that this domain possesses all the features of the parent molecule that enable it to accumulate rapidly and selectively in the nucleus.

Nucleoplasmin, a *Xenopus* Oocyte Nuclear Protein

Nucleoplasmin is the most abundant protein in the *X. laevis* oocyte nucleus, constituting approximately 10% of the nuclear protein (Krohne & Franke, 1980; Mills *et al.*, 1980). Each oocyte nucleus contains 190–250 ng of nucleoplasmin. The protein was discovered in cell extracts by its ability to mediate nucleosome assembly *in vitro* at physiological ionic strength (Laskey *et al.*, 1978).

Nucleoplasmin behaves as a pentamer in solution (Earnshaw *et al.*, 1980), each

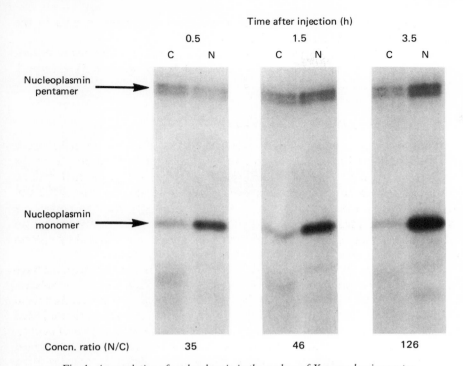

Fig. 1. *Accumulation of nucleoplasmin in the nucleus of Xenopus laevis oocytes*

Radiolabelled nucleoplasmin was injected into the cytoplasm of *X. laevis* oocytes and the oocytes were maintained at 20 °C in modified Barth's saline. Nuclei (N) and cytoplasms (C) were manually separated at the indicated times. N/C, Nuclear to cytoplasmic concentration ratio.

subunit having an apparent molecular weight of 33000. The protein is acidic with an isoelectric point of about 5 and is phosphorylated.

Migration of Purified Nucleoplasmin into the Nucleus

The experimental strategy used to examine the migration of purified nucleoplasmin into the nucleus is similar to that used by Bonner, namely microinjection of radiolabelled nucleoplasmin into the cytoplasm of *Xenopus* oocytes and analysis by SDS–polyacrylamide gel electrophoresis of the separated nuclei and cytoplasms. The ability of purified nucleoplasmin to accumulate in the nucleus was first demonstrated by Mills *et al.* (1980). In the experiments described here the rate of accumulation in the nucleus was examined by looking at the subcellular distribution of the injected nucleoplasmin soon after microinjection.

The result is shown in Fig. 1. The injected nucleoplasmin can be seen to migrate rapidly from the cytoplasm into the nucleus, where it becomes highly concentrated. The nuclear/cytoplasmic concentration ratios were calculated assuming that the cytoplasmic volume (including yolk) is approximately 25 times the nuclear volume (Bonner, 1978). Therefore when the nucleus and cytoplasm

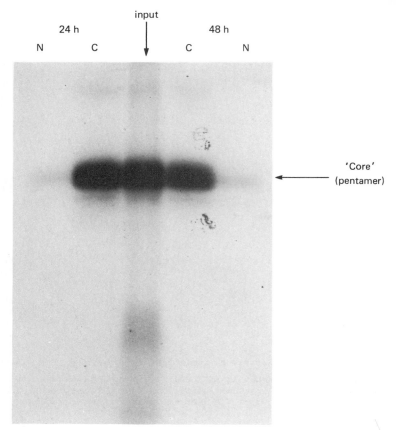

Fig. 2. *Failure of the trypsin resistant nucleoplasmin core to enter the nucleus*

Radiolabelled trypsin resistant nucleoplasmin core was injected into the cytoplasm of *X. laevis* oocytes and the oocytes were maintained at 20 °C in modified Barth's saline. Nuclei (N) and cytoplasms (C) were manually separated at the indicated times. Input: injected material.

contain the same amount of nucleoplasmin the concentration is 25 times higher in the nucleus than in the cytoplasm. It should be noted that the nucleoplasmin pentamer is unusually resistant to denaturation by SDS and runs partly as pentamer and partly as monomer on SDS–polyacrylamide gels.

In this experiment, approximately 1–2 ng of radiolabelled protein is micro-injected into the oocyte cytoplasm. Therefore, since the oocyte nucleus contains 190–250 ng of endogenous nucleoplasmin this large protein (M_r 165000) can accumulate rapidly in the nucleus against a substantial concentration gradient.

Partial proteolysis of nucleoplasmin abolishes transport into the nucleus

Treatment of nucleoplasmin with trypsin (and a variety of other serine proteases) produces a relatively protease resistant 'core' molecule, which remains pentameric. The core monomer has an apparent molecular weight of

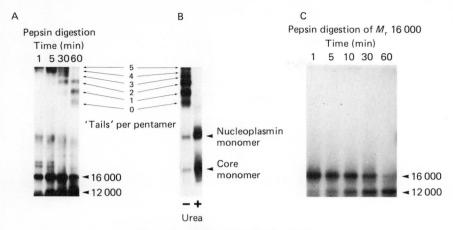

Fig. 3. *Pepsin digestion of nucleoplasmin*

(A) Time course of digestion of radiolabelled nucleoplasmin. (B) Dissociation of pepsin cleaved nucleoplasmin pentamers into nucleoplasmin monomers and core monomers. (C) Production of the M_r 12 000 tail fragment from the M_r 16 000 tail fragment by further pepsin cleavage.

23 000 and the 'tail' region of the pentamer which is removed by the protease is degraded to very small peptides.

When assayed for its ability to accumulate in the oocyte nucleus by micro-injection of a radiolabelled core preparation into the cytoplasm it was found that it could not accumulate in the nucleus but remained in the cytoplasm (Fig. 2).

The observed nuclear to cytoplasmic concentration ratio for the core was stable at 0.5 after 24 and 48 h post-injection compared with a value of 475 for intact nucleoplasmin 24 h post-injection. The implication of this observation is that the protease sensitive tail regions of the protein are responsible for the accumulation of nucleoplasmin in the nucleus. In order to establish that this is the case, it is necessary to prepare the tail region as an intact fragment and assay its ability to accumulate in the nucleus. This was achieved by pepsin digestion of nucleoplasmin (Dingwall *et al.*, 1982).

Pepsin digestion of nucleoplasm

The digestion of nucleoplasmin with pepsin is shown in Fig. 3A. Pepsin cleaves the subunits of the pentamer at a unique site to liberate a M_r 16 000 tail fragment. The pentamer therefore successively loses a M_r 16 000 tail fragment from each subunit during digestion. The pentamers that lack various numbers of tails are resolved on an SDS–polyacrylamide gel as a ladder of bands.

The cleaved pentamers can be dissociated into intact and cleaved monomeric subunits by treatment with urea (Fig. 3B).

The M_r 16 000 tail fragment is further cleaved by pepsin to produce a M_r 12 000 tail fragment (Fig. 3A, 3C).

That the M_r 12 000 tail fragment encompasses the tail region removed by trypsin digestion is indicated by digestion of the trypsin resistant core with pepsin. This treatment results in the production of the characteristic pepsin core

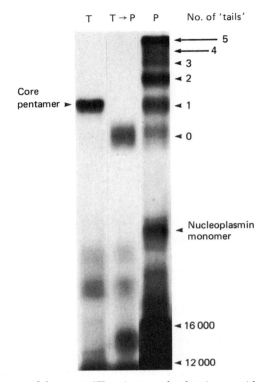

Fig. 4. *Cleavage of the trypsin (T) resistant nucleoplasmin core with pepsin (P)*

Lane T, the trypsin resistant nucleoplasmin core; lane T → P, trypsin resistant nucleoplasmin core digested with pepsin; lane P, partial pepsin digest of nucleoplasmin.

(Fig. 4), as seen in the digestion of intact nucleoplasmin by pepsin. In addition, the M_r 12000 tail fragment is extremely sensitive to trypsin digestion.

Accumulation of pepsin fragments in the nucleus

The ability of the various fragments of nucleoplasmin produced by pepsin digestion to accumulate in the nucleus was assayed by microinjection of a radiolabelled digest into *Xenopus* oocytes. At various times after injection the nuclei were isolated and the separated cytoplasms and nuclei were analysed as described previously. The results are shown in Fig. 5. Lane b shows the material that was microinjected. The sample contained no detectable intact pentamers and the M_r 16000 tail fragment had been largely converted to the M_r 12000 fragment.

Fig. 5 shows that a cleaved pentamer that has only one intact monomer (i.e. one tail per pentamer) can still accumulate to the same extent as intact nucleoplasmin. The pentameric core that completely lacks tails is not detected in the nucleus even after 48 h but remains stably in the cytoplasm, confirming the observation with cores generated by trypsin cleavage.

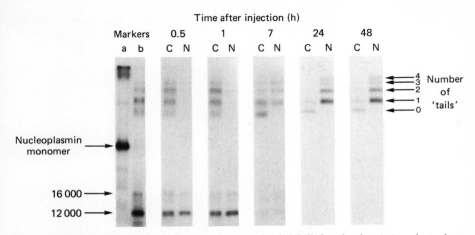

Fig. 5. *Migration of the products of pepsin cleavage of radiolabelled nucleoplasmin into the nucleus*

The material shown in lane b was microinjected into the cytoplasm of *X. laevis* oocytes. The oocytes were maintained at 20 °C in modified Barth's saline. Nuclei (N) and cytoplasms (C) were manually separated at the indicated times.

Table 1. *Accumulation of the M_r 12000 tail fragment in the oocyte nucleus compared with the extent of its degradation*

Densitometer traces of the autoradiograph shown in Fig. 5 were used to determine the nuclear to cytoplasmic concentration ratio and the amount of tail fragment in the oocyte. To eliminate variation between samples, all values for the M_r 12000 tail fragment were normalized by reference to the total amount of radioactivity in the stable pentameric bands in the sample.

Time (h) post-injection	Concentration ratio N/C	Amount (%) of input M_r 12000 fragment remaining in oocyte
0.5	21	91
1	37	85
7	32	17

The dissociated tail fragment rapidly accumulates in the nucleus

Fig. 5 also shows that the tail fragment of apparent M_r 12000 that is produced by prolonged digestion of nucleoplasmin rapidly accumulates in the nucleus.

This fragment is unstable in the oocyte and is degraded, but we can exclude the possibility that the accumulation of this fragment in the nucleus is only apparent, being due to preferential degradation in the cytoplasm. Densitometric traces of the gels shown in Fig. 5 reveal that 1 h after injection, when 85% of the injected tail fragment is still intact, it has accumulated 37-fold in the nucleus (Table 1). This small amount of degradation is therefore insufficient to account for the marked accumulation in the nucleus.

To confirm that the M_r 12000 tail fragment alone is sufficient for migration it was purified from a pepsin digest of nucleoplasmin, radiolabelled and injected

Time after injection (h)

0.5   1.6   5.5

C   N   C   N   C   N   a

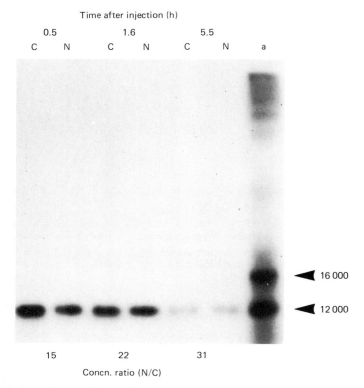

◀ 16 000

◀ 12 000

15   22   31

Concn. ratio (N/C)

Fig. 6. *Accumulation of the purified M_r 12000 tail fragment in the nucleus*

Pure, radiolabelled M_r 12000 tail fragment was microinjected into the cytoplasm of *X. laevis* oocytes. The oocytes were maintained at 20 °C in modified Barth's saline. Nuclei (N) and cytoplasms (C) were separated manually at the indicated times. Lane a, marker M_r 16000 and M_r 12000 fragments. N/C, Nuclear to cytoplasmic concentration ratio.

into the cytoplasm of *X. laevis* oocytes to assay its ability to migrate into the nucleus. The results are shown in Fig. 6. The observed nuclear to cytoplasmic concentration ratios show that the M_r 12000 tail fragment of nucleoplasmin accumulates rapidly in the nucleus. This fragment is therefore both necessary and sufficient for accumulation in the nucleus. The features of nucleoplasmin that enable it to accumulate in the nucleus are thus located within a discrete polypeptide domain.

The tail fragment is a signal for entry rather than a domain conferring binding within the nucleus

Early studies of nucleo–cytoplasmic exchange involved the microinjection of exogenous proteins, and dextrans of known molecular dimensions into oocytes (reviewed by Bonner, 1978). These studies indicated that molecules up to a certain size can enter the nucleus by a simple diffusion process and that the pores through which this process occurs have an approximate diameter of 9 nm.

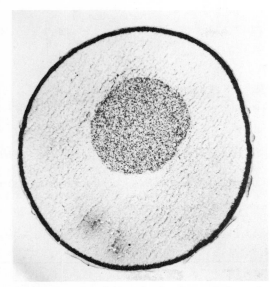

Fig. 7. *Retention of the nucleoplasmin core in the nucleus*

Radiolabelled trypsin resistant nucleoplasmin core was microinjected directly into the nucleus of *X. laevis* oocytes. The oocytes were maintained at 20 °C in modified Barth's saline for 24 h. They were then fixed, embedded, sectioned and autoradiographed at 4 °C.

The oocyte nucleus contains large stock piles of proteins, e.g. histones and nucleoplasmin for use in early development. For example the histone content of the oocyte nucleus is sufficient for the DNA of 10000 diploid nuclei. Since the amount of DNA in the oocyte nucleus is only 6 pg then the histones must be retained in a form other than chromatin. These facts lead to the suggestion (Bonner, 1978) that proteins are accumulated in the oocyte nucleus by free diffusion, followed by selective retention of nuclear proteins by binding to insoluble components of the nucleus.

The opposing view is that only nuclear proteins are allowed to enter the nucleus, i.e. there is selective entry of nuclear proteins into the nucleus.

Even though nucleoplasmin is much larger than exogenous proteins that have been seen to enter the nucleus by diffusion, the data presented so far can be interpreted in terms of either mechanism. The tail could either confer selective retention of nucleoplasmin in the nucleus, hence the 'core' molecule can diffuse into the nucleus but cannot be retained. Or the tail region is the signal for entry and thus the 'core' molecule cannot enter as it lacks the signal.

These two alternative mechanisms can be distinguished by microinjection of the 'core' molecule directly into the oocyte nucleus. If the tail region confers selective retention, we would predict that the 'core' molecules would diffuse back into the cytoplasm, since they lack tails. The observed result was that the core molecules remained in the nucleus (Fig. 7) after microinjection, hence the tail region is not necessary for retention but is necessary for entry. This result therefore argues against a mechanism of accumulation involving passive

Time after injection (h)

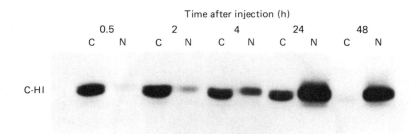

Fig. 8. *Accumulation of the C-terminal domain fragment (C-H1) of histone H1 in the oocyte nucleus*

Radiolabelled *C*-terminal domain fragment (C-H1) of calf thymus was injected into the cytoplasm of *X. laevis* oocytes. The oocytes were maintained at 20 °C in modified Barth's saline. Nuclei (N) and cytoplasms (C) were separated at the indicated times.

Table 2. *Amino acid composition (mol/100 mol) of the C-terminal domain fragment of histone H1 (C-H1)*

	CH-1
Asx	0.2
Thr	4.8
Ser	2.7
Gix	1.2
Pro	10.0
Gly	4.4
Ala	30.9
Val	2.5
Met	–
Ile	–
Leu	0.1
Tyr	–
Phe	–
His	–
Lys	43.2
Arg	–

diffusion and selective retention and favours a mechanism involving selective entry.

Sequence requirements for entry into the nucleus

Histone H1. Bonner (1975a) has shown that purified histone H1 from *Xenopus* erythrocytes accumulates in the oocyte nucleus. We have confirmed this result using histone H1 from calf thymus (Dingwall & Allan, 1984) and in addition we have shown that a proteolytic fragment encompassing the isolated *C*-terminal domain of the protein can accumulate in the nucleus (Fig. 8). This fragment has an extremely unusual amino acid composition (Table 2) in which lysine, proline, and alanine constitute about 75 of the 89 amino acids in this fragment. The remaining 14 amino acids are not clustered but are scattered throughout the fragment, therefore any primary sequence that specifies nuclear entry would seem necessarily to involve the amino acids lysine, proline and alanine.

Yeast MAT α2 protein. Corroborative evidence for the above suggestion comes from the investigation of protein accumulation in yeast nuclei. Hall *et al.* (1984) have used recombinant DNA techniques to fuse small pieces of the yeast MAT α2 protein (a nuclear protein) to *Escherichia coli β*-galactosidase and then monitored the ability of the fusion proteins to accumulate in the yeast nucleus. In this way the smallest piece of MAT α2 protein that will cause the fusion protein to accumulate in the nucleus is 13 amino acids derived from the *N*-terminus of MAT α2. By sequence comparison regions of homology have been found in the MAT α1 protein and the histone H2B, H4 and H2A. In these regions of homology the amino acids lysine, proline and alanine occur frequently. The overall feature of the homologous region is a group of three hydrophobic residues, one of which is proline, flanked on each side by a basic residue, predominantly lysine.

SV40 large T antigen. In cells infected with SV40 a virus encoded large T antigen is produced which accumulates in the cell nucleus. In 1969 Butel *et al.* found an SV40 mutant in which the large T antigen failed to accumulate in the nucleus. This mutant has now been characterized (Lanford & Butel, 1984) and shown to be the substitution of a positively charged lysine residue at position 128 by a neutral asparagine residue. This lysine residue lies in the sequence Pro-Lys-Lys-Lys[128]-Arg-Lys-Val-Glu-Asp-Pro.

In an independent study, Kalderon *et al.* (1984) generated a series of mutants in the large T antigen gene in SV40 by mixed oligonucleotide mutagenesis. They demonstrated that conversion of lysine-128 to threonine abolishes transport of large T antigen into the nucleus. Mutation of lysine 127 did not affect nuclear accumulation, and mutation of residues 129, 130 or 131 produced T antigen molecules which were present in both the nucleus and cytoplasm. These authors also produced deletions in the large T antigen gene in the vicinity of lysine-128 and showed that interruption of the tract of basic amino acids in this region abolishes entry of large T antigen into the nucleus.

Summary

The ability of nuclear proteins to enter and accumulate in the nucleus is a feature of the mature protein. The signal for nuclear entry is not a feature of the overall shape of the protein but resides in a discrete polypeptide domain. The exact nature of the primary sequence that specifies nuclear entry is not clear but several lines of evidence point to the specific involvement of lysine, proline and perhaps alanine.

References

Bonner, W. M. (1975*a*) *J. Cell Biol.* **64**, 421–430
Bonner, W. M. (1975*b*) *J. Cell Biol.* **64**, 431–437

Bonner, W. M. (1978) in *The Cell Nucleus 6* (Busch, H. ed.), part C, pp. 97–148, Academic Press, New York

Butel, J. S., Geuntzel, M. J. & Rapp, I. (1969) *J. Virol.* **4**, 632–641

Dingwall, C. & Allan, J. (1984) *EMBO J.* **3**, 1933–1937

Dingwall, C., Sharnick, S. V. & Laskey, R. A. (1982) *Cell* **30**, 449–458

Earnshaw, W. C., Honda, B. M., Laskey, R. A. & Thomas, J. O. (1980) *Cell* **21**, 373–383

Gurdon, J. B. (1970) *Proc. Roy. Soc. Lond. B* **176**, 303–314

Hall, M. N., Hereford, L. & Horskowitz, I. (1984) *Cell* **36**, 1057–1065

Kalderon, D., Richardson, W. D., Markham, A. F. & Smith, A. E. (1984) *Nature (London)* **311**, 33–38

Kreil, G. (1981) *Annu. Rev. Biochem.* **50**, 317–348

Krohne, G. & Franke, W. (1980) *Proc. Natl. Acad. Sci. U.S.A.* **77**, 1034–1038

Lanford, R. E. & Butel, J. S. (1984) *Cell* **37**, 801–813

Laskey, R. A., Honda, B. M., Mills, A. D. & Finch, J. T. (1978) *Nature (London)* **275**, 416–420

Mills, A. D., Laskey, R. A., Black, P. & De Robertis, E. M. (1980) *J. Mol. Biol.* **139**, 561–568

Biochem. Soc. Symp. **50**, 205–220
Printed in Great Britain

Ionic Signal Transduction in Growth Factor Action

S. W. DE LAAT, W. H. MOOLENAAR, L. H. K. DEFIZE, J. BOONSTRA* and
P. T. VAN DER SAAG

*Hubrecht Laboratory, International Embryological Institute, Uppsalalaan 8, 3584 CT Utrecht, The Netherlands, and *Department of Molecular Cell Biology, State University Utrecht, Padualaan 8, 3584 CH Utrecht, The Netherlands*

Synopsis

Growth factors are polypeptides which exert their mitogenic action through binding to specific high-affinity receptor molecules on the cell surface of target cells. This interaction leads to the rapid activation of a receptor-linked signal transduction system, involving the stimulation of an intrinsic receptor tyrosine phosphokinase activity, the breakdown of inositol lipids, and the production of ionic signals. In this contribution we have analysed the nature and origin of the ionic signals, and we have applied monoclonal antibodies against the receptor for epidermal growth factor (EGF), as well as tumour-promoting phorbol esters, to dissociate the early cellular responses to growth factors. Evidence is presented that the ionic signals are coupled to the breakdown of inositol lipids. The hydrolysis of phosphatidylinositol 4,5-bisphosphate (PIP2) would lead to the production of 1,2-diacylglycerol (DG) and inositol triphosphate (IP3). DG production results in the stimulation of protein kinase C, which causes the activation of Na^+/H^+ exchange by increasing its affinity for cytoplasmic H^+. Consequently, a rise in cytoplasmic pH is observed. This response can be mimicked by the tumour promoter TPA, which can replace DG in activating the protein kinase C. Independently, IP3 production leads to the rapid mobilization of Ca^{2+} from intracellular stores. Monoclonal antibodies against the EGF receptor differed in their ability to evoke EGF-like responses upon binding to the EGF receptor. Of the three anti-EGF receptor IgGs tested, one was directed against the EGF-binding domain (2E9), and the others (2D11 and 2G5) were directed against sugar moieties not involved in EGF binding. Receptor tyrosine kinase activity could be stimulated by 2E9 as well as 2D11, but not by 2G5. Only 2D11 induced morphological changes similar to EGF. None of the antibodies was able to trigger the production of ionic signals, which implies probably that antibody binding to the receptor, even when they bind to the EGF binding domain, is an insufficient stimulus for the breakdown of inositol lipids. Most importantly, these monoclonal antibodies were also not able to induce DNA synthesis in quiescent human fibroblasts, not even after cross-linking of the EGF receptors by a second antibody. It may thus be concluded that the stimulation of the intrinsic receptor tyrosine phosphokinase activity can be dissociated from other early responses, and that none of the identified early responses is a sufficient trigger for the mitogenic action of growth factors.

Introduction

Polypeptide growth factors play an essential role in the regulation of cellular growth and differentiation. Generally, growth factors exert their effects at nanomolar concentrations. Their action is initiated by binding to specific high-affinity receptor molecules localized in the plasma membrane. This interaction triggers a cascade of biochemical and physiological responses in the target cell, which ultimately lead to overt changes in cellular behaviour, such as the initiation of proliferation or differentiation. Interestingly, the earliest receptor-mediated responses are very similar for the various growth factors, suggesting that common molecular pathways are utilized in reaching diverse biological end points in different cell types (de Laat et al., 1983, 1985). These early responses are detectable within seconds or minutes of growth factor binding, and include the stimulation of tyrosine phosphokinase activity of the receptor itself (Cohen et al., 1982; Ek & Heldin, 1982), the breakdown of inositol-phospholipid (Sawyer & Cohen, 1981; Habenicht et al., 1981; Berridge & Irvine, 1984), the mobilization of intracellular Ca^{2+} (Moolenaar et al., 1984a) and the activation of Na^+/H^+ exchange, resulting in a rise in cytoplasmic pH (pH_i) (Moolenaar et al., 1983, 1984b; Cassel et al., 1983; L'Allemain et al., 1984). How these early responses lead to the stimulation of DNA replication and cell division is as yet unknown.

Epidermal growth factor (EGF) and platelet-derived growth factor (PDGF) are among the most intensively studied growth factors. The importance of the growth regulatory function of these molecules was emphasized by the recent discovery that growth factors or their receptor show a striking structural homology with certain oncogene products: the sis oncogen encodes for a product homologous to PDGF (Deuel et al., 1983; Doolittle et al., 1983; Robbins et al., 1983; Waterfield et al., 1983; Chiu et al., 1984), and the erb-B oncogene product is homologous to the intracellular domain of the EGF receptor (Downward et al., 1984; Ullrich et al., 1984). Furthermore, there is evidence that other oncogene products exert their transforming action by interference with certain growth factor receptor-mediated responses (Berridge & Irvine, 1984).

Inappropriate expression of cellular oncogenes is thought to be involved in the initiation and maintenance of malignant growth. Knowledge of the mode of action of polypeptide growth factor such as EGF and PDGF is undoubtedly of importance for our understanding of the mechanisms of cancer, in particular in view of these recent findings. In this contribution we will focus on the generation of ionic signals by growth factors and we will describe some new experimental approaches to assess the interrelationship between the various early responses and their biological relevance, using monoclonal antibodies against the human EGF receptor and tumour-promoting phorbol esters as partial activators of the response machinery.

Ionic Signal Transduction

Activation of Na⁺/H⁺ exchange and cytoplasmic alkalinization

Alterations in monovalent cation fluxes have been implicated in the action of mitogens since the observation that the activation of quiescent mouse fibroblasts leads to a rapid stimulation of the Na^+,K^+-ATPase (Na^+,K^+ pump) (Rozengurt & Heppel, 1975). Since then it has been established that this is a general response to various growth factors in a variety of cells (de Laat *et al.*, 1983, 1985). For example, EGF stimulates the activity of the Na^+,K^+ pump in human fibroblasts (Moolenaar *et al.*, 1982*a*) neuroblastoma cells (Mummery *et al.*, 1983), rat phaeochromocytoma cells (Boonstra *et al.*, 1983) and other cell types (Rozengurt & Heppel, 1975; Fehlman *et al.*, 1981). This was, however, shown to be a secondary response. Smith & Rozengurt (1978*a*) were the first to demonstrate that the mitogenic stimulation of quiescent mouse fibroblasts leads to an amiloride-sensitive increase in the rate of Li^+ uptake. It has now been established that the observed stimulation of the Na^+,K^+ pump is a direct consequence of a prior increase in Na^+ influx and serves merely for the extrusion of entered Na^+ (Smith & Rozengurt, 1978*b*; Mendoza *et al.*, 1980; Moolenaar *et al.*, 1981*a*, *b*), this Na^+ influx being mediated largely by growth factor-activated, amiloride-sensitive Na^+/H^+ exchange (Moolenaar *et al.*, 1981*a*, *b*, 1982*a*; Boonstra *et al.*, 1983).

In recent years it has become clear that Na^+/H^+ exchange has a primary role in the regulation of pH_i in mammalian cells (Moolenaar *et al.*, 1981*a*, 1984*c*). Its functioning can be assessed by monitoring the kinetics of pH_i recovery after an acute acid load. This can be achieved by the use of intracellularly trapped fluorescent pH_i indicators or by pH-sensitive microelectrodes, as illustrated in Fig. 1. This pH_i recovery follows an exponential time course and is, at least in human fibroblasts, mediated entirely by an electroneutral amiloride-sensitive Na^+/H^+ exchanger in the plasma membrane, which is driven by the transmembrane Na^+ and H^+ gradients. The evidence hereto can be summarized as follows (Moolenaar *et al.*, 1983, 1984*c*):

(1) pH_i recovery is accompanied by Na^+ uptake and H^+ efflux, approximately with a 1:1 stoichiometry;
(2) pH_i recovery, Na^+ influx and H^+ efflux are reversibly blocked by amiloride;
(3) the rate of pH_i recovery depends critically on the external Na^+ concentration (half-maximal at ~ 30 mM-Na^+), but not on external anions;
(4) pH_i recovery is insensitive to alterations in membrane potential;
(5) the Na^+/H^+ exchanger can act in a reversed mode by the choice of appropriate Na^+ gradients.

The exponential recovery of pH_i after an acute acid load facilitates the kinetic analysis of the Na^+/H^+ exchange process. It implies that the rate of Na^+/H^+ exchange, proportional to $d(pH_i)/dt$, is linearly dependent on pH_i. Under normal ionic conditions the rate of Na^+/H^+ exchange is virtually zero in quiescent cells, although the system is not in thermodynamic equilibrium. In fact, the physiological transmembrane Na^+ and H^+ gradients provide sufficient energy to raise pH_i by nearly 1 pH unit. Our results are consistent with the concept that the pH_i sensitivity of the system reflects an allosteric activation by

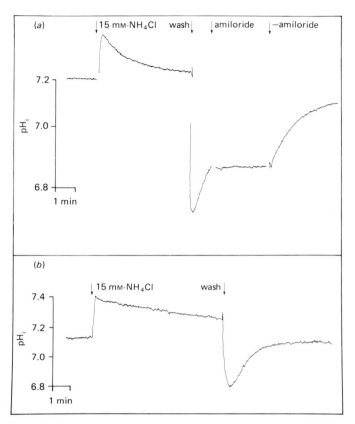

Fig. 1. *Continuous recordings of pH$_i$ recovery from an acid load induced by a pre-pulse with*
15 mM-NH$_4$Cl

(a) pH$_i$ was recorded fluorimetrically in human fibroblasts, using bis-(carboxyethyl)-carboxy-
fluorescein (BCECF) as an intracellularly trapped pH$_i$ indicator. This experiment shows in
addition the reversible inhibition of pH$_i$ recovery by 1 mM-amiloride (see Moolenaar *et al.*, 1983,
1984*b*). (b) pH$_i$ was monitored electrophysiologically, in N1E-115 neuroblastoma cells, with ultra-
fine pH-sensitive micro-electrodes.

cytoplasmic H$^+$ at a modifier site, distinct from the internal H$^+$ transport site
(Aronson *et al.*, 1982).

A possible clue to the mechanism of activation of the Na$^+$/H$^+$ exchanger by
growth factors was given by comparing the kinetics of pH$_i$ recovery in quiescent
fibroblasts before and after growth factor addition (Moolenaar *et al.*, 1983).
These experiments demonstrated that growth factors act by increasing the
apparent affinity of the Na$^+$/H$^+$ exchanger for cytoplasmic H$^+$. Consequently,
a rise of pH$_i$ might result. Initial indications that mitogenic stimulation of
quiescent fibroblasts indeed induces an amiloride-sensitive rise in pH
of ~ 0.15 pH unit came from conventional weak-acid distribution studies
(Moolenaar *et al.*, 1982*b*; Schuldiner & Rozengurt, 1982). Direct evidence for a
growth factor-induced increase in pH$_i$ was obtained with a trapped fluorescent
pH$_i$ indicator as well as from pH-sensitive microelectrode recordings (Moolenaar

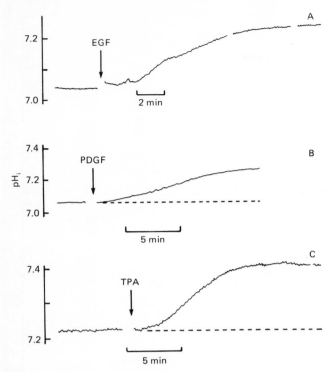

Fig. 2. *Effect of EGF (50 ng/ml), PDGF (40 ng/ml) and TPA (50 ng/ml) on pH$_i$ of quiescent human fibroblasts*

pH$_i$ was monitored by the fluorescent pH$_i$ indicator BCECF. Data are from Moolenaar *et al.* (1983, 1984*b*).

et al., 1983, 1984*b*). Some typical examples of pH$_i$ shifts after addition of EGF or PDGF to human fibroblasts are given in Fig. 2. An increase in pH$_i$ is usually detectable within 30 s of growth factor addition and is completed after 10–15 min. In the continuous presence of the growth factor the elevated pH$_i$ is maintained for at least 2 h, whereas pH$_i$ only slowly decays after washout of the growth factors. The activation of the system is apparently a rapid process, in contrast to the deactivation. The involvement of Na$^+$/H$^+$ exchange in the growth factor-induced cytoplasmic alkalinization is evidenced by its sensitivity for the external Na$^+$ concentration and its inhibition by amiloride. Interestingly, insulin has no effect on pH$_i$ by itself but it markedly potentiates the effects of EGF and PDGF.

The observations summarized here make it clear that the activation of Na$^+$/H$^+$ exchange, and consequently a cytoplasmic alkalinization, is one of the earliest detectable responses to growth factor–receptor interaction. Furthermore it is a general response evoked by a variety of mitogens in normal and transformed mammalian cells (Pouysségur *et al.*, 1982; Moolenaar, 1982*b*, 1983, 1984*b*; Cassel *et al.*, 1983; Rothenberg *et al.*, 1983; Schuldiner & Rozengurt, 1982). This suggests that pH$_i$ acts as a signal to modulate pH-sensitive processes

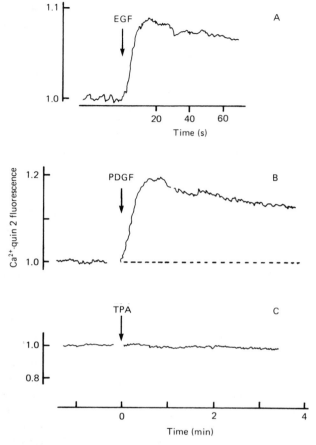

Fig. 3. *Effect of EGF (50 ng/ml), PDGF (40 ng/ml) and TPA (50 ng/ml) on Ca$_i^{2+}$ mobilization in quiescent human fibroblasts*

Ca^{2+} mobilization was monitored by the fluorescent Ca^{2+} indicator quin 2. Data are from Moolenaar *et al.* (1984*a*, *b*).

in the cell, and that in this way cytoplasmic alkalinization is directly involved in the mitogenic activation process.

Ca^{2+} mobilization

Cytoplasmic free Ca^{2+} (Ca$_i^{2+}$) has been implicated as a second messenger in the action of various hormones. Relatively little was known, however, about the role of Ca$_i^{2+}$ in the action of growth factors, due mainly to the lack of reliable methods to determine dynamic Ca$_i^{2+}$ changes. The availability of the fluorescent Ca$_i^{2+}$ indicator quin 2 has recently provided a means to monitor such changes. With this method it was established, for example, that lectin-stimulation of lymphocytes results in a rapid twofold increase in Ca$_i^{2+}$ (Tsien *et al.*, 1982). We have applied quin 2 to analyse the possible effects of EGF and PDGF on Ca$_i^{2+}$

in quiescent human fibroblasts (Moolenaar et al., 1984b). As shown in Fig. 3, EGF and PDGF cause a 20–30% increase in quin 2 fluorescence without a detectable lag period, a maximum being reached within 20–40 s. This represents a two- to three-fold increase in Ca_i^{2+}. The fluorescence then slowly declines to a new steady-state level about 30% above the normal basal level of 100–150 nM. The growth factor induced rise in Ca_i^{2+} was independent of the presence of extracellular Ca^{2+} and could thus not result from a change in Ca^{2+} influx. We concluded therefore that growth factor–receptor interaction leads to the mobilization of Ca^{2+} from intracellular stores (Moolenaar et al., 1984a). As such this Ca_i^{2+} rise is the fastest response to growth factors described so far, which suggests that Ca_i^{2+} acts as a primary trigger in mitogenic activation.

Mechanism of activation of ionic signals

Now that the nature of the early ionic signals is known it remains to be established by what molecular mechanism(s) growth factors activate the Na^+/H^+ exchanger and mobilize Ca_i^{2+}, and whether these events are causally linked. As stated in the Introduction, there are at least two other types of responses detectable at a comparable time scale. These are the stimulation of tyrosine phosphokinase activity of the receptor itself and the breakdown of plasma membrane inositol lipids, in particular the hydrolysis of phosphatidyl-inositol 4,5-bisphosphate (PIP2). As reviewed recently by Berridge & Irvine (1984), PIP2 hydrolysis is a very general response to a wide range of external stimuli in all kinds of cells. Phosphodiesterase hydrolysis of PIP2 yields 1,2-diacylglycerol (DG) and inositol triphosphate (IP3), and both products are thought to act as second messengers: DG is a potent activator of protein kinase C, which phosphorylates proteins at serine and threonine residues (Nishizuka, 1984); IP3 is an effective mobilizer of Ca_i^{2+}, probably from stores such as the endoplasmic reticulum (Berridge & Irvine, 1984). Interestingly, certain phorbol-ester tumour promoters, such as 12-O-tetradecanoyl-phorbol-13-acetate (TPA), that bear a structural similarity to DG can mimic DG in stimulating the protein kinase C (Nishizuka, 1984). We have made use of this feature to analyse the molecular mechanism underlying the production of ionic signals. As before, a fluorescent pH_i indicator as well as pH-sensitive microelectrodes were used to monitor now possible effects of TPA on pH_i in a number of cell types, including human fibroblasts (Moolenaar et al., 1984a). TPA, but also synthetic diacylglycerol, mimic fully the action of growth factors in activating Na^+/H^+ exchange and elevating pH_i, as shown in Fig. 2. Quin 2 monitoring of Ca_i^{2+} demonstrated, however, that these stimulants of protein kinase C have no effect on cytoplasmic free Ca_i^{2+} (Fig. 3). These results indicate that growth factor-induced activation of Na^+/H^+ exchange results through protein kinase C stimulation, by the production of diacylglycerol upon PIP2 hydrolysis. Possibly, protein kinase C directly phosphorylates the Na^+/H^+ exchanger. In addition, these data demonstrate that an early rise in Ca_i^{2+} is not essential for the activation of Na^+/H^+ exchange (Moolenaar et al., 1984a). As illustrated in Fig. 4, it is thus possible to dissociate the early responses to growth factors by partial activation of the signal transduction system, in this case by using TPA or synthetic

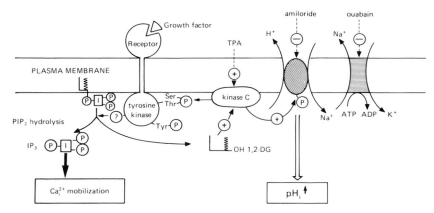

Fig. 4. *Schematic representation of the early responses caused by growth factor–receptor interaction*

For explanation see the text.

diacylglycerol as specific activators of protein kinase C. How the PIP2-mediated ionic responses are coupled to the stimulation of the tyrosine phosphokinase activity of the receptor itself, and what the relevance of the early response is for the ultimate biological effects of growth factors, remains to be established.

Dissociation of EGF Receptor-Mediated Responses by Monoclonal Antibodies

Production and characterization of monoclonal antibodies

A novel approach in the analysis of cause and effect among the various cellular responses to growth factors is the use of monoclonal antibodies as partial activators of the receptor. We have recently employed this strategy for a further characterization of the signal transduction system of the EGF receptor. Monoclonal antibodies against different antigenic determinants of the EGF receptor have been raised in several laboratories, usually by immunization with human epidermoid carcinoma A431 cells or plasma membrane preparations of these cells (Schreiber *et al.*, 1981; Waterfield *et al.*, 1982; Kawamoto *et al.*, 1983; Richert *et al.*, 1983; Gregoriou & Rees, 1984). A431 cells provide an excellent material for immunization since these cells express $2-3 \times 10^6$ EGF receptors per cell on their cell surface. We have raised a series of monoclonal antibodies against the extracellular domain of the human EGF receptor by immunization of Balb/c mice with a plasma membrane preparation of A431 cells. Hybridoma cultures were obtained as described (Herzenberg *et al.*, 1978) and culture supernatants were screened by a number of assays with increasing specificity. Three out of 200 hybridoma cultures, designated as 2E9, 2D11 and 2G5, were selected for subcloning by limiting dilution, and were characterized further. All three produced IgG antibodies, which were able to precipitate a functional EGF receptor tyrosine phosphokinase (M_r 170000) from solubilized A431 plasma membranes (Fig. 5). Scatchard analysis of the binding of iodinated antibody to intact A431 cells revealed a similar number of 2E9 binding sites as compared

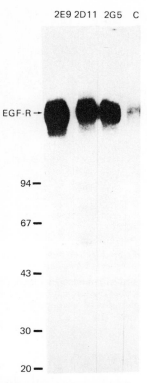

Fig. 5. *Immunoprecipitation of functional EGF receptor by monoclonal antibodies against the EGF receptor*

Plasma membranes from A431 cells were incubated with EGF and [γ-^{32}P]ATP, after which immunoprecipitation was performed, by using protein A–Sepharose.

with EGF (2–3×10^6 receptors per cell). In contrast, the number of binding sites for 2D11 and 2G5 was 5–10 times higher, which suggests the presence of common antigenic sites for 2D11 and 2G5 on the EGF receptor and other membrane constituents or the occurrence of multiple antigenic sites per receptor molecule. These binding studies also revealed that 2E9, and not 2D11 or 2G5, is able to compete effectively with EGF for binding to the receptor and vice versa. This indicates that 2E9 recognizes an antigenic determinant at, or very close to, the EGF binding domain of the receptor. The nature of the different antigenic determinants was further identified by immunoprecipitation of solubilized A431 cells, after [^{35}S]methionine labelling in the presence or absence of tunicamycin. After tunicamycin treatment 2E9 was still able to precipitate a single protein with M_r 130000, which is equal to that of the unglycosylated EGF receptor (Decker, 1984). It can thus be concluded that 2E9 is directed against a peptide determinant. Under identical circumstances 2D11 and 2G5 failed to precipitate detectable amounts of protein, and in the absence of tunicamycin treatment these antibiotics were able to precipitate multiple proteins. 2D11 and 2G5 are thus directed against sugar moieties of the EGF receptor which probably are common for multiple proteins on the cell surface of A431 cells.

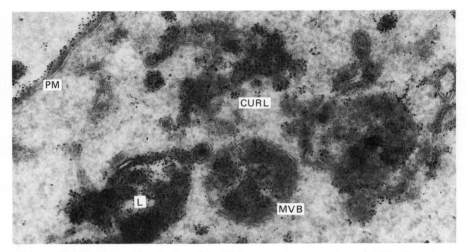

Fig. 6. *Identification and localization of the EGF receptor in cryo-sections of A431 cells, using gold-immune labelling*

A431 cells were cryo-sectioned and labelled with monoclonal IgG 2E9, rabbit-anti-mouse IgG and protein A–gold complex (8 nm), as described by Boonstra *et al.* (1985). Note the dense labelling on the plasma membrane (PM), compartment of uncoupling of receptor and ligand (CURL), multi-vesicular bodies (MVB) and lysosomes (L).

Gold-immune labelling of the EGF receptor

Various lines of evidence have indicated that EGF binding induces a clustering of EGF–receptor complexes, which are rapidly internalized via clathrin-coated pits and vesicles. It is believed that these complexes, like other receptor molecules (Geuze *et al.*, 1984), are transferred via so-called multi-vesicular bodies to lysosomes, where degradation occurs (for review see Carpenter, 1983). Clearly, the exact nature of this process as well as the ultimate fate of the ligand and receptor remain to be established.

A direct use of available (monoclonal) antibodies against the EGF receptor is given by their applicability for visualization of the receptor molecule. Cryo-ultramicrotomy followed by gold-immune labelling enables high-resolution localization of the receptor at any time during the endocytotic routing.

We have applied this method to localize EGF receptor in cryo-sections of cultured A431 cells, using the monoclonal antibodies 2E9, 2D11 and 2G5 (Boonstra *et al.*, 1985). After a consecutive labelling with anti-EGF receptor antibody, rabbit-anti-mouse antibody and protein A-colloidal gold complex, a high number of gold particles was observed at the external side of the plasma membrane (Fig. 6). In particular cases it was possible to identify gold particle-containing coated pits (Boonstra *et al.*, 1985). Intracellularly, gold particles were observed on vesicular and tubular membrane structures, most likely representing the compartment previously called CURL (Geuze, 1984), which is probably involved in the sorting-out process of receptor and ligand. Furthermore, multi-vesicular bodies and structures resembling lysosomes were heavily labelled with gold particles (Fig. 6).

These results demonstrate the applicability of cryo-ultramicrotomy immuno-gold labelling in the localization and identification of EGF receptor on an ultrastructural level. In addition, however, the results indicate an interesting phenomenon as well. The cells shown in Fig. 6 were not treated with EGF at any time, and therefore the intracellular labelling of CURL, multi-vesicular bodies and lysosomes indicate that A431 cells internalize EGF receptor in the absence of binding of EGF. This phenomenon would also explain the absence of a mitogenic response of A431 cells towards EGF (Boonstra et al., 1985). Further studies along these lines are under way.

Partial activation of the EGF receptor by monoclonal antibodies

As a first step in analysing the effects of monoclonal antibodies on the signal transduction system of the EGF receptor, we tested our antibodies 2E9, 2D11 and 2G5 for their capacity to mimic EGF in stimulating the tyrosine phospho-kinase activity of the receptor in plasma membranes of A431 cells, human fibroblasts and HeLa cells, as well as in intact A431 cells.

Tyrosine phosphokinase activity in membrane preparations can be assayed simply by using peptide substrates, containing tyrosine but no serine or threonine, such as angiotensin I or II, or synthetic peptides (Pike et al., 1982). With angiotensin I as a substrate, 2E9 was able to stimulate tyrosine phospho-kinase activity in plasma membranes of A431 cells, in contrast to 2D11 and 2G5. However, 2E9 was less effective than EGF itself. Under the assay conditions used the maximal stimulation obtained for EGF was eightfold, whereas 2E9 caused a 3.5-fold stimulation. Interestingly, the maximal stimulation by EGF could be reduced by low concentration of 2E9 (sixfold stimulation at 0.5 μM-2E9), whereas at saturation concentrations of the antibody (2.8 μM) EGF had no effect. These results demonstrate that the stimulation of tyrosine phosphokinase activity is dependent on the antigenic site to which the antibodies bind, and they support the notion that EGF and 2E9 compete for the same binding domain (see above). In addition, they show that the nature of the external ligand is not essential for the stimulation of the internal enzymic activity of the receptor, although quantitative differences in the degree of stimulation are found.

A more conventional way to monitor the stimulation of kinase activity of the receptor is the determination of phosphorylation of endogenous proteins in plasma membrane preparations (Mummery et al., 1985). We have applied this method for A431 cells (see Fig. 7), human fibroblasts and HeLa cells. As expected, EGF enhanced the autophosphorylation of the EGF receptor (M_r 170000) as well as the phosphorylation of a number of other membrane-associated proteins, most prominently an M_r 24000 and a 36000 protein, in all three cell types. 2E9 and 2D11 mimicked EGF in its stimulation of phosphorylation of the EGF receptor, but interesting differences are found in their capacity to induce the phosphorylation of other proteins, such as the M_r 24000 and 36000 proteins. It should be noted in particular that the phosphorylation of the latter proteins is not enhanced by 2E9. The antibody 2G5 had no effects on endogenous protein phosphorylation in plasma membrane preparations (not shown). Taken together, these experiments show that there is only a partial correlation between the

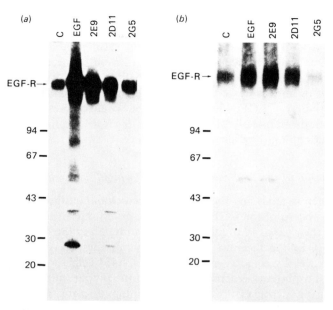

Fig. 7. *Stimulation of tyrosine phosphokinase activity of the EGF receptor in (a) solubilized plasma membranes of A431 cells and (b) intact A431 cells, by EGF and anti-EGF receptor monoclonal antibodies*

(a) Endogenous phosphorylation pattern of solubilized plasma membranes from A431 cells. Membranes were incubated with EGF or antibody for 15 min at 20 °C, after which [γ-^{32}P]ATP was added for 10 min at 0 °C. (b) Immunoprecipitation of EGF receptor from intact A431 cells, after EGF or antibody induced receptor phosphorylation. Cells were labelled with 0.5 mCi of ^{32}P for 14 h, after which EGF or antibody was added for 10 or 30 min respectively. The cells were thus lysed in a 1% NP-40 containing buffer, and immunoprecipitation was carried out using a polyclonal antibody directed against a peptide sequence of the cytoplasmic domain of the EGF receptor (kindly provided by J. Schlessinger).

antibody effects on the phosphorylation of exogenous peptides and endogenous membrane proteins. 2E9 is effective under both conditions, but 2D11 only in the latter assay. In view of the other effects of the antibodies, it could be important that the receptor kinase shows substrate specificity depending on the nature of the activating ligand.

Direct evidence for the stimulation of kinase activity of the receptor under physiological conditions came from the determination of receptor phosphorylation in intact cells, as induced by EGF or antibody binding. A431 cells were metabolically labelled with ^{32}P at 37 °C for 14 h, after which EGF or antibody was added for 10 or 30 min respectively. Subsequently the cells were lysed and solubilized, and the EGF receptor was immunoprecipitated by an antibody directed against the cytoplasmic domain of the receptor (kindly provided by J. Schlessinger). As shown in Fig. 7, EGF and the antibodies 2E9 and 2D11 were capable of inducing phosphorylation of the EGF receptor, whereas 2G5 was not.

The above results lead to the conclusion that the kinase activity, located on the cytoplasmic domain of the EGF receptor, can be stimulated by extracellular

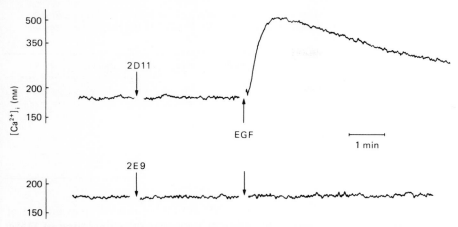

Fig. 8. *Lack of Ca$_i^{2+}$ mobilization in A431 cells upon binding of monoclonal anti-EGF receptor antibody*

Note the inhibition of growth factor-induced Ca$_i^{2+}$mobilization by the EGF competing antibody 2E9.

ligands other than EGF, although not by all ligands (e.g. 2G5). Stimulation can be directed via the EGF binding domain (EGF and the anti-peptide antibody 2E9), but also binding to other domains can be a sufficient trigger (the anti-sugar antibody 2D11). The results indicate furthermore that the kinase activity as well as the substrate specificity of the enzyme are determined by the spatial conformation of the receptor molecule, as its conformation is probably influenced by the nature of the ligand.

The differential ability of monoclonal antibodies to stimulate the kinase activity of the EGF receptor opened the possibility to analyse the involvement of this enzymatic activity in other signal transduction events. Using the same techniques as before, we found that none of the monoclonal antibodies was able to induce cytoplasmic alkalinization (not shown) or Ca$_i^{2+}$ mobilization in A431 cells (Fig. 8), even after secondary addition of a cross-linking rabbit-anti-mouse IgC. On the other hand, the EGF competing antibody 2E9 completely inhibited the effect of EGF on Ca$_i^{2+}$ mobilization (Fig. 8) as would be expected from a competitive ligand. The stimulation of tyrosine phosphokinase activity of the EGF receptor and clustering of the receptors are apparently insufficient triggers for the induction of the ionic signals.

Other rapid cellular changes in EGF treated A431 cells are cytoskeletal alterations leading to the rapid rounding up of cells in low Ca^{2+} media (Chinkers *et al.*, 1981). These morphological effects of EGF could be mimicked by 2D11 and by TPA, but not by 2E9 or 2G5.

Clearly, the cytoskeletal alterations are not well understood, as there is no simple correlation with any of the other early responses (see Table 1).

Finally, the mitogenic properties of the antibodies were examined in quiescent human fibroblasts. None of them, nor TPA, was able to induce DNA synthesis, not even in the presence of a cross-linking rabbit-anti-mouse IgG. These results

Table 1. *Comparison of the effects of EGF, monoclonal antibodies directed against the extracellular domain of the EGF receptor (2E9, 2D11 and 2G5), and the tumour-promoting phorbol ester TPA*

	EGF	2E9	2D11	2G5	TPA
Precipitation of EGF receptor	−	+	+	+	−
EGF binding competition	+	+	−	−	−
Stimulation of tyrosine kinase	+	+	+	−	−
Na^+/H^+ exchange, $pH_i\uparrow$	+	−	−	−	+
Ca_i^{2+} mobilization	+	−	−	−	−
Cytoskeletal organization	+	−	+	−	+
DNA synthesis	+	−	−	−	−

are in disagreement with the proposal that antibody-induced cross-linking of EGF receptors is a necessary and sufficient signal for the stimulation of DNA synthesis in human fibroblasts (Schreiber *et al.*, 1983).

Conclusions

In this contribution we have summarized recent results on the characterization of the growth factor–receptor signal transduction system, emphasizing the nature and origin of ionic signals. As an important tool for the dissection of the complex early responses to growth factor–receptor interaction we have studied the effects of partial activation of the receptor by monoclonal antibodies, and of by-passing the receptor by tumour-promoting phorbol esters. The main results are summarized in Table 1. If one accepts the view that the ionic signals are generated by the hydrolysis of PIP2 (see Fig. 4), i.e. Ca_i^{2+} mobilization by the production of IP3 and activation of Na^+/H^+ exchange by 1,2-diacylglycerol stimulation of protein kinase C, our results provide evidence for the following conclusions:

(1) cross-linking of the EGF receptors and/or stimulation of the intrinsic EGF receptor tyrosine phosphokinase does not necessarily lead to inositol lipid breakdown, cytoskeletal alterations or the induction of DNA synthesis;

(2) the control of the EGF receptor kinase activity is a complex one: substrate specificity and enzyme activity depend on the nature of the ligand–receptor interaction, and are thus probably determined by the spatial configuration;

(3) the nature of the triggering of PIP2 hydrolysis is as yet unknown, and the resulting ionic signals are necessary but insufficient signals for the ultimate mitogenic effect of the growth factor.

Clearly, the interaction between a growth factor and its receptor induces a pleiotropic response in the target cell. Although the knowledge about this signal transduction system is evolving rapidly, the relevance of most, if not all of the receptor-linked early events for the mitogenic action remains to be proven. In particular the intermediate events are as yet largely obscure, and future studies should certainly involve the characterization of the nature and role of the poorly defined substrates of the receptor kinase and protein kinase C.

Research carried out at the Hubrecht Laboratory was supported in part by the Netherlands Cancer Foundation (Koningin Wilhelmina Fonds).

References

Aronson, P. S., Nee, J. & Suhm, M. (1982) *Nature (London)* **299**, 161–163

Berridge, M. J. & Irvine, R. F. (1984) *Nature (London)* **312**, 315–321

Boonstra, J., Moolenaar, W. H., Harrison, P. H., Moed, P., van der Saag, P. T. & de Laat, S. W. (1983) *J. Cell Biol.* **97**, 92–98

Boonstra, J., van Maurik, P. A. M., Defize, L. H. K., de Laat, S. W., Leunissen, J. L. M. & Verkley, A. J. (1985) *Eur. J. Cell Biol.* **36**, 209–216

Carpenter, G. (1983) *Mol. Cell. Endocrinol.* **31**, 1–19

Cassel, D., Rothenberg, P., Zhuang, Y.-X., Deuel, T. F. & Glaser, L. (1983) *Proc. Natl. Acad. Sci. U.S.A.* **80**, 6224–6228

Chinkers, M., McKanna, J. A. & Cohen, S. (1981) *J. Cell Biol.* **88**, 422–429

Chiu, L. M., Reddy, E. P., Givol, D., Robbins, K. C., Tronick, S. R. & Aaronson, S. A. (1984) *Cell* **37**, 123–129

Cohen, S., Ushiro, H., Stoschek, C. & Chinkers, M. (1982) *J. Biol. Chem.* **257**, 1423–1431

Decker, S. (1984) *Mol. Cell Biol.* **4**, 571–575

de Laat, S. W., Boonstra, J., Moolenaar, W. H., Mummery, C. L., van der Saag, P. T. & van Zoelen, E. J. J. (1983) in *Development in Mammals* (Johnson, M. H., ed.), vol. 5, pp. 33–106, Elsevier Science Publishers, Amsterdam

de Laat, S. W., Boonstra, J., Moolenaar, W. H. & Mummery, C. L. (1985) in *Growth and Maturation Factors* (Guroff, G., ed.), vol. 3, pp. 219–279, John Wiley, New York

Deuel, T. F., Huang, J. S., Huang, S. S., Stroobant, P. & Waterfield, M. D. (1983) *Science* **221**, 1348–1350

Doolittle, R. F., Hunkapillar, H. W., Hood, L. E., Devare, S. G., Robbins, K. C., Aaronson, S. A. & Antoniades, H. N. (1983) *Science* **221**, 275–277

Downward, J., Yarden, Y., Mayes, E., Scarce, G., Totty, N., Stockwell, P., Ullrich, A., Schlessinger, J. & Waterfield, M. D. (1984) *Nature (London)* **307**, 521–527

Ek, B. & Heldin, C.-H. (1982) *J. Biol. Chem.* **257**, 10486–10492

Fehlman, M., Canivet, B. & Freycket, P. (1981) *Biochem. Biophys. Res. Commun.* **100**, 254–260

Geuze, H. J., Slot, J. W., Strous, G. J. A. M., Peppard, J., von Figura, K., Hasilik, A. & Schwartz, A. L. (1984) *Cell* **37**, 195–204

Gregoriou, M. & Rees, A. (1984) *EMBO J.* **3**, 929–937

Habenicht, A. J. R., Glomset, J. A., King, W. C., Nist, C., Mitchell, C. D. & Ross, R. (1981) *J. Biol. Chem.* **256**, 12329–12335

Herzenberg, L., Herzenberg, L. & Mistein, C. (1978) in *Handbook of Experimental Immunology* (Weiz, D. M., ed.), pp. 25.1–25.7, Blackwell, Oxford

Kawamoto, T., Sato, J. P., Le, A., Polikoff, J., Sato, G. H. & Mendelsohn, J. (1983) *Proc. Natl. Acad. Sci. U.S.A.* **80**, 1337–1341

L'Allemain, G., Franchi, A., Cragoe, E. & Pousségur, J. (1984) *J. Biol. Chem.* **259**, 4313–4319

Mendoza, S. A., Wigglesworth, N. M., Pohjanpelto, P. & Rozengurt, E. (1980) *J. Cell. Physiol.* **103**, 17–27

Moolenaar, W. H., Boonstra, J., van der Saag, P. T. & de Laat, S. W. (1981 *a*) *J. Biol. Chem.* **256**, 12883–12887

Moolenaar, W. H., Mummery, C. L., van der Saag, P. T. & de Laat, S. W. (1981 *b*) *Cell* **23**, 789–798

Moolenaar, W. H., Yarden, Y., de Laat, S. W. & Schlessinger, J. (1982 *a*) *J. Biol. Chem.* **257**, 8502–8506

Moolenaar, W. H., de Laat, S. W., Mummery, C. L. & van der Saag, P. T. (1982 *b*) in *Ions, Cell Proliferation and Cancer* (McKeehan, W. L. & Boynton, A. L., eds.), pp. 152–162, Academic Press, New York

Moolenaar, W. H., Tsien, R. Y., van der Saag, P. T. & de Laat, S. W. (1983) *Nature (London)* **304**, 645–648

Moolenaar, W. H., Tertoolen, L. G. J. & de Laat, S. W. (1984 *a*) *J. Biol. Chem.* **259**, 8066–8069

Moolenaar, W. H., Tertoolen, L. G. J. & de Laat, S. W. (1984 *b*) *Nature (London)* **312**, 371–373

Moolenaar, W. H., Tertoolen, L. G. J. & de Laat, S. W. (1984 *c*) *J. Biol. Chem.* **259**, 7563–7569

Mummery, C. L., van der Saag, P. T. & de Laat, S. W. (1983) *J. Cell. Biochem.* **21**, 63–75

Mummery, C. L., Feijen, A., van der Saag, P. T., van den Brink, C. E. & de Laat, S. W. (1985) *Dev. Biol.* in the press

Nishizuka, Y. (1984) *Nature (London)* **308**, 693–697

Pike, L. J., Marquardt, H., Todaro, G. J., Gallis, B., Casnellie, J. E., Bornstein, P. & Krebs, E. G. (1982) *J. Biol. Chem.* **257,** 14628–14631

Pouysségur, J., Chambardt, J. C., Franchi, A., Paris, S. & van Obberghen-Schilling, E. (1982) *Proc. Natl. Acad. Sci. U.S.A.* **79,** 3935–3939

Richert, N., Willingham, M. D. & Pastan, I. (1983) *J. Biol. Chem.* **258,** 8902–8907

Robbins, K. C., Antoniades, H. N., Devare, S. G., Hunkapillar, M. W. & Aaronson, S. A. (1983) *Nature (London)* **305,** 605–609

Rothenberg, P., Glaser, L., Schlessinger, P. & Cassel, D. (1983) *J. Biol. Chem.* **258,** 12644–12653

Rozengurt, E. & Heppel, L. A. (1975) *Proc. Natl. Acad. Sci. U.S.A.* **72,** 4492–4495

Sawyer, S. T. & Cohen, S. (1981) *Biochemistry* **20,** 6280–6286

Schreiber, A. B., Lax, I., Yarden, Y., Eshhar, Z. & Schlessinger, J. (1981) *Proc. Natl. Acad. Sci. U.S.A.* **78,** 7535–7539

Schreiber, A. B., Libermann, T., Lax, I., Yarden, Y. & Schlessinger, J. (1983) *J. Biol. Chem.* **258,** 846–853

Schuldiner, S. & Rozengurt, E. (1982) *Proc. Natl. Acad. Sci. U.S.A.* **79,** 7778–7782

Smith, J. B. & Rozengurt, E. (1978a) *J. Cell. Physiol.* **97,** 441–450

Smith, J. B. & Rozengurt, E. (1978b) *Proc. Natl. Acad. Sci. U.S.A.* **75,** 5560–5564

Tsien, R. Y., Pozzan, T. & Rink, T. J. (1982) *Nature (London)* **295,** 68–71

Ullrich, A., Caussens, L., Hayflick, J. S., Dull, T. J., Gray, A., Tam, A. L. S., Lee, J., Yarden, Y., Libermann, T. A., Schlessinger, J., Downward, J., Mayes, E. L. V., Whittle, N., Waterfield, M. D. & Seebing, P. H. (1984) *Nature (London)* **309,** 418–425

Waterfield, M. D., Mayes, E., Stroobant, P., Bennet, P., Young, S., Goodfellow, P., Banting, G. & Ozanne, B. (1982) *J. Cell. Biochem.* **20,** 115–127

Waterfield, M. D., Scrace, G. T., Whittle, N., Stroobant, P., Johnson, A., Wasteson, A., Westermark, B., Heldin, C.-H., Huang, J. S. & Deuel, T. F. (1983) *Nature (London)* **304,** 35–39

Biochem. Soc. Symp. **50**, 221–233
Printed in Great Britain

Membrane Damage by Channel-Forming Proteins: Staphylococcal α-Toxin, Streptolysin-O and the C5b-9 Complement Complex

SUCHARIT BHAKDI* and JØRGEN TRANUM-JENSEN†

* *Institute of Medical Microbiology, University of Giessen, Schubertstrasse 1, D-6300 Giessen, West Germany, and* † *Anatomy Institute C, University of Copenhagen, Blegdamsvej 3C, DK-2200 Copenhagen N, Denmark*

Synopsis

One mechanism through which cells can be damaged involves insertion of alien proteins into the membrane bilayer and the formation of hydrophilic transmembrane pores. Three examples for this process are discussed, namely membrane damage by staphylococcal α-toxin, streptolysin-O, and by the terminal C5b-9 complement complex. Common to all is the principle of a transition of the proteins from a water-soluble state to an amphiphilic state, occurring through the appearance or exposure of apolar surfaces during oligomerization of the protein molecules into supramolecular aggregates. The resulting complexes or protein oligomers insert spontaneously into the target lipid bilayer and assume properties akin to those of integral membrane proteins. The protein channels can be isolated from membranes after their solubilization by mild detergents and characterized on a bio-immunochemical and ultra-structural level.

Introduction

The complement C5b-9 complex was the first membrane-damaging channel to be isolated and characterized (Bhakdi *et al.*, 1976, 1978; Tranum-Jensen *et al.*, 1978; Bhakdi & Tranum-Jensen, 1978), and the resulting contention that pore formation by the complex in a target membrane constitutes the primary mechanism of complement cytolysis (Mayer, 1972; Bhakdi & Tranum-Jensen, 1978) is now generally accepted (Mayer *et al.*, 1981; Bhakdi & Tranum-Jensen, 1983; Tschopp *et al.*, 1982*a*, *b*). Subsequently, the recognition emerged that membrane damage by channel formation was quite a widespread phenomenon apparently governed by distinct and partially repetitive patterns of events. The present paper will review recent data obtained in our laboratory on three channel-formers, namely staphylococcal α-toxin, streptolysin-O and the terminal C5b-9 complement complex, and attempts will be made to focus on the numerous analogies existing amongst these three membrane-damaging protein systems.

General Features of Channel-Formers

The native channel-forming proteins are produced as hydrophilic polypeptides that undergo an irreversible transition from a hydrophilic to an amphiphilic state when they bind to and insert into the target lipid bilayers. Exposure of hydrophobic surfaces occurs as the proteins oligomerize into supramolecular aggregates on the cell surface, possibly via conformational changes leading to unfolding of the molecules with the exposure of initially hidden apolar domains. Membrane insertion *per se* is probably a spontaneous process driven primarily by the energetically favoured hydrophobic–hydrophobic interactions between these domains and membrane lipid. Generation of functional pores may result either from the embedment of completely circularized channel structures within the bilayer, or through the insertion of partially circularized structures. In some cases, functional pore formation is not accompanied by the appearance of an ultrastructurally recognizable channel at all.

The channels may constitute a homogeneous population of structures, as appears to be the case with staphylococcal α-toxin. They may, however, also exhibit marked heterogeneity due to the presence of varying numbers of monomeric subunits contained within the individual polymers (e.g. streptolysin-O). In all events, the offending proteins probably represent the sole constituents of the channels; membrane proteins and lipids have not been found to participate in the formation of the pore structures. Our present belief is that these membrane constituents are passively forced aside in the lateral membrane plane of eukaryotic cells during the entry of the alien proteins.

With regard to the initial membrane binding process, two broad categories of pore-formers may be differentiated. The first category comprises molecules that bind with high affinity to specific substrates on the target membranes. Classic examples are the thiol-activated bacterial toxins including streptolysin-O, which bind initially to surface-exposed membrane cholesterol. The second category comprises molecules that may not require specific binders for initial membrane attachment. One of the best studied toxins that probably belongs to this group is staphylococcal α-toxin. When specific binders are present, attack by the channel-former will probably be generally efficient, regardless of cell target. In contrast, many 'unspecific' membrane factors (e.g. surface charge) may influence the efficacy in binding of the 'receptor-less' toxins and the susceptibilities of different cell species towards the cytotoxic effects of such pore-formers thus will tend to vary widely.

Protein channels have been functionally characterized with the use of many experimental systems, including measurement of release of intracellularly trapped markers (Thelestam & Möllby, 1975; Giavedoni et al., 1979; Füssle et al., 1981; Buckingham & Duncan, 1983), and conductance measurements across planar lipid bilayers (Michaels et al., 1976). The different experimental approaches have all yielded data supporting the concept of transmembrane pore formation. Minor discrepancies arising with regard to the true dimensions of the channels probably stem from the specific experimental conditions selected by the different groups of investigators (for discussion, see Bhakdi & Tranum-Jensen, 1983).

All the oligomerized protein channels studied in our laboratory have been found to be remarkably stable. They withstand the action of very high concentrations of non-ionic detergent and deoxycholate, and also resist destruction by proteases at neutral pH (Tranum-Jensen et al., 1978; Füssle et al., 1981; Bhakdi et al., 1985). All three protein channels to be discussed were initially isolated from membranes after lysis of erythrocytes with unpurified toxin or complement components. High concentrations of deoxycholate were used to solubilize the membranes quantitatively, and a single centrifugation of membrane solubilizates through sucrose density gradients led to satisfactory purification of both C5b-9 (Bhakdi & Tranum-Jensen, 1982; Bhakdi et al., 1983b) and streptolysin-O (Bhakdi et al., 1985). The isolated channels have regularly been found to be very immunogenic, and antisera raised against the complexes have proven to be useful tools in immunological studies of the oligomerized as well as the native proteins (e.g. Bhakdi et al., 1978).

Staphylococcal α-Toxin

Staphylococcal α-toxin is considered one of the major factors of staphylococcal pathogenicity (McCartney & Arbuthnott, 1978). The toxin is secreted as a hydrophilic, 3.3 S monomer of M_r 34000 (Bhakdi et al., 1981) with an isoelectric point of approximately 8.5. The toxin monomers self-associate on a target membrane to form homogeneous, ring-structured 11–12 S hexamers (M_r 200000; Bhakdi et al., 1981) that bind lipid and detergent. The embedment of these rings in the membrane generates functional holes in the bilayers of resealed erythrocyte ghosts whose effective diameters appear to be 2–3 nm (Füssle et al., 1981). Electron microscopy reveals that the hexamers indeed display a central pore of these dimensions (Fig. 1). The external diameter of the hexamers measures 8.5–10 nm and the channel extends approximately 4 nm from the membrane surface into the extramembranous space. The volume of the extramembranously oriented portion of the complex can thus be estimated to be 200–250 nm³, which corresponds to a mass of 160–200 kDa. These estimates indicate that the intramembranous domain of the toxin rings may comprise only a small part of the total mass; the walls of the pore within the membrane have not been directly visualized by electron microscopy and may in fact be much thinner than those forming the walls of the externally oriented cylinder. Indeed, the amino acid sequence of α-toxin deduced from the nucleotide sequence of the cloned gene does not reveal the presence of any remarkably long stretches of apolar residues (Gray & Kehoe, 1984). This serves to underline the difficulties that may be encountered in future attempts to construct molecular models for the organization of the polypeptide chains in the amphiphilic protein complexes. Photolabelling may become a valuable aid for eventually identifying the membrane-embedded regions of the molecule. As is to be expected from the lipid-binding studies, it has indeed been found that membrane-inserted α-toxin is labelled by hydrophobic probes (Thelestam et al., 1983).

Triggering of the hexamerization process is not recognizably dependent on the presence of a biochemically defined membrane substrate (Freer et al., 1968; Buckelew & Colaccico, 1971). Spontaneous hexamer formation occurs through

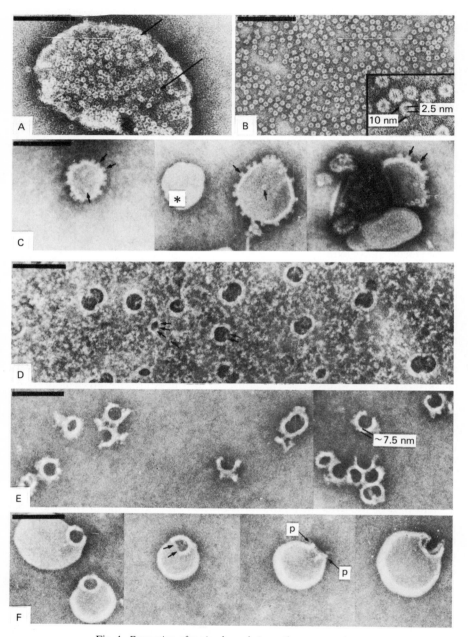

Fig. 1. *Formation of toxin channels in erythrocyte membranes*

(A)Fragment of rabbit erythrocyte lysed with *Staphylococcus aureus* α-toxin; 10 nm ring-shaped structures are seen over the membrane (arrows). (B) Isolated α-toxin hexamers in detergent solution. (C) Lecithin liposomes carrying re-incorporated α-toxin hexamers, seen as stubs along the edge of the liposomal membrane and as rings over the membrane (arrows). (D) Erythrocyte membrane lysed by streptolysin-O, showing numerous 25–100 nm long and approximately 7.5 nm broad, curved rods. Most rods are approximately semicircular, often joined in pairs at their ends. Dense accumulations of stain are seen at the concave side of the rods. When these do not form closed heating of the toxin (Arbuthnott *et al.*, 1967) and by exposure to deoxycholate detergent (Bhakdi *et al.*, 1981). Hexamers also form upon contact of the toxin with human plasma low-density lipoprotein (LDL), probably through non-specific triggering of the oligomerization process by lipid contained in the LDL particle (Bhakdi *et al.*, 1983a). As with all other channel-formers, toxin oligomers are haemolytically inactive since they aggregate in aqueous solution. Binding and hexamer formation of α-toxin on LDL thus causes toxin inactivation and may represent a significant non-immune defence mechanism of the host towards this toxin.

The efficiency of toxin-binding to cell membranes is generally low: at toxin concentrations of 0.03–3 μM (1–30 μg/ml), less than 5% of toxin offered to a surplus of erythrocytes becomes membrane-bound. We believe that this is due to the absence of specific membrane binders for the toxin. As toxin levels are raised, overall binding will increase to reach final levels of 50–70%. Thus, binding of the toxin to cells in this concentration range does not exhibit characteristics of a simple ligand–receptor interaction (Bhakdi *et al.*, 1984a).

Since specific binders are probably absent, the initial diffusion and attachment of the toxin to the cell surface can probably be influenced by quite unspecific factors, including surface charge. This is possibly why different cell species display a very wide variation in overall sensitivity towards cytolytic and cytotoxic toxin action. For example, human erythrocytes are lysed at toxin concentrations approximately 100-fold higher than those required to lyse rabbit erythrocytes (McCartney & Arbuthnott, 1978). However, we have noted a marked pH-dependence in sensitivity of human erythrocytes towards lytic toxin action. At pH 5–5.5, these cells are lysed by toxin levels approximately fivefold lower than required at neutral pH. We have suggested that this derives from neutralization of negative charges on the membrane surface which otherwise restrict diffusion of the positively charged toxin molecules to the lipidic surface (Bhakdi *et al.*, 1984a).

Streptolysin-O

This toxin is secreted by group A β-haemolytic streptococci and belongs with at least 14 other bacterial toxins to the group of thiol-activated cytolysins (Alouf, 1980; Bernheimer, 1974; Smyth & Duncan, 1978). Toxins of this group are reversibly inactivated by atmospheric oxygen and primarily bind to membrane cholesterol. The initial binding process appears to be temperature independent and also occurs at 0 °C. At elevated temperatures, oligomerization of the toxin molecules into high molecular weight aggregates takes place, probably following lateral diffusion and collision of the toxin–cholesterol complexes in the membrane plane. Very large, curved rod structures are thus generated that insert into the

profiles, the stain deposit is partly bordered by a 'free' edge of the erythrocyte membrane (arrows). (E) Isolated streptolysin-O oligomers. (F) Purified streptolysin-O complexes re-incorporated into cholesterol-free lecithin liposomes. The toxin oligomers form holes in the liposomes. Part of the circumference of such holes appears bordered by a 'free' edge of liposomal membrane (unlabelled arrows). p: lesion seen in profile. Scale bars indicate 100 nm in all frames. Sodium silicotungstate was used as negative stain in frames B–F, and uranyl acetate in A.

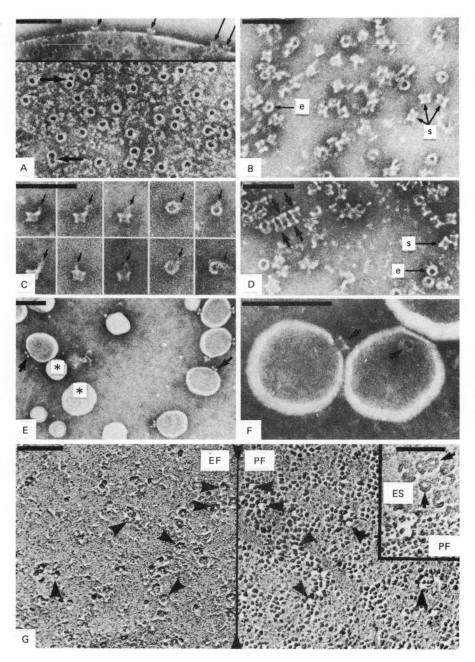

Fig. 2. *Structure of cytolytic C5b-9 complement complexes*

(A) Complement-lysed erythrocytes: C5b-9(m) complexes are seen as numerous circular lesions over the membrane together with some 'twinned' forms (bold arrows), and as 10 nm high cylindrical projections along the bent edge of the ghost membrane (arrows). (B) Isolated C5b-9(m) complexes in detergent solution. The complex has the basic structure of a 15 nm high, thin-walled cylinder, rimmed by an annulus at one end. The cylinder is seen in various levels of tilt between side views bilayer to create water-filled channels (Duncan & Schlegel, 1975; Alouf, 1980; Smyth & Duncan, 1978; Buckingham & Duncan, 1983; Bhakdi et al., 1985).

We have recently isolated streptolysin-O from bacterial culture supernatants and characterized the native toxin as a water-soluble, 3.9 S molecule of M_r 69000 (Bhakdi et al., 1984b). When incubated with target erythrocytes, toxin oligomers form that exhibit marked heterogeneity. A broad array of structures ranging from C-shaped curved rods to fully closed, single and fused rings is thus observed (Duncan & Schlegel, 1975; Bhakdi et al., 1985; Fig. 1). The complexes generate very large, 30–35 nm wide defects in the bilayer. Functional studies have also revealed the existence of large pores across such membranes (Buckingham & Duncan, 1983). The protein channels have been isolated from deoxycholate-solubilized erythrocyte membranes and they can be particularly well visualized after their re-incorporation into unilamellar bilayers of phosphatidylcholine (Fig. 1; Bhakdi et al., 1985). When C-shaped polymers become membrane-incorporated in this system, the membrane pores appear to be lined only in part by the protein (Fig. 1; Bhakdi et al., 1985), and what appars as a free edge of liposomal membrane extends between the ends of a given rod. These observations are in full accordance with the observation of similar, apparently free edges of membrane in toxin lesions on erythrocytes (Fig. 1). When seen in profile, the rods appear only slightly elevated above the membrane surface. It is thus apparent that the bulk of the oligomers lie embedded within the lipid bilayer. The structural data indicate that insertion of frankly 'incomplete' channels consisting of C-shaped toxin oligomers already appears to create trans-membrane pores, possibly because of repelling of lipid by the hydrophilic concave sides of the membrane-inserted complexes. It will obviously be of interest eventually to determine the orientation and molecular packing of lipid molecules in such membrane regions.

Although the C-shaped and circular toxin structures have generally been thought to represent toxin–cholesterol complexes, we have no evidence that the isolated complexes contain constituents other than the toxin molecules themselves (Bhakdi et al., 1985). The fact that such extensively delipidated toxin polymers can re-associate with phosphatidylcholine without any requirement for cholesterol underlines the contention that the sterol, although important for initial toxin binding, is not itself involved in formation of the channel. Once exposed,

---

(s) and axial projection (e). (C) Selection of C5b-9(m) complexes exhibiting a small appendage (arrows), often seen on the annulus. (D) 'Poly-C9' formed by prolonged incubation of purified human C9 in detergent-free buffer solution at 37 °C, as described by Tschopp et al. (1982a). Occasionally, small ordered arrays of cylinders are seen (arrows), associated at the putative apolar terminus opposite the annulus. (E) and (F) C5b-9(m) complexes (arrows) re-incorporated into phosphatidylcholine liposomes. Vesicles that escaped incorporation of a complex (asterisks) are characteristically impermeable to the stain. (G) Complementary freeze–etch replicas of complement-lysed sheep erythrocyte. Fracture E-faces (left frame) exhibit numerous ring-shaped structures, representing the intramembranous portion of C5b-9(m) cylindrical complexes. The rings are complementary to circular defects in the lipid plateau of the inner membrane leaflet (PF-face). A number of complementary lesions are labelled by arrowheads. Inset at G (upper right) shows C5b-9(m) annuli (arrows) on the etched true outer surface (ES) of a proteolytically stripped ghost membrane. 25° rotary shadowing with Pt. Scale bars indicate 100 nm in frames A–G. Sodium silicotungstate was used for negative staining in frames A–F. [Figs. 1 and 2 are reproduced from Bhakdi & Tranum-Jensen (1984a) with permission of The Royal Society, London.]

the apolar regions of the protein complexes exhibit no recognizable specificity with respect to the type of lipid that can be bound (Bhakdi *et al.*, 1985).

Volume calculations based on electron microscopy indicate a volume of approximately 6000 nm³ for a typical closed toxin ring, corresponding to a molecular mass of about 5 MDa. Since the molecular weight of native toxin is 69 000, we estimate that such closed ring structures harbour 70–80 molecules of toxin monomers. Haemolytic titrations indeed indicate that the presence of 70–120 molecules of toxin per erythrocyte is sufficient to generate a mean of one functional (haemolytic) lesion per cell (Bhakdi *et al.*, 1984*b*; Alouf, 1980). The collective data thus indicate that one or very few oligomers bound to an erythrocyte will cause lysis. It is noteworthy that our structural studies were conducted with high toxin doses (approximately 50 µg of toxin/10⁹ cells). The structures of lesions induced by low, just lytic doses has not been studied. In these cases, it is possible that lesion formation follows the same basic pattern, but that the oligomers contain fewer toxin molecules. It is also possible that some functional lesions may not be visible in the electron microscope.

Terminal C5b-9 Complement Complex

The C5b-9(m) complex (the suffix 'm' is used to denote the membrane location of the molecule) forms on a membrane after initiation of both classical and alternative complement pathways whenever the complement cascade is able to over-run the regulatory mechanisms that operate at the level of C3b (Fearon, 1979; Kazatchkine & Nydegger, 1982). C5 is thereby cleaved to its active derivative C5b, and formation of C5b-9 occurs through subsequent complexing of this molecule with the C6–C9 proteins. No proteolytic cleavage step of functional significance has been identified in the terminal reaction after the generation of C5b.

The first appearance of hydrophobic domains occurs at the stage of C5b-7 assembly. Thus, whereas C5b-6 complexes bind reversibly to membranes, C5b-7 becomes firmly embedded within the bilayer (Hammer *et al.*, 1975) and can be labelled by apolar, membrane-restricted photolabels (Hu *et al.*, 1981). Insertion of C5b-7 does not, however, create instability or leakiness of erythrocyte membranes. The complex possesses one binding site for the β-chain of C8 (Monahan & Sodetz, 1981). Binding of this component generates the tetrameric C5b-8 complex, and functional pores appear under specific experimental conditions. The molecular basis for the generation of these C5b-8 channels is unclear. Present data indicate that both C5b-7 and C5b-8 complexes exhibit a pronounced tendency to aggregate through lateral diffusion in the membrane; this property appears to be suppressed in the presence of C9. We suspect that the formation of functional C5b-8 channels is the consequence of such C5b-8 aggregation. Thus, sheep erythrocytes laden with low densities of C5b-8 lyse only to a very small extent, but lysis becomes pronounced as the surface density of C5b-8 is raised. At the same time, fluffy aggregates of protein tufts, probably representing C5b-8 complexes, are observed in the electron microscope. Upon post-treatment with C9, the ensuing circular C5b-9 rings or lesions are also seen in primarily aggregated states on the membranes (Bhakdi & Tranum-Jensen, 1984*b*).

The functional dimensions of the C5b-8 channels have been estimated to be in the order of 0.9 nm (Ramm *et al.*, 1982*b*). It should, however, be considered that these channels may represent *in vitro* artifacts arising through secondary C5b-8 aggregation, and that monomeric C5b-8 complexes might not primarily create channels at all.

The C5b-8 complex serves as the binding substrate of C9. We have obtained evidence that controlled binding of the last complement component can be achieved at low temperature. Thus, incubation of C5b-8 cells with an excess of C9 at 0 °C leads to formation of C5b-9 complexes that contain an average stochiometry of $C5b-8_1C9_1$ as determined by densitometry of SDS–polyacrylamide gel electrophoresis scans of the isolated protein. Binding of one molecule of C9 to C5b-8 at low temperature causes no appreciable cell lysis if the cells are kept in the cold. However, if cells are washed at 4 °C (to remove non-bound C9) and re-incubated in saline at 37 °C, haemolysis occurs. Thus, raising the temperature to 37 °C induces the formation of a functional pore, probably through conformational changes that take place in the terminal complement components leading to insertion of the C9 molecule into the bilayer. Of particular interest is the fact that the $C5b-8_1C9_1$ pore is stable and of intermediate size (1–3 nm effective diameter, as estimated by osmotic protection experiments). These cells are effectively protected from lysis by dextran 4 (mean M_r 4000), but not by sucrose. If the dextran is removed and the cells are re-suspended in saline, lysis occurs. In the electron microscope, the cell membranes carry no circular lesions and, indeed, no clear micromorphological indications for the presence of the moderately large channels can be obtained. These findings demonstrate that the binding of an average of only one C9 molecule to C5b-8 complexes suffices to generate stable channels, which are, however, not detectable by electron microscopy.

$C5b-8_1C9_1$ complexes can bind further C9 molecules at elevated temperature. In this reaction, C9–C9 oligomerization takes place as a rapid autocatalytic process without any recognizable specific requirements (e.g. metal ions). It appears that the polymerization process is sensitive to disturbances and, once interrupted, the process cannot re-continue. As a consequence, a very heterogeneous population of C5b-9 complexes is generated, comprising complexes with various numbers of C9 molecules. At a certain, as yet undefined, ratio of C9:C8 molecules, channel structures become visible by electron microscopy. A fully circularized complement lesion probably exhibits the molecular stoichiometry $C5b-8_1C9_{12-16}$ (Tschopp, 1984), but the majority of complexes forming on target erythrocytes probably contain only four to eight C9 molecules when the cells are treated with very high doses of complement (Bhakdi & Tranum-Jensen, 1984*b*). In any event, there is now general consensus that the unit lesion is a monomer with respect to C5b-8 (Bhakdi & Tranum-Jensen, 1981; Tschopp, 1984; Ramm *et al.*, 1982*a*), and not a dimer as widely proposed previously (e.g. Podack & Müller-Eberhard, 1981). It is also clear that functional lesion heterogeneity cannot arise from the parallel presence of monomeric and dimeric species of C5b-9, as advanced by Podack *et al.* (1982).

Circularized C5b-9 complexes have been characterized by electron microscopy as hollow protein cylinders rimmed by an annulus at one terminus and harbouring an internal pore of approximately 10 nm diameter (Tranum-Jensen *et al.*, 1978). The cylinders are vertically oriented to the membrane plane and are seen as the typical ring structures that were originally described on complement-lysed cells (Borsos *et al.*, 1964; Humphrey & Dourmashkin, 1969). The complexes can be isolated in detergent solution (Bhakdi *et al.*, 1976; Bhakdi & Tranum-Jensen, 1982) and re-incorporated into unilamellar liposomes (Bhakdi & Tranum-Jensen, 1978). These studies have revealed that 4–5 nm of the thin-walled portion of the cylinder distal to the annulus carry the lipid-binding sites which insert into the membrane (Fig. 2). Frank interruptions in the continuity of the bilayer are observed at the sites of protein insertion, and stain deposits are seen traversing the internal diameter of the channel to enter the liposomes (Tranum-Jensen *et al.*, 1978; Bhakdi & Tranum-Jensen, 1978). The micromorphology of such protein cylinders embedded within the bilayer of target erythrocytes has additionally been studied by freeze–fracture electron microscopy (Tranum-Jensen & Bhakdi, 1983). The intramembrane portion of the C5b-9 cylinder is observed on EF-faces as an elevated, circular structure. In non-etched fractures, the complex appears as a solid stub, whereas in etched fractures a central pit is viewed, confirming the existence of a central, water-filled pore in the molecule. Complementary replicas (Fig. 2) show that each EF-face ring corresponds to a hole in the lipid plateau of the PF-face. Etched fractures of proteolytically stripped membranes reveal the extramembranous annulus of the C5b-9 cylinder on ES-faces, as well as putative internal openings of the pores on PS-faces. The dimensions of EF-face rings and ES-face annuli fully conform with expectations derived from negatively stained preparations. The collective data have provided strong support for the concept that the hollow C5b-9 cylindrical complex penetrates into the inner leaflet of the target erythrocyte membrane bilayer to form a stable transmembrane protein channel. Recent studies have provided evidence that C9 epitopes indeed become exposed on the cytoplasmic membrane surface of erythrocytes after C5b-9 insertion (Morgan *et al.*, 1984). Thus, it is likely that the C5b-9 channel spans the entire membrane width, as originally expected (Tranum-Jensen *et al.*, 1978; Bhakdi & Tranum-Jensen, 1978). The major portion of the intramembrane domain is probably built by C9 molecules (Hu *et al.*, 1981). However, other terminal components also appear to contribute towards formation of the membrane-embedded region (Bhakdi & Tranum-Jensen, 1979; Bhakdi *et al.*, 1980; Steckel *et al.*, 1983).

Much attention has recently been drawn to the fact that purified C9 will, under specific conditions *in vitro*, polymerize in the absence of a target membrane to form cylindrical structures typical of the C5b-9 channels (Tschopp *et al.*, 1982*a*; Podack & Tschopp, 1984). From these studies, the conclusions have been drawn that (a) pore formation by C5b-9 is primarily the consequence of insertion of 'poly-C9' (i.e. circularized C9 polymers) into the membrane and (b) C5b-8 does not significantly contribute towards formation of the intramembranous, channel-forming portion of the C5b-9 complex. C5 and C6 antigenic determinants have indeed been localized to a stalk structure projecting from one side of the C9-polymer in the C5b-9 complex (Tschopp *et al.*, 1982*b*). Whereas it is clear that the poly-C9 data do identify the major part of the C5b-9 channel as C9, we disagree with the generalization that generation of the pores is a function elicited solely by 'poly-C9'. The latter term applies to circularized

C9-polymers that resist dissociation by SDS (Podack & Tschopp, 1984) and harbour 12–16 molecules of C9. As discussed above, most C5b-9 complexes forming on erythrocyte membranes do not contain such completely circularized C9-polymers, yet it is generally accepted that they generate functional channels. It is also probable that heterogeneity in channel sizes derives from the parallel existence of complexes carrying lower numbers of C9 molecules (Boyle *et al.*, 1981). Finally, since C5b-8_1C9$_1$ complexes (containing no polymerized C9) still generate stable, intermediate-sized channels, it is obvious that this function is not invariably fulfilled by the poly-C9 complexes. We adhere to our earlier proposal that polypeptide chains derived from other terminal components (C5b-C8) do play a significant part in channel-formation (Bhakdi & Tranum-Jensen, 1979; Bhakdi *et al.*, 1980), and the participation of these components should not be ignored.

Conclusions

A relatively simple chain of events can be envisaged to govern the process of membrane penetration by channel-forming proteins. The initial reaction involves binding of the protomers to the membrane target, and this may be mediated by specific binders, or may take place as the result of simple diffusion of the protein molecules to the cell surfaces. Oligomer formation, arising through lateral membrane diffusion of protomers (e.g. streptolysin-O) or through binding of native monomers from the fluid-phase (e.g. C9) leads to unfolding of the proteins at the lipidic surface. It is conceivable that the crucial step in this process is the formation of dimers. Conformational changes then lead to the exposure of hydrophobic domains in the molecules, and to the triggering of an autocatalytic reaction that causes rapid formation of the oligomeric aggregates. Membrane insertion may be a spontaneous process driven by the energetically favoured hydrophobic–hydrophobic protein–lipid interactions. With regard to the fate of membrane lipid and protein at the channel sites, two basic processes may be envisaged. These membrane constituents may be expelled into the extramembranous phase, or they could be laterally displaced by the channels. Recent data from our laboratory clearly support the latter contention in the case of erythrocyte targets (unpublished data). Repelling of lipid (and membrane proteins) may be initiated by the insertion of 'incomplete', non-circularized oligomeric structures due to the exposure of hydrophilic surfaces within the lipid plane, such as appears to occur with the majority of streptolysin-O complexes. Complete circularization is thus probably not even necessary for generation of functional transmembrane pores. These considerations pertain also to C5b-9 complexes that do not contain 'poly-C9' rings, but which nevertheless clearly form pores. The generation of lesion heterogeneity in C5b-9 complexes through differential binding of C9 to C5b-8 exhibits obvious analogies to the heterogeneity encountered in streptolysin-O channels.

The concept of membrane damage by channel-formers is thus now well documented and many other protein systems will probably be found to operate similarly. A recent finding of wide general interest is that channel-formation also appears to operate in lymphocyte-mediated cytotoxicity (Simone & Henkart,

1980; Henkart *et al.*, 1984). It will now be of interest to investigate cell biological and pathophysiological aspects of these processes. Obvious questions relate to the fate of membrane-bound channels, and to secondary effects possibly elicited by the proteins.

We thank Margit Pohl and Marion Muhly for outstanding technical assistance. Our studies were supported by the Deutsche Forschungsgemeinschaft (Bh2/1-5, Bh 2/2 and SFB 47).

References

Alouf, J. E. (1980) *Pharmac. Ther.* **11**, 661–717
Arbuthnott, J. P., Freer, J. H. & Bernheimer, A. W. (1967) *J. Bacteriol.* **94**, 1170–1177
Bernheimer, A. W. (1974) *Biochim. Biophys. Acta* **344**, 27–50
Bhakdi, S. & Tranum-Jensen, J. (1978) *Proc. Natl. Acad. Sci. U.S.A.* **75**, 5655–5659
Bhakdi, S. & Tranum-Jensen, J. (1979) *Proc. Natl. Acad. Sci. U.S.A.* **76**, 5872–5876
Bhakdi, S. & Tranum-Jensen, J. (1981) *Proc. Natl. Acad. Sci. U.S.A.* **78**, 1818–1822
Bhakdi, S. & Tranum-Jensen, J. (1982) *J. Cell. Biol.* **94**, 755–759
Bhakdi, S. & Tranum-Jensen, J. (1983) *Biochim. Biophys. Acta* **737**, 343–372
Bhakdi, S. & Tranum-Jensen, J. (1984a) *Proc. R. Soc. London Ser B* **306**, 311–324
Bhakdi, S. & Tranum-Jensen, J. (1984b) *J. Immunol.* **133**, 1453–1463
Bhakdi, S., Ey, P. & Bhakdi-Lehnen, B. (1976) *Biochim. Biophys. Acta* **413**, 445–456
Bhakdi, S., Bjerrum, O. J., Bhakdi-Lehnen, B. & Tranum-Jensen, J. (1978) *J. Immunol.* **121**, 2526–2532
Bhakdi, S., Tranum-Jensen, J. & Klump, O. (1980) *J. Immunol.* **124**, 2451–2457
Bhakdi, S., Füssle, R. & Tranum-Jensen, J. (1981) *Proc. Natl. Acad. Sci. U.S.A.* **78**, 5475–5479
Bhakdi, S., Füssle, R., Utermann, G. & Tranum-Jensen, J. (1983a) *J. Biol. Chem.* **258**, 5899–5904
Bhakdi, S., Muhly, M. & Roth, M. (1983b) *Meth. Enzymol.* **93**, 409–420
Bhakdi, S., Muhly, M. & Füssle, R. (1984a) *Infect. Immun.* **46**, 318–323
Bhakdi, S., Roth, M., Sziegoleit, A. & Tranum-Jensen, J. (1984b) *Infect Immun.* **46**, 394–400
Bhakdi, S., Tranum-Jensen, J. & Sziegoleit, A. (1985). *Infect. Immun.* **47**, 52–60
Borsos, T., Dourmashkin, R. R. & Humphrey, J. H. (1964) *Nature (London)* **202**, 251–254
Boyle, M. D. P., Gee, A. P. & Borsos, T. (1981) *Clin. Immunol. Immunopathol.* **20**, 287–295
Buckelew, A. R. & Colaccico, G. (1971) *Biochim. Biophys. Acta* **233**, 7–16
Buckingham, L. & Duncan, J. L. (1983) *Biochim. Biophys. Acta* **729**, 115–122
Duncan, J. L. & Schlegel, R. (1975) *J. Cell Biol.* **67**, 160–173
Fearon, D. T. (1979) *Proc. Natl. Acad. Sci. U.S.A.* **76**, 5867–5871
Freer, J. H., Arbuthnott, J. P. & Bernheimer, A. W. (1968) *J. Bacteriol.* **95**, 1153–1168
Füssle, R., Bhakdi, S., Sziegoleit, A., Tranum-Jensen, J., Kranz, T. & Wellensiek, H. J. (1981) *J. Cell Biol.* **91**, 83–94
Giavedoni, E. B., Chow, Y. M. & Dalmasso, A. P. (1979) *J. Immunol.* **122**, 240–245
Gray, G. S. & Kehoe, M. (1984) *Infect. Immun.* **46**, 615–618
Hammer, C. H., Nicholson, A. & Mayer, M. M. (1975) *Proc. Natl. Acad. Sci. U.S.A.* **72**, 5076–5079
Henkart, P. A., Millard, P. J., Reynolds, C. W. & Henkart, M. P. (1984) *J. Exp. Med.* **160**, 75–93
Hu, V., Esser, A. F., Podack, E. R. & Wisnieski, B. J. (1981) *J. Immunol.* **127**, 380–386
Humphrey, J. H. & Dourmashkin, R. R. (1969) *Adv. Immunol.* **11**, 75–115
Kazatchkine, J. & Nydegger, U. E. (1982) *Prog. Allergy* **30**, 193–234
Mayer, M. M. (1972) *Proc. Natl. Acad. Sci. U.S.A.* **69**, 2954–2959
Mayer, M. M., Michaels, D. W., Ramm, L. E., Whitlow, M. B., Willoughby, J. B. & Shin, M. L. (1981) *Crit. Rev. Immunol.* **2**, 133–165
McCartney, C. & Arbuthnott, J. P. (1978) Mode of action of membrane-damaging toxins produced by staphylococci, in *Bacterial Toxins and Cell Membranes* (Jeljaszewicz, J. & Wadström, T., eds.), pp. 89–122, Academic Press, New York
Michaels, D. W., Abramovitz, A. S., Hammer, C. H. & Mayer, M. M. (1976). *Proc. Natl. Acad. Sci. U.S.A.* **73**, 2852–2856
Monahan, J. B. & Sodetz, J. M. (1981) *J. Biol. Chem.* **256**, 3258–3262
Morgan, B. P., Luzio, J. P. & Campbell, A. K. (1984) *Biochem. Biophys. Res. Comm.* **118**, 616–622
Podack, E. R. & Müller-Eberhard, H. J. (1981) *J. Biol. Chem.* **256**, 3145–3148
Podack, E. R. & Tschopp, J. (1984) *Mol. Immunol.* **21**, 589–603
Podack, E. R., Tschopp, J. & Müller-Eberhard, H. J. (1982) *J. Exp. Med.* **156**, 268–282
Ramm, L. E., Whitlow, M. B. & Mayer, M. M. (1982a) *Proc. Natl. Acad. Sci. U.S.A.* **79**, 4751–4755

Ramm, L. E., Whitlow, M. B. & Mayer, M. M. (1982b) *J. Immunol.* **123**, 1143–1146
Simone, C. B. & Henkart, P. (1980) *J. Immunol.* **124**, 954–963
Smyth, C. J. & Duncan, J. L. (1978) in *Bacterial Toxins and Cell Membranes* (Jeljaszewicz, J. & Wadström, T. eds.), pp. 129–183, Academic Press, New York
Steckel, E. W., Welbaum, B. E. & Sodetz, J. M. (1983) *J. Biol. Chem.* **258**, 4318–4324
Thelestam, M. & Möllby, R. (1975) *Infect. Immun.* **12**, 225–232
Thelestam, M., Jolivet-Reynaud, C. & Alouf, J. E. (1983) *Biochem. Biophys. Res. Commun.* **111**, 444–449
Tranum-Jensen, J. & Bhakdi, S. (1983) *J. Cell Biol.* **97**, 618–626
Tranum-Jensen, J., Bhakdi, S., Bhakdi-Lehnen, B., Bjerrum, O. J. & Speth, V. (1978) *Scand. J. Immunol.* **7**, 45–56
Tschopp, J. (1984) *J. Biol. Chem.* **259**, 7857–7863
Tschopp, J., Müller-Eberhard, H. J. & Podack, E. R. (1982a) *Nature (London)* **298**, 534–537
Tschopp, J., Podack, E. R. & Müller-Eberhard, H. J. (1982b) *Proc. Natl. Acad. Sci. U.S.A.* **79**, 7474–7478

Biochem. Soc. Symp. **50**, 235–246
Printed in Great Britain

Transmembrane Channel-Formation by Five Complement Proteins

HANS J. MÜLLER-EBERHARD

*Department of Immunology, Research Institute of Scripps Clinic, La Jolla, California 92037,
U.S.A.*

Synopsis

Five serum proteins act in concert to form the membrane attack complex (MAC) of complement. The precursor proteins, C5, C6, C7, C8 and C9, are hydrophilic glycoproteins with molecular weights ranging from 70000 to 180000. When C5 is cleaved by the serine protease C5 convertase, nascent C5b is produced which forms together with C6 a soluble and stable bimolecular complex (C5b,6). Upon binding of C5b,6 to C7 a trimolecular complex (C5b-7) is formed, which expresses a metastable membrane binding site. Membrane-bound C5b-7 constitutes the receptor for C8 and the tetramolecular C5b-8 complex binds and polymerizes C9. During the assembly process the proteins undergo hydrophilic–amphiphilic transition and the end product consists of C5b-8 (M_r approx. 550000) and of tubular poly C9 (M_r approx. 1100000). The functional channel size varies but its maximal diameter is approximately 10 nm. C9 polymerization appears to involve initial reversible associations of several C9 molecules, which leads to temperature dependent, constrained unfolding. Unfolded C9 monomers then associate laterally with each other and polymerization terminates with closure of the circular structure, which consists of 12–18 C9 monomers. Amino acid composition and sequence indicate that the *N*-terminal half of the single chain C9 molecule is hydrophilic and the *C*-terminal half rather hydrophobic. Phospholipid binding and insertion into membranes are functions of the *C*-terminal portion of the molecule.

Introduction

Membrane damage by complement occurs as a consequence of activation of either the classical or the alternative pathway on the surface of a cell. The channel-former is the membrane attack complex (MAC). It constitutes a supramolecular organization that is composed of approximately 20 protein molecules and has a molecular weight of about 1.7 million. The complex has five precursors, C5, C6, C7, C8 and C9, which are hydrophilic glycoproteins with molecular weights ranging from 70000 to 180000. When C5 is cleaved by C5 convertase and nascent C5b (5b*) is produced, self-assembly of the MAC ensues: C5b* and C6 form a soluble and stable bimolecular complex which binds to C7 and induces it to express a metastable site through which the nascent complex

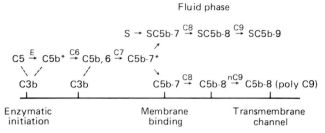

Fig. 1. *Schematic representation of assembly of the MAC and its control by S-protein*
Asterisks denote metastable forms of C5b and C5b-7 respectively.

(C5b-7*) can bind to membranes. Membrane-bound C5b-7 constitutes the receptor for C8 and the C5b-8 complex binds and polymerizes C9. During the assembly process the proteins undergo hydrophilic–amphiphilic transition. Thus, metastable Cb5-7 expresses hydrophobic phospholipid binding sites. The lipid binding capacity increases on addition of C8 and again on addition of C9. The fully assembled MAC contains one molecule each of C5b, C6, C7 and C8 and multiple molecules of C9. The end-product consists of the tetramolecular C5b-8 complex (M_r approx. 550 000) and of tubular poly C9 (M_r approx. 1 100 000). The available evidence suggests that poly C9 represents the major complement transmembrane channel and C5b-8 the unit that catalyses C9 polymerization (Fig. 1).

Structural and functional membrane perturbation begins before the large poly C9 channel is formed. In absence of C9, the C5b-8 complex tends to aggregate in the target membrane and these large protein aggregates produce membrane leakiness. Aggregation also occurs in the presence of C9 when the amount of C9 is insufficient for poly C9 formation. Whereas the poly C9 channel is not necessary for complement dependent lysis of erythrocytes, it may be essential for the killing of nucleated cells.

In the electron microscope the MAC projects the image of a hollow cylinder and, inserted into a target membrane, it evokes the image of a 10 nm wide circular membrane hole which is surrounded by a 5 nm wide rim. These characteristic and well-known images of the MAC and the membrane lesions produced by complement are by and large due to the tubular poly C9 contained in the MAC.

MAC formation in the fluid phase is controlled by the S-protein, a M_r 80 000 plasma glycoprotein which functions as MAC inhibitor. It binds to the metastable binding site of C5b-7* and thereby prevents membrane attack. SC5b-7 remains in the fluid phase as a hydrophilic complex. Whereas the S-protein within the complex does not block the attachment of C8 and C9, it limits the number of C9 molecules to two or three and inhibits C9 polymerization (Fig. 1).

Some of the properties of the proteins of the membrane attack pathway are listed in Table 1.

Table I. *Properties of the proteins of the membrane attack pathway*

N.A., Information not available.

Protein	M_r	Subunits	M_r carbohydrate	Sedimentation coefficient (S)	Serum concn. (μg/ml)
C5	191 000	α: 115 000 β: 75 000	5200	8.7	70
C5b	180 000	a′: 104 000 β: 75 000	3000	7.5	—
C6	120 000	Single chain	4000	6.0	64
C7	110 000	Single chain	6000	5.6	56
C8	151 000	α: 64 000 β: 64 000 γ: 22 000	N.A.	8.0	55
C9	71 000	Single chain	5500	4.7	59
S-protein	83 000	Single chain	N.A.	4.0	505

C5 Activating Enzymes

The classical C5 convertase, C4b,2a,3b, consists of the Mg^{2+}-dependent bimolecular complex C4b,2a (M_r approx. 280 000) and C3b (M_r approx. 176 000). C4b is a three-chain fragment of C4, and C2a a single-chain fragment of C2. C2a is a serine protease as evidenced by inactivation with DFP (Medicus *et al.*, 1976) and cDNA sequence analysis (Bentley & Porter, 1984). Viewed by electron microscopy, it consists of two globular domains, one measuring 4 nm in diameter and the other 4.7 nm, which are connected by a thin 1 nm linker segment (Smith *et al.*, 1984). It is bound to C4b through the larger of the two domains, suggesting that the freely projecting domain bears the catalytic site. C2a is non-covalently linked to C4b and dissociates spontaneously with concomitant loss of enzymatic activity, the complex having a half-life at 37 °C of approximately 3 min. C3b and C4b are covalently linked to cell surface proteins through ester or amide bonds. Covalent attachment of C3b occurs when the internal thioester (Tack, 1984) of metastable C3b (Müller-Eberhard *et al.*, 1966) comes under stress and its carbonyl group reacts with a hydroxyl or an amino group of cell surface constituents (Law & Levine, 1977). Similarly, a covalent bond is established between C4b and a reactive surface when the internal thioester of metastable C4b enters into a transacylation reaction.

The C5 convertase of the alternative pathway is C3b,Bb,C3b. The catalytic site bearing subunit Bb (M_r 60 000) is a serine protease (Medicus *et al.*, 1976) which is derived from Factor B, the structure of which has been elucidated by sequencing of the protein (Mole *et al.*, 1983) and of cDNA (Campbell & Porter, 1983). The Mg^{2+}-dependent enzyme undergoes rapid decay–dissociation at 37 °C, its half-life being 2 min. By using Ni^{2+} instead of Mg^{2+} for enzyme formation, the half-life at 37 °C was increased to 11 min and at 4 °C to 32 h (Fishelson & Müller-Eberhard, 1982). The stability of C3b,Bb(Ni) was sufficient to allow demonstration of the enzyme complex by molecular transport experiments (Fishelson *et al.*, 1983), and with the radioisotope [63]Ni it was found that a single metal ion is incorporated into the complex, probably into the Bb subunit, and that this metal ion is resistant to chelation by EDTA. C3b,Bb has been visualized by electron microscopy and found to be similar in structure to C4b,2a, except that the two Bb domains are of equal size (Smith *et al.*, 1984).

The Metastable Membrane Binding Site of C5b-7

The assembly of C5b-7 commences with proteolytic activation of C5 by C5 convertase. At the target cell surface, C5 binds to one of the C3b molecules surrounding the C3 convertase complex. Suitably presented and modulated by C3b, C5 is cleaved and metastable C5b (C5b*) is created, the labile binding site of which has specificity for C6. If it collides with a molecule of C6, the bimolecular C5b,6 complex is formed (Lachmann & Thompson, 1970). C5b,6 remains loosely bound to C3b awaiting reaction with C7.

Upon binding of C7 to C5b,6 on the target cell surface, the trimolecular complex undergoes a hydrophilic–amphiphilic transition, in the course of which the metastable membrane binding site is generated. C5b-7* leaves the C3b holding position and transfers to the surface of the membrane. It overcomes the charge barrier of the membrane by ionic interactions probably involving its C5b,6 portion and then anchors itself firmly in the lipid bilayer by hydrophobic interactions, primarily through its C7 subunit. The ability of C5b-7 to anchor itself in the membrane is due to the acquisition of high affinity phospholipid binding sites during the hydrophilic–amphiphilic transition. Approximately 400 mol of lecithin was found to be firmly bound/mol of C5b-7 protein when the complex was formed in presence of sodium deoxycholate (DOC)–phospholipid mixed micelles and the DOC was removed (Podack *et al.*, 1979). C5b-7 does not compromise membrane function, as evidenced by lack of leakiness of lipid vesicles or erythrocytes. The life-time of the transient binding site of C5b-7* has been estimated to be less than 10 ms (Podack *et al.*, 1978).

Bound to single bilayer phospholipid vesicles of 20–40 nm diameter, C5b-7 predominantly appeared as a V-shaped two-leaflet structure on electron microscopy, which extends approx. 21 nm above the vesicle surface and is associated with the membrane through an approx. 4 nm long stalk (Fig. 2) (Preissner *et al.*, 1985*a*). This structure is thought to be the dimer of C5b-7, because occasionally a single leaflet structure is visualized on the smallest vesicles, which is likely to be the monomer. When C5b-7* is formed in the fluid phase, self-aggregation occurs and its cytolytic activity is lost. Soluble protein micelles are formed which exhibit a sedimentation coefficient of approximately 36 S (Preissner *et al.*, 1985*a*). Polymerization of C5b-7* is prevented by deoxycholate (DOC) micelles which preserve the complex in monomeric form, DOC–C5b-7 having a sedimentation coefficient of approximately 16 S (Podack & Müller-Eberhard, 1978). In the electron microscope the aggregates are imaged as flower-like structures in which the C5b-7 monomers remain clearly distinguishable (Fig. 2) (Preissner *et al.*, 1985*a*). Three to six leaflet-like structures can be identified per aggregate, each approx. 20 nm long, 10 nm wide and terminating in a pedicle at the centre of the flower, indicating that the pedicellar regions are the sites of hydrophobic interactions.

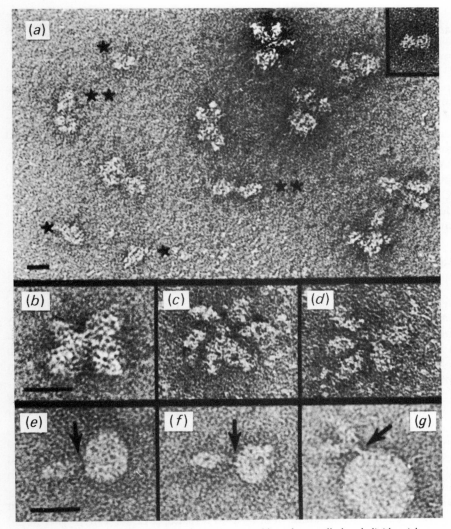

Fig. 2. *Electron microscopy of C5b-7 in solution and bound to small phospholipid vesicles*

(*a*) C5b-7 5 min after start of reaction between 5b,6 (inset) and C7 at 0 °C; *, monomeric, **, dimeric complexes. (*b–d*) C5b-7 protein micelles after completion of the reaction at 30 °C. (*e–g*) C5b-7 bound to phospholipid vesicles in monomeric (*e*) and dimeric (*f, g*) form via short stalk (arrows). Scale bars: 25 nm. (Preissner *et al.*, 1985 *a*.)

Recent studies (Preissner *et al.*, 1985 *b*) have provided strong evidence that it is primarily activated C7 that confers amphiphilicity on C5b-7*. It was found that isolated C7, but not C5 or C6, has the propensity to undergo hydrophilic–amphiphilic transition, to aggregate and to incorporate itself into lipid bilayers. Treatment of isolated C7 with 1 % DOC at 37 °C for 10 min resulted in complete loss of haemolytic activity and dimerization of the protein. The modified C7 has M_r 230000, which is twice that of native C7; it is more negatively charged than native C7 and it contains 18% more β-pleated sheet structure (Preissner et al., 1985b). The dimer possesses hydrophobic surface domains, as evidenced by its ability to bind approximately 82 mol of DOC/mol of dimer and by its prompt precipitation from solution on removal of the detergent.

Taken together, the available information suggests that, in the physiological pathway, C5b,6 acts on C7 much in the manner DOC does: it converts C7 to an activated, amphiphilic state in which it expresses polar and apolar binding sites which are concealed in the native molecule. The binding sites to be distinguished are these: native C7 has a binding site for C5b,6, C5b,6-bound and activated C7 has multiple phospholipid binding sites, which constitute the hydrophobic membrane binding site of C5b-7*, and it has a largely polar protein binding site by means of which C5b-7 can dimerize or possibly react with C8. C7 thus emerges as a highly versatile molecule in the MAC assembly and as the subunit that supplies the initial anchor of the forming complex in the target membrane.

Structure and Function of the Tetramolecular C5b-8 Complex

Native C8 has no detectable affinity for membranes and is therefore entirely dependent on the mediating function of C5b-7, which determines the site of membrane attack and serves as C8 receptor. Membrane-bound C5b-7 not only binds C8, but it also affords its insertion into the membrane. It thus has at least two distinct functions with respect to C8 and the question arises as to which of its three subunits are involved in ligand binding and in the subsequent processing of C8.

Before fusing into the tight MAC complex, the five precursor proteins already express some stereochemical affinity for each other. Reversible interactions occur between C5 and C6, C5 and C7, C5 and C8 and C8 and C9 (Kolb et al., 1973). Accordingly, the binding site for C8 on the receptor may be situated on its C5b subunit, whereas the recognition site on C8 for the receptor has been located to the β-chain of C8 (Monahan & Sodetz, 1980, 1981). This chain was shown to bind to C5b-7 in the absence of the α-γ subunit and with the same affinity as native C8. The α-γ subunit did not attach to C5b-7 without the mediation of the β-chain (Monahan & Sodetz, 1981; Tschopp et al., 1981). This initial contact probably brings a second site of the receptor into play, which enables C8 to undergo the necessary conformational rearrangement to allow the α-γ subunit to penetrate into the hydrophobic core of the lipid bilayer (Podack et al., 1981).

Electron microscopy of C5b-8 bound to small phospholipid vesicles visualized monomeric C5b-8 as a 25 nm long and 5 and 14 nm wide rod-like structure with a rather polymorphic appearance (Tschopp et al., 1982a). The C5b,6 subunit was detected at the end of the rod, distal to the lipid binding site, by using avidin-coated gold probes recognizing biotinyl C5b,6.

There are two distinct ways in which C5b-8 acts on C9, namely binding and polymerization; both reactions proceed irrespective of whether C5b-8 is membrane-bound or free in the fluid phase. Binding of C9 occurs via the C8 portion of C5b-8, as evidenced by the facts that C9 cannot bind to C5b-7, that binding to C5b-8 is inhibited by anti-C8 and that C8 and C9 interact in free solution (Kolb *et al.*, 1973). Soluble C8 binds only one molecule of C9 even when C9 is in considerable molar excess. This fact is in contrast to the ability of C5b-8 to bind multiple C9 molecules, which, of course, is due to C9 polymerization.

The penetration of C5b-8 into the membrane and its influence on the order of the bilayer were measured on planar lipid bilayers doped with spin-labels, by electron paramagnetic resonance spectroscopy (Esser *et al.*, 1979). These studies showed that whereas C5b-7 interacted strongly with the ionic part of the bilayer and penetrated only slightly into the hydrophobic region, C5b-8 penetrated into the hydrophobic phase more deeply. Insertion caused an increase in disorder of the membrane lipids, which means that the fatty acyl chains adopted a wider distribution of angles with respect to each other than they would in a normal bilayer structure. This reorientation of the ordered bilayer lipids into domains more micellar in nature is due to strong binding of phospholipid molecules to the inserted polypeptides of C5b-8. Direct lipid binding studies employing radiolabelled lecithin showed that approximately 800 mol of phospholipid can be bound per mol of C5b-8, which is about twice as much as can be bound by C5b-7 (Podack *et al.*, 1979). Studies designed to determine the parts of C5b-8 most intimately in contact with the interior of the membrane revealed the α-chain of C8 (Steckel *et al.*, 1983) to be primarily interacting with the hydrocarbon phase.

The evidence for the existence of a functional transmembrane channel produced by C5b-8 is the following: (a) sheep erythrocytes bearing C1-8 released haemoglobin on incubation at 37 °C (Tamura *et al.*, 1972). The reaction was slow and did not reach an endpoint within 8 h. Release of ^{86}Rb from these cells, however, was rapid and approached completion within 1 h (Gee *et al.*, 1980). (b) Treatment of resealed erythrocyte ghosts containing trapped [^{14}C]sucrose with C5b-8 resulted in slow sucrose release, which neared a plateau after 7–9 h (Ramm *et al.*, 1982*a*). (c) Liposomes containing entrapped glucose (Haxby *et al.*, 1969) or ^{86}Rb (Shin *et al.*, 1978) released marker when reacted with C1-8 or C5b-8. (d) A modest but definite increase in ion permeability of planar lecithin bilayers was observed after treatment of the membrane with C5b-8 (Michaels *et al.*, 1976).

The size of the C5b-8 channel was explored by performing kinetic sieving experiments with resealed erythrocyte ghosts, with sucrose and inulin as markers (Ramm *et al.*, 1982*a*). C5b-8 readily released sucrose, which has a molecular diameter of 0.9 nm, but not inulin, which has a diameter of 3 nm (Ramm *et al.*, 1982*b*). Inhibition of EAC1-8 lysis by sugars correlated well with their molecular weight (Kitamura & Kazuyoshi, 1981). Conductance changes across black lipid membranes suggested a channel diameter of 1.6 nm compared with 2.5 nm for C5b-9 (Michaels *et al.*, 1978). In the membrane of nucleated cells (guinea-pig hepatoma) the Cb5-8 complex caused rapid ^{86}Rb release, but did not cause cell death (Boyle *et al.*, 1976).

Recently, the size of the C5b-8 channel was explored by using the liposome swelling assay (Zalman & Müller-Eberhard, 1985). The radius of the pore formed by C5b-8 assembled on liposomes was estimated to be between 1.1 and 1.5 nm. Solute flux could be inhibited by certain monoclonal anti-C8 and slightly by one anti-C7, suggesting a channel composed of C8 with possible involvement of C7.

Structure and Function of the C5b-7 Complex

The MAC is heterogeneous with respect to size and composition. Heterogeneity is due to variation in C9 content and secondary aggregation of C5b-9. Nevertheless, the basic composition of the MAC is described by the formula $C5b_1, C6_1, C7_1, C8_1, C9_n$, where n can vary between 1 and 18. Hence, the minimum molecular weight of the MAC ranges between 660000 and 1850000.

There is now very little doubt that the electron microscopic image of the ring-like or cylinder-like membrane lesion caused by complement is evoked primarily by poly C9 (Podack et al., 1982). It is not known how C5b-8 causes C9 polymerization. C8 in solution has one C9 binding site and C8-C9 association is reversible ($K_a \simeq 10^7$ M^{-1}) (Podack et al., 1982), whereas cell-bound C8 can mediate the binding of many C9 molecules and that association is virtually irreversible ($K_a \simeq 10^{11}$ M^{-1}) (Podack et al., 1978). Binding of a molecule of C9 to C5b-8 might facilitate high affinity C9–C9 interaction and the energy derived from that interaction might effect constrained unfolding of monomeric C9 (long axis $\simeq 8$ nm) to the poly C9 subunit (long axis $\simeq 16$ nm), with exposure of hydrophobic regions that insert themselves into the membrane. Polymerization of the C9 oligomers continues until closure of the tubule occurs or the supply of monomeric C9 is exhausted. It is the C-terminal region of the unfolded C9 that inserts into the hydrocarbon phase of the membrane (Ishida et al., 1982).

The poly C9 tubule of the membrane-bound MAC extends 12 nm above the surface of the membrane; it has an inner diameter of 11 nm and it terminates at its upper, hydrophilic end in a 3 nm thick annulus which has an outer diameter of 21 nm (Tschopp et al., 1982b; Bhakdi & Tranum-Jensen, 1983). The C5b-8 subunit appears firmly attached to the poly C9 tubule and extends 16 to 18 nm above its annulus as a 5–14 nm wide elongated structure (Tschopp et al., 1982a). The total length of the structure is thus 28 to 30 nm above the surface of the membrane.

The structure of the MAC varies greatly with the supply of C9, i.e. with the molar C9/C5b-8 ratio and also with the topographical distribution of C5b-8 on the cell surface. When the MAC was assembled on vesicles or erythrocytes and the molar C9/C5b-8 ratio was 1 or 3, no poly C9 was detectable by SDS–polyacrylamide gel electrophoresis (Tschopp et al., 1985) and no complement membrane lesions were seen on electron microscopy (Podack et al., 1982). Instead, large protein aggregates were observed, which probably represented a C5b-9 network maintained by C9–C9 interactions. The poly C9 content of the MAC was found to be directly proportional to the molar C9/C5b-8 ratio. When this ratio was 6 or 12, the percentage of C9 present as poly C9 was 35 and 72, respectively (Tschopp et al., 1985). At these ratios discrete ring structures were observed on the cell surface by electron microscopy, which appeared unaggregated and well separated from each other (Podack et al., 1982). Examination of DOC-solubilized MAC for poly C9 content after molecular sieve chromato-

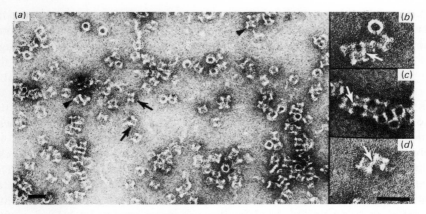

Fig. 3. *Ultrastructure of poly C9 made from isolated monomeric C9 by prolonged incubation at 37 °C*

Poly C9 is visualized as a stain-filled tubule with an inner diameter of 11 nm (top views are imaged as rings) and a length of 16 nm (side views, black arrows). The annulus at one end of the tubule is approx. 3 nm thick and has an outer diameter of 21 nm (black arrowhead). The other end of the tubules consists of a 4 nm long hydrophobic segment through which the tubules form side-by-side aggregates with overlapping areas (white arrows). Bars: 40 nm. (Tschopp *et al.* 1982*b*.)

graphy, in the presence of DOC, detected the greatest amount in the relatively low molecular weight unaggregated material (Tschopp *et al.*, 1985).

It thus appears that even within the lipid bilayer C5b-9 will either aggregate to larger clusters when the C9 multiplicity is low, or form poly C9-containing monomers when the C9 multiplicity is high. Obviously, SDS-resistant poly C9 cannot be regarded an obligatory constituent of the MAC and C5b-9 does not necessarily manifest itself as the typical ultrastructural membrane lesion. The functional significance of oligomeric C9 as subunit of the MAC requires investigation.

The MAC has a greater phospholipid binding capacity than its precursor complexes. It possesses high affinity binding sites for 1400 mol of lecithin/mol of complex (Podack *et al.*, 1979). It also binds other phospholipids but not cholesterol.

The size of the functional transmembrane pore produced by C5b-9 is highly variable and dependent on the experimental conditions. The smallest C5b-9 pore was found to have a diameter of 0.7 nm and the largest of more than 5.5 nm (Sims & Lauf, 1978; Boyle *et al.*, 1979; Ramm *et al.*, 1983). The size increases with C5b-9 density on a target membrane and with the C9/C5b-8 ratio (Boyle *et al.*, 1979). At very high complement dose the maximum diameter of the pore was 11 nm with human complement and 7 nm with guinea-pig complement.

Spontaneous Poly C9 Formation

Although C9 is a water soluble glycoprotein, it has the propensity to aggregate in isolated form and then to undergo a hydrophilic–amphiphilic transition, which results in cyclic polymerization. The outcome of this process is a tubular structure which has a hydrophilic annulus of 3 nm thickness at one end and which is lipophilic at the other end. The height of the tubule is 16 nm, the inner diameter 10 nm and the wall thickness approximately 2 nm (Fig. 3). Through the lipophilic end the forming tubule is capable of inserting itself into the membrane of phospholipid vesicles and to render them leaky. The hydrophobic portion of the poly C9 tubule corresponds to the C-terminal region of its subunits. By the liposome swelling assay, large pores could be detected when the liposomes were prepared with preformed poly C9. If preformed poly C9 was added to preformed liposomes, no functional pore was formed. However, if the poly C9 was allowed to assemble in the presence of preformed liposomes, the forming poly C9 inserted into the bilayer and caused channel-formation. The channels were of the same size as those seen with the MAC (Zalman & Müller-Eberhard, 1985). Twelve to 18 molecules of C9 are incorporated into tubular poly C9, which is also defined as an approximately 1.1 MDa complex that is resistant to dissociation by SDS and reducing agents. Poly C9 possesses more β-pleated sheet structure than native, monomeric C9 and it expresses antigenic determinants that are not detectable on its precursor (Podack et al., 1982; Tschopp et al., 1982b).

Because the dimensions of native C9 are 8×5.5 nm and because the length of the poly C9 cylinder is 16 nm, it is assumed that C9 during polymerization undergoes constrained unfolding. Unfolding is preceded by an initial reversible association of several C9 molecules, which can be demonstrated by ultra-centrifugation under appropriate conditions. Spontaneous poly C9 formation is strongly temperature dependent and does not occur at 15 °C. Examination of C9 at 15 °C by analytical ultracentrifugation revealed self-association that was metal ion and protein concentration dependent, ionic and reversible (Chiu et al., 1985). Since conditions promoting reversible C9–C9 interaction at 15 °C allowed poly C9 formation at 37 °C, reversible oligomerization appears a prerequisite for tubular poly C9 formation. Tubular poly C9 could be completely dissociated to monomeric C9 by treatment with succinic anhydride in the presence of 6 M-guanidine isothiocyanate for 1 h at 20 °C (Chiu et al., 1985).

The remarkable phenomenon of poly C9 formation raises mechanistic questions, especially whether enzymes are involved in spontaneous C9 poly-merization and how C5b-8 mediates it. Zinc was particularly effective in that it permitted at a concentration of 50 μM rapid polymerization even in the presence of 150 mM-NaCl (Chiu & Müller-Eberhard, 1984; Tschopp, 1984). The poly C9 rapidly generated under optimal conditions had the typical ultrastructural appearance. Gleaning from these observations what the specific function of C5b-8 might be, one might say that C5b-8 induces the initial association of several C9 molecules, probably two, under conditions (0.15 M-NaCl, 0.01 M-EDTA, pH 7.4) that would not allow this reaction to occur in its absence. C5b-8 reduces the activation energy of C9 polymerization and greatly enhances the rate of polymerization, thus acting as a catalyst. Finally, whereas poly-merizing C9 can attack lipid vesicles in the absence of C5b-8, it cannot lyse biological membranes unless C5b-8 is present on their surface.

It is of considerable interest that killer lymphocytes produce and insert into the membrane of target cells tubular structures that resemble poly C9. This was first reported by Dourmashkin *et al.* (1980), who examined cells after lysis by the antibody dependent cellular cytotoxicity reaction, and who found the human lymphocyte derived cylinders to have an inner diameter of 16 nm. Subsequently, similar tubular structures were found to be produced by cloned mouse natural killer cells and cytolytic T cells (Podack & Dennert, 1983). The precursor proteins which are contained in the cytoplasmic dense granules of these cells differ in size from C9.

Control of MAC-Formation

The S-protein of plasma (Kolb & Müller-Eberhard, 1975) competes with membrane lipids for the metastable binding site of C5b-7, and by binding to the complex it prevents its attachment to the cell surface. The inhibition constant, K_i, is 39 μg/ml, which is less than one-tenth of the plasma concentration of the protein (Podack & Müller-Eberhard, 1979). Its function appears to be to protect cells adjacent to the site of complement activation from accidental attack. The resultant SC5b-7 complex contains three molecules of S-protein (M_r 80000), has a sedimentation coefficient of 18.5 S and M_r 668000. The hydrophilic complex binds C8 and three molecules of C9 to form SC5b-8 and SC5b-9 (M_r 1030000) (Boyle *et al.*, 1976). All three complexes contain neoantigens that are not detectable in the precursor proteins and that are distinct from the neoantigens of poly C9. In addition to blocking the membrane binding site, S-protein also prevents polymerization of C9 (Podack *et al.*, 1984).

References

Bentley, D. R. & Porter, R. R. (1984) *Proc. Natl. Acad. Sci. U.S.A.* **81**, 1212–1215
Bhakdi, S. & Tranum-Jensen, J. (1983) *Biochim. Biophys. Acta* **737**, 343–372
Boyle, M. D. P., Ohanian, S. H. & Borsos, T. (1976) *J. Immunol.* **117**, 1346–1350
Boyle, M. D. P., Gee, A. P. & Borsos, T. (1979) *J. Immunol.* **123**, 77–82
Campbell, R. & Porter, R. R. (1983) *Proc. Natl. Acad. Sci. U.S.A.* **80**, 4464–4468
Chiu, F. J. & Müller-Eberhard, H. J. (1984) *Fed. Proc. Fed. Am. Soc. Exp. Biol.* **43**,1449
Chiu, F. J., Ziccardi, R. J. & Müller-Eberhard, H. J. (1985) *Fed. Proc. Fed. Am. Soc. Exp. Biol.* **44**, 1874
Dourmashkin, R. R., Deteix, P., Simone, C. B. & Henkart, P. (1980) *Clin. Exp. Immunol.* **42**, 554–560
Esser, A. F., Kolb, W. P., Podack, E. R. & Müller-Eberhard, H. J. (1979) *Proc. Natl. Acad. Sci. U.S.A.* **76**, 1410–1414
Fishelson, Z. & Müller-Eberhard, H. J. (1982) *J. Immunol.* **129**, 2603–2607
Fishelson, Z., Pangburn, M. K. & Müller-Eberhard, H. J. (1983) *J. Biol. Chem.* **258**, 7411–7415
Gee, A. P., Boyle, M. D. P. & Borsos, T. (1980) *J. Immunol.* **124**, 1905–1910
Haxby, J. A., Götze, O., Müller-Eberhard, H. J. & Kinsky, S. C. (1969) *Proc. Natl. Acad. Sci. U.S.A.* **64**, 290–295
Ishida, B., Wisnieski, B. J., Lavine, C. H. & Esser, A. F. (1982) *J. Biol. Chem.* **257**, 10551–10553
Kitamura, H. & Kazuyoshi, N. (1981) *Mol. Immunol.* **18**, 985–990
Kolb, W. P. & Müller-Eberhard, H. J. (1975) *J. Exp. Med.* **141**, 724–735
Kolb, W. P., Haxby, J. A., Arroyave, C. M. & Müller-Eberhard, H. J. (1973) *J. Exp. Med.* **138**, 428–437
Lachmann, P. J. & Thompson, R. A. (1970) *J. Exp. Med.* **131**, 643–657
Law, S. K. & Levine, R. P. (1977) *Proc. Natl. Acad. Sci. U.S.A.* **74**, 2701–2705
Medicus, R. G., Götze, O. & Müller-Eberhard, H. J. (1976) *Scand. J. Immunol.* **5**, 1049–1055
Michaels, D. W., Abramovitz, A. S., Hammer, C. H. & Mayer, M. M. (1976) *Proc. Natl. Acad. Sci. U.S.A.* **73**, 2852–2856
Michaels, D. W., Abramovitz, A. S., Hammer, C. H. & Mayer, M. M. (1978) *J. Immunol.* **120**, 1785

Mole, J. E., Woods, D., Colten, H. R. & Anderson, J. K. (1983) *Immunobiology* **164**, 279–280
Monahan, J. B. & Sodetz, J. M. (1980) *J. Biol. Chem.* **255**, 10579–10582
Monahan, J. B. & Sodetz, J. M. (1981) *J. Biol. Chem.* **256**, 3258–3263
Müller-Eberhard, H. J., Dalmasso, A. P. & Calcott, M. A. (1966) *J. Exp. Med.* **123**, 33–54
Podack, E. R. & Dennert, G. (1983) *Nature (London)* **302**, 442–445
Podack, E. R. & Müller-Eberhard, H. J. (1978) *J. Immunol.* **12.**, 1025–1030
Podack, E. R. & Müller-Eberhard, H. J. (1979) *J. Biol. Chem.* **254**, 9908–9914
Podack, E. R., Biesecker, G., Kolb, W. P. & Müller-Eberhard, H. J. (1978) *J. Immunol.* **121**, 484–490
Podack, E. R., Biesecker, G. & Müller-Eberhard, H. J. (1979) *Proc. Natl. Acad. Sci. U.S.A.* **76**, 897–901
Podack, E. R., Stoffel, W., Esser, A. F. & Müller-Eberhard, H. J. (1981) *Proc. Natl. Acad. Sci. U.S.A.* **78**, 4544–4548
Podack, E. R., Tschopp, J. & Müller-Eberhard, H. J. (1982) *J. Exp. Med.* **156**, 268–282
Podack, E. R., Preissner, K. T. & Müller-Eberhard, H. J. (1984) *Acta Pathol. Microbiol. Immunol. Scand., Sect. C*, **92**, *Suppl. 284*, 89–96
Preissner, K. T., Podack, E. R. & Müller-Eberhard, H. J. (1985*a*) *J. Immunol.*, in the press
Preissner, K. T., Podack, E. R. & Müller-Eberhard, H. J. (1985*b*) *J. Immunol.*, in the press
Ramm, L. E., Whitlow, M. B. & Mayer, M. M. (1982*a*) *J. Immunol.* **129**, 1143–1146
Ramm, L. E., Whitlow, M. B. & Mayer, M. M. (1982*b*) *Proc. Natl. Acad. Sci. U.S.A.* **79**, 4751–4755
Ramm, L. E., Whitlow, M. B. & Mayer, M. M. (1983) *Mol. Immunol.* **20**, 155–160
Shin, M. L., Paznekas, W. A. & Mayer, M. M. (1978) *J. Immunol.* **120**, 1996–2002
Sims, P. J. & Lauf, P. K. (1978) *Proc. Natl. Acad. Sci. U.S.A.* **75**, 5669–5673
Smith, C. A., Vogel, C.-W. & Müller-Eberhard, H. J. (1984) *J. Exp. Med.* **159**, 324–329
Steckel, E. W., Welbaum, B. E. & Sodetz, J. M. (1983) *J. Biol. Chem.* **258**, 4318–4324
Tack, B. F. (1984) *Springer Semin. Immunopathol.* **6**, 259–282
Tamura, N., Shimada, A. & Chang, S. (1972) *Immunology* **22**, 131–140
Tschopp, J. (1984) *J. Biol. Chem.* **259**, 10569–10573
Tschopp, J., Esser, A. F., Spira, T. J. & Müller-Eberhard, H. J. (1981) *J. Exp. Med.* **154**, 1599–1607
Tchopp, J., Podack, E. R. & Müller-Eberhard, H. J. (1982*a*) *Proc. Natl. Acad. Sci. U.S.A.* **79**, 7474–7478
Tschopp, J., Müller-Eberhard, H. J. & Podack, E. R. (1982*b*) *Nature (London)* **298**, 534–538
Tschopp, J., Podack, E. R. & Müller-Eberhard, H. J. (1985) *J. Immunol.* **134**, 495–499
Zalman, L. S. & Müller-Eberhard, H. J. (1985) *Fed. Proc. Fed. Am. Soc. Exp. Biol.* **44**, 551

Biochem. Soc. Symp. **50**, 247–264
Printed in Great Britain

Cell Damage by Viruses, Toxins and Complement: Common Features of Pore-Formation and its Inhibition by Ca^{2+}

C. A. PASTERNAK, G. M. ALDER, C. L. BASHFORD, C. D. BUCKLEY,
K. J. MICKLEM* and K. PATEL

*Department of Biochemistry, St George's Hospital Medical School, London SW17 0RE and
Department of Biochemistry, University of Oxford, Oxford OX1 3QU

Synopsis

Haemolytic paramyxoviruses interact with cells in the following way: a potentially leaky viral envelope fuses with the plasma membrane, creating a hydrophilic pore of approximately 1 nm in diameter; this allows ions and low molecular weight compounds, but not proteins, to leak into and out of cells. Other viruses act similarly if the pH is reduced to 5. Leakage (measured by collapse of membrane potential, by movement of monovalent cations and by loss of phosphorylated intermediates from cells) is prevented by extracellular Ca^{2+}. Ca^{2+} does not affect binding or fusion of virus to cells. It inhibits leakage as well as preventing it, and it aids in the recovery (i.e. the restoration of non-leakiness) of cells. Certain 'anti-Ca^{2+}' drugs have an opposite effect. Experiments with the bee venom protein melittin, with the α-toxin of *Staphylococcus aureus* and with activated complement, show that the lesions produced by these agents, too, are sensitive to extracellular Ca^{2+} and to 'anti-Ca^{2+}' drugs. The mechanisms of these effects are discussed.

Introduction

Many cytotoxic agents act by introducing hydrophilic pores into the plasma membrane of susceptible cells. The ensuing cell damage depends on the size of the pore: pores with an effective diameter around 0.9 nm will leak ions such as Na$^+$, K$^+$, Ca^{2+} and Cl$^-$ and molecules such as glucose, sucrose and amino acids; pores with an effective diameter around 1–1.5 nm will, in addition, leak phosphorylated intermediates of metabolism such as nucleotides, sugar phosphates and phosphorylcholine; pores larger than this may leak cytoplasmic proteins such as lactate dehydrogenase.

The first action of many pore-forming agents is to bind to a receptor at the plasma membrane. Susceptibility of different cell types to such an agent is obviously dependent on the presence of the receptor, and cells bearing many receptors are likely to suffer more damage than cells bearing only a few receptors. Human red cells possess receptors for several cytotoxic agents and since pore-formation in such cells generally leads to haemolysis [because red cells are particularly sensitive to ionically-induced osmotic swelling (Knutton et al., 1976)], haemolysis, which is easy to detect by haemoglobin leakage, has proved to be a simple and useful assay for the pore-forming ability of certain viruses, toxins and complement. In a recent study we showed that haemolysis of sheep or human red cells by Sendai virus, influenza virus at pH 5.3, the α-toxin of *Staphylococcus aureus* or the later components of activated complement is sensitive to inhibition by Ca^{2+} (Bashford *et al.*, 1984). Although the concentration of Ca^{2+} necessary to prevent haemolysis was outside the physiological range (> 10 mM-Ca^{2+} required to inhibit haemolysis caused by each reagent by 50%), Ca^{2+} was more effective than Mg^{2+} in every case except for α-toxin, against which Mg^{2+} was as effective as Ca^{2+}. When the ability of Ca^{2+} and Mg^{2+} to prevent leakage induced by these agents from Daudi cells was examined, it was again found that Ca^{2+} was more effective than Mg^{2+}. With these cells, however, Ca^{2+} was active at lower concentrations, suggesting that in this instance extracellular Ca^{2+} might play a role in regulating pore-formation and the ensuing cell damage *in vivo* (Bashford *et al.*, 1984). In any case, the amount of Ca^{2+} required to prevent leakage depends on the amount of pore-forming agent present (e.g. Micklem *et al.*, 1985).

Another type of cell in which pore-formation, at least by Sendai virus, is known to be sensitive to 1 mM (or lower) concentration of Ca^{2+} (Impraim *et al.*, 1979) is the Lettré cell (a type of Ehrlich ascites tumour cell; Lettré *et al.*, 1972). This cell type is easy to grow by intraperitonial passage in mice, with a yield of $> 10^9$ cells/mouse; the resulting ascites fluid is virtually devoid of polymorphs, macrophages or other contaminating cells (Mehta *et al.*, 1985). Virally-induced permeability changes in Lettré cells have been extensively studied. Although the precise mechanism of pore-formation is not yet clear, the events leading up to it are, and have been recently reviewed (Pasternak, 1984a, b).

Fusion of a potentially leaky viral envelope with the plasma membrane follows binding to the cell surface. Fusion is triggered by a glycoprotein which is active at pH 7 in the case of paramyxoviruses, but which requires pH < 6 (in order to protonate certain amino acid residues and to cause a conformational change; Sato *et al.*, 1983; Skehel *et al.*, 1985) in the case of other enveloped viruses. As fusion, which is the most temperature-sensitive step in the induction of pores (Micklem *et al.*, 1985), proceeds, a critical or threshold amount of damage has to occur before leakage commences (Patel & Pasternak, 1985): that is, leakage shows a characteristic lag, whereas fusion does not (Micklem *et al.*, 1985). Once threshold has been attained, leakage commences in a pH-independent manner (Patel & Pasternak, 1985). At the same time as leakage is being induced, repair of the lesion takes place (Micklem *et al.*, 1985). Thus the outcome of a virus–cell interaction is a balance between induction of pores and their repair.

Ca^{2+} appears to prevent leakage for a number of reasons, which may have a common cause. Although the concentration of Ca^{2+} required to prevent leakage depends on the virus/cell ratio, Ca^{2+} does not prevent binding of virus to cells (Wyke *et al.*, 1980), nor does it affect fusion between viral envelope and cells (Micklem *et al.*, 1984a). Ca^{2+} does, however, lengthen the lag to onset of leakage and it decreases the extent of eventual leakage: that is, it raises threshold (Micklem *et al.*, 1984a); recovery is promoted. To a certain extent, Ca^{2+} is also

Table 1. *Pore-forming agents used in this study*

References: [a]Knutton (1978), Pasternak (1984a, b); [b]Wyke et al. (1980); [c]Huang et al. (1981), Lenard & Miller (1981); [d]Patel & Pasternak (1983, 1985); [e]Dawson et al. (1978), Tosteson & Tosteson (1981), Dufton et al. (1984); [f]Bhakdi & Tranum-Jensen (1984); [g]Füssle et al. (1981); [h]Tschopp et al. (1982), Bhakdi & Tranum-Jensen (1984); [i]Ramm & Mayer (1980); [j]Humphrey & Dourmashkin (1969), Tschopp et al. (1982); [k]Dufton et al. (1984).

Agent	Nature	Suggested mechanism of pore-formation	Effective pore size (diameter)
Sendai virus	Glycoproteins embedded in membrane surrounding ~ 100 nm diam. particle	Insertion of potentially leaky membrane by fusion with cell plasma membrane[a]	~ 1 nm[b]
Influenza virus at pH 5.5	Similar to above	As above[e]	? ~ 1 nm[d]
Bee venom toxin melittin	M_r 3000 cationic protein	Cross-linking of intrinsic plasma membrane protein and/or insertion of melittin (? tetramer) into membrane[e]	?
Staphylococcus aureus α-toxin	M_r 34000 protein	Generation of hydrophobic regions by hexamer formation followed by partial insertion into membrane[f]	? 2–3 nm[g]
Activated complement (C5b-9n)	$M_r > 10^6$ protein aggregate	Deposition of C5b-8 aggregate on plasma membrane and generation of ~ 0.85 nm pore, followed by polymerization of C9 (generating hydrophobic regions) and sequential addition to C5b-8 aggregates to increase pore size[h]	? 5–7 nm[i] to 10–11 nm[j]
Polylysine	M_r 4000–15000 cationic protein	Cross-linking of intrinsic plasma membrane proteins[k]	?

able to 'close' existing pores, and EGTA is able to 're-open' them (Impraim *et al.*, 1980; Patel & Pasternak, 1985). Drugs that block Ca^{2+} channels in excitable cells, such as verapamil, D600 or prenylamine, and drugs that inhibit calmodulin-requiring reactions, such as trifluoperazine and calmidazolium (R24571), have an action opposite to that of Ca^{2+} in that lag to onset of leakage is shortened and extent of leakage is increased; at concentrations of Ca^{2+} at which virally-induced leakage is entirely prevented, these drugs induce leakage from virally treated cells. They are particularly effective at lower temperatures, although they do not appear to affect membrane fusion (Micklem *et al.*, 1984*b*). Although these data might suggest an involvement of calmodulin in virally-induced permeability changes, such a conclusion must await more direct evidence.

With these data on virally-mediated permeability changes to hand, we decided to measure the effect of other cytotoxic agents on Lettré cells, in order to determine (a) what the common features of pore-formation are, and (b) whether Ca^{2+} blocks leakage in a similar manner. The cytotoxic agents studied are listed in Table 1, together with some of their properties. The data to be presented show that despite obvious differences in the mechanisms by which induction of pores is achieved, the actions of these agents do have several features in common. Moreover in each case Ca^{2+} at physiological concentration is able to prevent leakage; the mechanism by which this occurs has been explored: as with virally-induced leakage, Ca^{2+} appears to act at a late stage of pore-formation and is able partially to close existing pores.

Methodology

Lettré cells, which were grown intraperitoneally in Swiss white mice as previously described (Impraim *et al.*, 1980), were used for most experiments. For some experiments with complement, human tonsil lymphocytes were used.

Experiments were carried out in Hepes-buffered saline [150 mM-NaCl, 5 mM-KCl, 5 mM-Hepes, 1 mM-$MgCl_2$, pH 7.4 (NaOH):HBS] or in Hepes-buffered mannitol [300 mM-mannitol, 10 mM-Hepes, 1 mM-$MgSO_4$, pH 7.4 (NaOH)], unless stated otherwise. Because cells were routinely diluted into medium from concentrated suspensions in Eagle's MEM medium with 3–10% fetal calf serum (Bashford *et al.*, 1985*a*), which contains approx. 2 mM-$CaCl_2$, the final Ca^{2+} concentration in incubation media to which no $CaCl_2$ was added was approx. 50 μM. Likewise the final monovalent ion concentration in mannitol medium was approx. 8 mM-Na^+ and approx. 100 μM-K^+ (total ionic strength approx. 0.01).

Sendai virus (Impraim *et al.*, 1980) and influenza virus (Patel & Pasternak, 1983) were grown as previously described. Purified phospholipase-free melittin was a gift from Dr. R. C. Hider, University of Essex; it was dissolved and kept in dimethylsulphoxide, since aqueous solutions were found to lose activity, whether stored frozen or not, *S. aureus* α-toxin, isolated from strain Wood 46 (NCTC 7121) by the method of McNiven *et al.* (1972) and kept as a suspension in ammonium sulphate at $-20\,^{\circ}$C, was donated by Dr. Joyce de Azavedo (Moyne Institute, Trinity College, Dublin). Polylysine (type VI, M_r 4000–15000;

Sigma, Poole, U.K.) was dissolved in HBS and kept at −20 °C. The antiserum used with Lettré cells was a rabbit anti-mouse brain/anti-thymocyte preparation donated by Dr. L. Hudson, St. George's Hospital Medical School; the antiserum used with tonsil lymphocytes was a rabbit anti-human lymphocyte preparation donated by Dr. A. P. Johnstone, St. George's Hospital Medical School.

Human serum (pre-adsorbed with Lettré cells) or guinea-pig serum were used as complement source; heating at 56 °C for 30 min abolished pore-forming activity (see Fig. 8). All cytotoxic agents were diluted in HBS before use; experiments in the absence of cytotoxic agent showed that residual dimethylsulphide or ammonium sulphate caused no leakage. Experiments with complement were carried out under conditions in which antiserum alone, or complement alone, caused little leakage.

Membrane potential was measured with the anionic probe oxonol-V as previously described (Bashford & Pasternak, 1984); the use of this probe to measure membrane potential has been critically assessed (Bashford *et al.*, 1985 *c*). Because Lettré cells depolarize at pH 5.5 (Bashford & Pasternak, 1984), membrane potential measurement during pore-formation by influenza virus could not be carried out. Nor was it possible to measure the effect of complement on membrane potential, since oxonol-V binds to serum proteins. Leakage of monovalent cations was assessed by measuring intracellular K^+ and Na^+ after spinning cells through oil (Bashford *et al.*, 1983); dilution in Hepes-buffered isotonic choline before spinning through oil was omitted, since extracellular choline partially exchanges with intracellular Na^+ (Bashford *et al.*, 1985 *a*). Leakage of phosphorylated metabolic intermediates was assessed by preincubating cells with [³H]choline and washing HBS, which effectively labels intracellular phosphorylcholine, and measuring the subsequent distribution of ³H between incubation medium and cells (Pasternak & Micklem, 1973; Impraim *et al.*, 1980). Leakage of lactate dehydrogenase was assessed similarly, by measuring the distribution of lactate dehydrogenase (Varley, 1969) between medium and cells.

Common Features of Action of Pore-Forming Agents

Sequential onset of leakage

A characteristic feature of virally-mediated leakage is the sequential manner in which changes occur: membrane potential is affected first, followed by changes in intracellular cation content, followed by leakage of phosphorylated metabolites, followed (at high virus/cell ratios) by leakage of soluble proteins (Bashford *et al.*, 1983; Bashford *et al.*, 1985 *a*). Although this is partly a reflection of the sensitivity of the respective measurements (very much fewer ions need to move to affect membrane potential than can be detected by chemical means), it is a real effect in that it is possible to observe ionic changes without metabolite leakage, or metabolite leakage without protein leakage (Poste & Pasternak, 1978; Pasternak, 1984 *a*). As shown in Fig. 1, a similar sequentiality of action occurs when α-toxin, melittin, complement or polylysine interact with Lettré cells. Note that polylysine is non-haemolytic in human cells at 20 times the concentration at which it causes leakage from Lettré cells (results not shown);

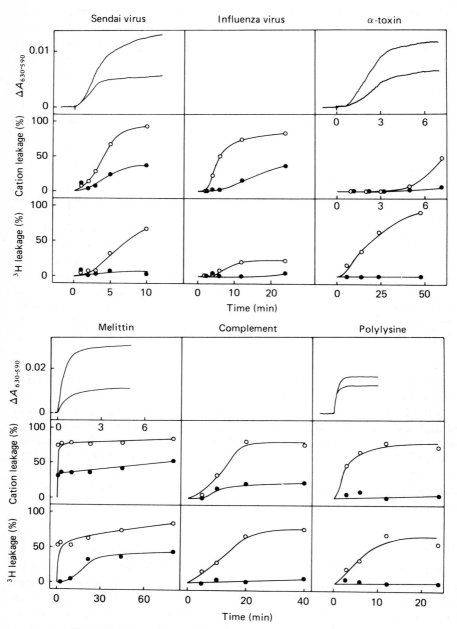

Fig. 1. *Membrane changes induced in Lettré cells by pore-forming agents*

Membrane potential is shown as $A_{630-590}$ (an increase of which indicates depolarization). Leakage of monovalent cations (calculated as $Na^+/Na^+ + K^+$ in cells) and [³H]phosphorylcholine are expressed as follows: 0% leakage is that before addition of agent and 100% leakage is that that would occur if cells were completely permeabilized. The lower curve in each case refers to changes induced in the presence of Ca^{2+} at the concentrations indicated below. Lettré cells ($2-5 \times 10^6$/ml) in HBS at 37 °C were exposed to agent as follows. Sendai virus: 14 haemagglutinating units (HAU)/ml; Ca^{2+} at 3 mM. Influenza virus: 2500 HAU/ml; Ca^{2+} at 1.5 mM; pH 5.5 buffer of Patel & Pasternak (1983). α-toxin: 2 μg/ml; Ca^{2+} at 10 mM. Melittin: 1 μM; Ca^{2+} at 2.5 mM. Complement: 1% human serum in the presence of 1.5% rabbit antiserum; Ca^{2+} at 2 mM. Polylysine: 20 μg/ml; Ca^{2+} at 5 mM. Reproduced from Pasternak *et al.* (1985) with permission.

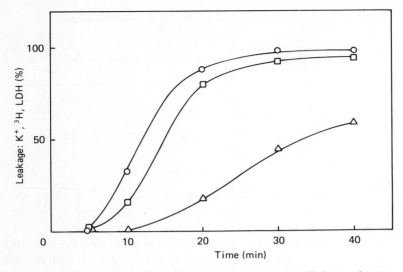

Fig. 2. *Sequential onset of leakage induced in Lettré cells by complement*

Lettré cells (4×10^6/ml) in HBS were incubated with complement (2% human serum) in the presence of rabbit antiserum (1.5%) at 37 °C and leakage of K^+ (○), [^{3}H]phosphorylcholine (□) and lactate dehydrogenase (LDH, △) was measured as described under Methodology. 0% leakage is taken as that from untreated cells; 100% leakage is taken as that that would occur if cells were completely permeabilized.

it does, however, share some properties (Dufton *et al.*, 1984) with melittin and certain basic peptides known to cause secretion from mast cells.

What is clear from Fig. 1 is the fact that in the presence of Ca^{2+}, leakage of phosphorylcholine is inhibited more than leakage of monovalent cations, which is inhibited more than membrane depolarization. In the case of complement-treated Lettré cells cytoplasmic protein begins to leak appreciably some time after the onset of leakage of phosphorylcholine, which follow that of monovalent cations (Fig. 2). As expected, Ca^{2+} has a greater effect on leakage of lactate dehydrogenase than on leakage of phosphorylcholine (results not shown).

Stimulation of leakage from cells in low ionic media by salt

We have recently observed that if Lettré cells are suspended in Hepes-buffered mannitol medium instead of in Hepes-buffered NaCl (HBS), Sendai virus-induced permeability changes are decreased; the rate of leakage can then be accelerated, in a Ca^{2+}-sensitive manner, by the addition of NaCl or KCl (Pasternak *et al.*, 1985). Although, in contrast to virus, strongly cationic pore-formers such as melittin (Pasternak *et al.*, 1985) or polylysine, and weakly cationic pore-formers such as α-toxin (pI approx. 8.5; Bhakdi & Tranum-Jensen, 1984), show a greater extent of leakage in mannitol medium than in HBS, especially at lower temperatures (Table 2), leakage in mannitol medium can nevertheless be accelerated by the addition of salt (Fig. 3); as might be expected, the effect is greater with Sendai virus than with the other agents (melittin > α-

Table 2. *Effect of low ionic strength on leakage induced by pore-formers in Lettré cells*

Lettré cells (5×10^6/ml) in HBS or in mannitol (low ionic strength) medium at various temperatures were exposed to pore-forming agent as follows: Sendai virus, 26 HAU/ml; melittin, 0.5 μM; α-toxin, 2 μg/ml; polylysine, 10 μg/ml. The extent of [^{3}H]phosphorylcholine leakage relative to that in HBS at 37 °C, taken as 100, is shown.

Pore-forming agent	Incubation temp.	HBS				Mannitol			
		37 °C	26 °C	16 °C	4 °C	37 °C	26 °C	16 °C	4 °C
Sendai virus		100	25	0	0	10	0	0	0
Melittin		100	62	45	88	102	117	121	118
S. aureus α-toxin		100	100	50	10	145	140	120	20
Polylysine		100	40	25	10	145	125	100	40

toxin > polylysine, which shows no stimulation). It is likely that the increased sensitivity of Lettré cells to cationic proteins in mannitol medium than in HBS reflects an increased binding of cationic protein to cells, whereas the sensitivity to subsequent addition of salt reflects an effect on the induction or function of pores. This is compatible with the observation that binding, at least of Sendai virus (Wyke *et al.*, 1980) or melittin (Pasternak *et al.*, 1985), is unaffected by Ca^{2+}, whereas the function of pores is sensitive to Ca^{2+} (Micklem *et al.*, 1984*a*, 1985; Patel & Pasternak, 1985; Figs. 8 and 9 below).

The difference between virus and the other agents in regard to the mechanism by which pores are induced is reflected in the temperature-sensitivity of leakage, regardless of the nature of the suspending medium (Table 2). Pore-formation by Sendai virus is dependent on membrane fusion (which has a Q_{10} of 4–7 between 37 and 7 °C; Micklem *et al.*, 1985) and therefore involves lipid–lipid interactions, whereas pore-formation by the other agents presumably involves only protein–lipid interactions.

Effect of Ca^{2+} and other divalent cations on leakage

Ca^{2+} and Mn^{2+} prevent virally-induced leakage at concentrations 10–100-fold lower than Mg^{2+} (Impraim *et al.*, 1979). A greater susceptibility to inhibition by Ca^{2+} compared with that by Mg^{2+} is seen with other pore-forming agents also. Fig. 4 illustrates this in the case of membrane depolarization of Lettré cells by melittin, and Figs. 2 and 3 of Bashford *et al.* (1984) show it for membrane depolarization, ion leakage and [^{3}H]phosphorylcholine leakage from Daudi cells by α-toxin and complement. Note, however, that not all agents that promote cation leakage are sensitive to Ca^{2+}: gramicidin, for example, which is a well-characterized (Harris & Pressman, 1967) Na^+/K^+ ionophore, is unaffected by Ca^{2+} at concentrations as high as 50 mM, nor does it cause leakage of phosphorylated metabolites (Pasternak *et al.*, 1985).

We recently came across the report by Boyle *et al.* (1979), in which a sequence of events for the induction of complement-mediated lysis of erythrocytes is proposed. Because that sequence is similar to the one which we propose for the

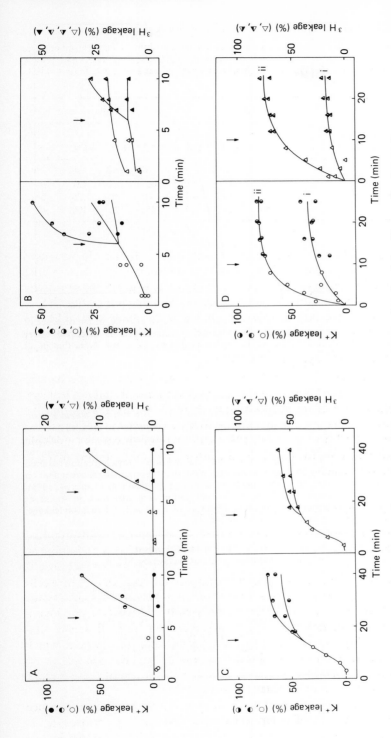

Fig. 3. *Effect of salt on leakage from Lettré cells in low ionic strength medium induced by pore-forming agents*

Lettré cells ($4-5 \times 10^6$/ml) were incubated in mannitol (low ionic strength) medium with pore-forming agent at 36 °C (α-toxin, 1 μg/ml, panel C), 35 °C (Sendai virus, 14 HAU/ml, panel A; melittin, 0.4 μM, panel B) or 22 °C [polylysine, 3 μg/ml (i) or 10 μg/ml (ii), panel D]; at the time indicated by the arrow, salt with or without Ca^{2+} was added as follows. Panel A: 20 mM-Na_2SO_4 (half-filled symbols); 20 mM-Na_2SO_4 plus 3.6 mM-$CaCl_2$ (filled symbols). Panel B: 20 mM-Na_2SO_4 (○, ▲); 80 mM-sucrose (●, ▲); 20 mM-Na_2SO_4 plus 3.6 mM-$CaCl_2$ (●, ▲). Panel C: 40 mM-NaCl (○, ▲); 80 mM-sucrose (●, ▲). Panel D: 40 mM-NaCl (●, ▲); 80 mM-sucrose (●, ▲). K$^+$ and [^{3}H]phosphorycholine leakage was measured as described under Methodology and expressed as described in the legend to Fig. 2.

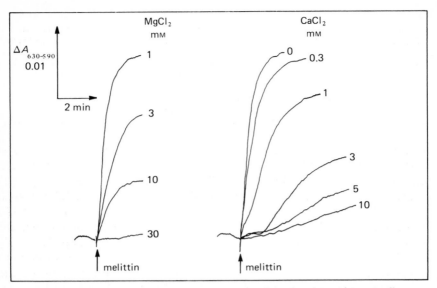

Fig. 4. *Effect of Mg^{2+} and Ca^{2+} on melittin-induced depolarization of Lettré cells*

Lettré cells (4×10^6/ml) were incubated in HBS with oxonol-V (2 mM) at 30 °C in the presence of $MgCl_2$ and $CaCl_2$ as indicated, and membrane potential was measured as described under Methodology. Melittin (0.2 μM) was added at the time indicated by the arrow. Note that in all experiments with $CaCl_2$, 1 mM-$MgCl_2$ was also present.

Table 3. *Inhibition of leakage by divalent cations*

Leakage from Lettré cells induced by Sendai virus, influenza virus (pH 5.5), melittin, *S. aureus* α-toxin or activated complement was assessed by measurement of membrane potential, intracellular Na^+ and K^+, and distribution of [^{3}H]phosphorylcholine between cells and medium. A number of different experiments were performed, in the presence of various amounts of Zn^{2+}, Ca^{2+} or Mg^{2+}. From the pooled results, illustrated in Fig. 9 of Pasternak *et al.* (1985), the amounts of divalent cation required to inhibit leakage by approximately 50% were calculated. Note that these values are dependent on cell: agonist ratio, temperature and other experiment-to-experiment variables (Micklem *et al.*, 1985); moreover they represent the mean of a range of divalent cation concentrations, in so far as [^{3}H]phosphorylcholine leakage is more sensitive to inhibition by divalent cation than is ion leakage, which is more sensitive than membrane potential (see Fig. 1).

Pore-forming agent	Concn. of divalent cation required to inhibit leakage by 50%		
	Zn^{2+}	Ca^{2+}	Mg^{2+}
Sendai virus	9.5×10^{-6} M	7.9×10^{-4} M	3.5×10^{-2} M
Influenza virus	$< 5.0 \times 10^{-5}$ M	3.6×10^{-4} M	2.9×10^{-3} M
Melittin	1.5×10^{-4} M	1.3×10^{-3} M	6.8×10^{-3} M
S. aureus α-toxin	3.2×10^{-4} M	7.6×10^{-3} M	7.6×10^{-3} M
Complement	6.0×10^{-5} M	7.4×10^{-4} M	5.2×10^{-3} M

Table 4. *Concentration of divalent cations in human serum*

The values quoted refer to the approximate mean concentration of ionized (non-bound) cation in healthy adults, with an indication of the range in this value in healthy donors. (From Diem & Lentner, 1970.)

	Zn^{2+}	Ca^{2+}	Mg^{2+}
Concentration	10^{-5} M	1.5×10^{-3} M	10^{-3} M
Range	$\pm 40\%$	$\pm 6-7\%$	$\pm 20-30\%$

induction of virally-mediated permeability changes [cf. Fig. 4 of Boyle *et al.* (1979) with Scheme 1 of Micklem & Pasternak (1977), or with Fig. 10 of Impraim *et al.* (1980)], and because Boyle *et al.* (1979) show that Zn^{2+} blocks haemolysis at a relatively early stage, we examined the effect of Zn^{2+} on permeability changes induced in Lettré cells by cytotoxic agents. The results (Bashford *et al.*, 1985*b*; Pasternak *et al.*, 1985) show that in every case Zn^{2+} is a more potent inhibitor of leakage than Ca^{2+}, and Ca^{2+} is a more potent inhibitor of leakage than Mg^{2+}. These data are expressed in terms of the concentrations required to inhibit a given amount of leakage by approximately 50% in Table 3. When these values are compared with the concentrations of Zn^{2+}, Ca^{2+} and Mg^{2+} found in adult human serum (Table 4), it is seen that the concentrations of Mg^{2+} required to inhibit leakage are well above that found in serum, and the same is true of Zn^{2+} (except for viruses). The concentrations of Ca^{2+} required to inhibit leakage, on the other hand, are near, or below, that of serum. Since the efficacy of divalent cations to inhibit leakage varies not only with agonist/cell ratio, but also from one cell type to another (Bashford *et al.*, 1984; Micklem *et al.*, 1984*a*), the results indicate that the extent of pathophysiological damage of pore-forming agents on particular cells may be influenced by the extracellular concentration of Ca^{2+}.

Nature of Inhibition by Divalent Cations

Pores through which monovalent cations and phosphorylated metabolites are able to leak are clearly likely to leak divalent cations also, and this has been shown to be so for virally- (Getz *et al.*, 1979; Impraim *et al.*, 1979) and complement- (Campbell *et al.*, 1981) induced pores. It is also true of pores induced by melittin, α-toxin and polylysine (see Fig. 6 below). In each case, as the extracellular Ca^{2+} concentration is raised it begins to inhibit its own entry. It might be thought, therefore, that Ca^{2+} inhibits leakage from the inside, although in other situations an increase in intracellular Ca^{2+} tends to make cells more, not less, leaky (e.g. Meech, 1976). In order to establish whether Ca^{2+} is acting extracellularly or intracellularly, the effect of A23187 on the ability of extracellularly added Ca^{2+} to inhibit leakage was assessed. Fig. 5 shows that in no case does A23187 potentiate the inhibitory action of Ca^{2+}; this is particularly clear for α-toxin (panel D). On the contrary, A23187 tends to make cells more, not less, leaky (cation leakage with Sendai virus, panel A, and melittin, panel C). Note that in each case the onset of leakage is sequential, in that cation leakage (solid line) precedes phosphorylcholine leakage (broken line). That A23187 is indeed affecting intracellular Ca^{2+}, by increasing its entry from

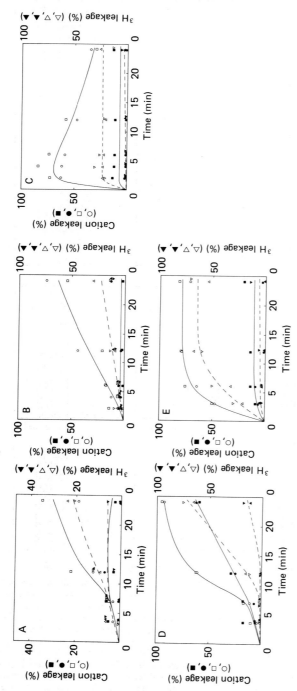

Fig. 5. *Effect of Ca²⁺ and A23187 on membrane changes induced in Lettré cells by pore-forming agents*

Lettré cells (5×10^6/ml) in HBS (except for influenza virus experiments) at 36 °C were incubated in the absence (open symbols) or presence (closed symbols) of 5 mm-CaCl₂, with (□, ■, ▽, ▼) or without (○, ●, △, ▲) 1 μM-A23187. Leakage of monovalent cations and of [³H]phosphorylcholine was measured as described under Methodology and expressed as described in the legend to Fig. 2. Pore-forming agent was added as follows. Panel A: Sendai virus (7 HAU/ml). Panel B: influenza virus (2500 HAU/ml; pH 5.5 buffer of Patel & Pasternak, 1983). Panel C: melittin (0.3 mM). Panel D: α-toxin (3 μg/ml). Panel E: Polylysine (10 μg/ml). Note the 'recovery' of cation content, presumably through action of the Na⁺ pump, at this relatively low dose of melittin.

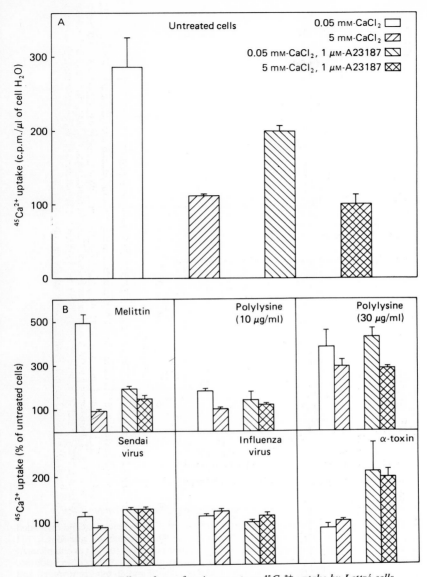

Fig. 6. *Effect of pore-forming agents on* $^{45}Ca^{2+}$ *uptake by Lettré cells*

During the experiment illustrated in Fig. 5, $^{45}CaCl_2$ (approx. 1 μCi/ml) was added at the same time as pore-forming agent, 5 mM-CaCl$_2$ or 1 μM-A23187 and ^{45}Ca associated with the cell pellet was measured; cell water was calculated from the Na$^+$/K$^+$ content of cell pellets on the assumption that Na$^+$ plus K$^+$ is 150 mM (Bashford *et al.*, 1985*a*). Because uptake of $^{45}Ca^{2+}$ is rapid, and in certain cases $^{45}Ca^{2+}$ actually leaks out during the time course of the experiment (25 min), the ^{45}Ca content of cell pellets is expressed as the mean over the 25 min incubation period. Bars indicate the S.E.M. Panel A: untreated cells. Panel B: cells treated with pore-forming agent as indicated in the legend to Fig. 5. The effect of increasing polylysine concentration from 10 to 30 μg/ml may be noted. In experiments such as this, in which uptake (of $^{45}Ca^{2+}$) and leakage (of [^{3}H]phosphorylcholine) are measured simultaneously, the effect of pore-forming agents, and its inhibition by Ca^{2+}, are particularly clear-cut (Poste & Pasternak, 1978).

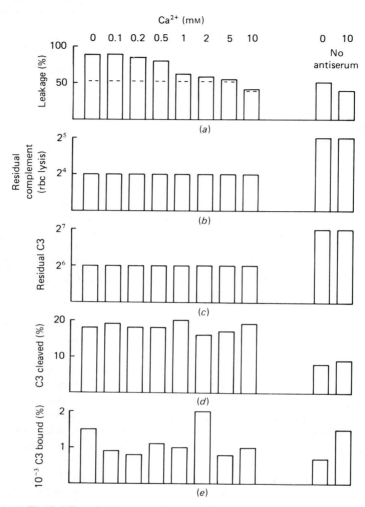

Fig. 7. *Effect of Ca²⁺ on complement-induced leakage from tonsil lymphocytes*

Human tonsil lymphocytes (2×10^7/ml), preincubated with [³H]choline, were washed and incubated in medium containing 150 mM-NaCl, 5 mM-KCl, 10 mM-Hepes, 5 mM-glucose, 0.1 mM-MgCl₂, pH 7.4 (NaOH) at 37 °C with anti-lymphocyte serum (0.7%), guinea-pig serum (0.3%) as complement source and CaCl₂ as indicated. One set of duplicate incubation mixtures contained, in addition, 10 μCi of ¹²⁵I-labelled C3 (1 μCi/μg). Samples were taken at 1, 3, 10 and 30 min, centrifuged and processed as follows. The supernatants from non-¹²⁵I-containing samples were assayed for total residual complement by incubation at 37 °C with antibody-coated sheep erythrocytes, and haemolysis was measured after 1 h. They were assayed for residual C3 with C1,4-coated sheep erythrocytes as described by Borsos & Rapp (1967). The cell pellets from these samples were assayed for ³H and leakage was expressed as the percentage of ³H in cells at 0 time. The portion above the dotted line corresponds to the amount of antiserum-dependent leakage (calculated from the without-antiserum controls at 0 and 10 mM-Ca²⁺). The supernatants from ¹²⁵I-containing samples were assayed for % C3 cleared by solubilization in polyacrylamide gel electrophoresis (PAGE) sample buffer and analysis by SDS–PAGE on a 6.5% gel. Cleavage of C3 was assessed by locating the labelled C3 fragments by autoradiography and measuring the ¹²⁵I in sections of the gel trace corresponding to the α and α' bands of C3 and C3b. Cleavage is expressed as $100 \, \alpha'/(\alpha' \times \alpha)$. The cell pellets from these samples were washed six times in medium and the cell-associated ¹²⁵I was measured. Binding is expressed as the percentage of total ¹²⁵I that is in the washed pellets. All values quoted are the means over the four time points.

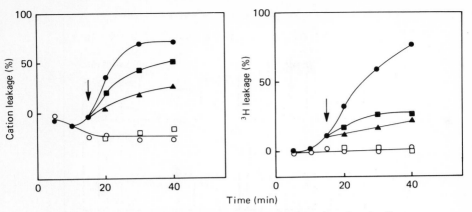

Fig. 8. *Effect of Ca²⁺ on complement-induced leakage from Lettré cells*

Lettré cells (5×10^6/ml) in HBS were incubated with complement (1% human serum; ●) or heat-inactivated complement (○) in the presence of rabbit antiserum (1.5%) at 37 °C; at the time indicated by the arrows a portion of cells was chilled to 4 °C and incubation continued in the absence (■, cells with complement; □, cells with heated complement) or presence (▲, cells with complement) of 2 mM-CaCl₂. Leakage of monovalent cations and of [³H]phosphorylcholine was measured as described under Methodology and is expressed as in the legend to Fig. 2.

the extracellular medium and/or its release from intracellular stores, is shown in Fig. 6. Although $^{45}Ca^{2+}$ data are difficult to interpret, the addition of A23187 clearly decreases the amount of $^{45}Ca^{2+}$ associated with untreated cells at 50 μM external Ca^{2+}, and has relatively less effect at 5 mM external Ca^{2+} (Fig. 6, upper panel); the same is true of agonist-treated cells, and this effect is so marked for melittin, polylysine (10 or 30 μg/ml) and Sendai virus, that it is discernible even when the data are expressed relative to untreated cells (Fig. 6, lower panel). The conclusion that the inhibitory action of Ca^{2+} is extracellular is compatible with the observation that at sufficiently high Ca^{2+} no leakage is ever manifest, and that EGTA, which does not penetrate such cells, is able to relieve this inhibition (Impraim *et al.*, 1980; Micklem *et al.*, 1984*a*; Fig. 9 over page).

In the case of complement, several stages leading to the formation of a lytic C5b-9 complex are known to be sensitive to extracellular divalent cation (Müller-Eberhard, 1975), and it might be thought that this is the mechanism by which Zn^{2+} or Ca^{2+} affect pore-formation. Thus the activation of C1 requires Ca^{2+}, C3 convertase requires Mg^{2+} and when Mg^{2+} is limiting, is inhibited by high concentrations of Ca^{2+} (A. W. Dodds, personal communication), and divalent metal ions (including Zn^{2+}; Tschopp, 1984) under certain conditions polymerize C9 and thus effectively prevent the latter from interacting with a membrane-associated C5b-8 complex (Chiu & Müller-Eberhard, 1984). That none of these mechanisms is involved in the inhibition of complement-induced leakage from tonsil lymphocytes by up to 10 mM-Ca^{2+} is indicated in Fig. 7. The upper panel (*a*) shows that 1 mM-Ca^{2+} reduces complement-specific leakage by more than 70%. Under these conditions, the ability of the residual complement proteins to lyse sheep erythrocytes in the presence of appropriate antiserum is unaffected by concentrations of Ca^{2+} as high as 10 mM (panel *b* of Fig. 7); the same is true of complement-mediated leakage from Lettré cells (results not

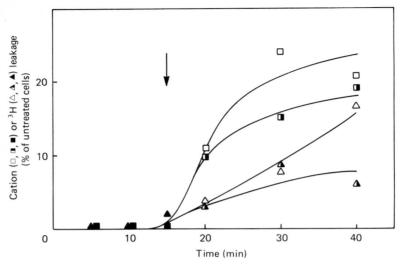

Fig. 9. *Effect of EGTA on complement-induced leakage from Lettré cells*

Lettré cells (5×10^6/ml) were incubated with complement (1% human serum) in th presence of rabbit antiserum (1.5%) and 2 mM-CaCl$_2$ (closed symbols) at 37 °C; at the time indicated by the arrow cells were chilled to 4 °C and incubation was continued in the absence (half-filled symbols) or presence (open symbols) of 4 mM-EGTA. Leakage of monovalent cations and of [³H]phosphorylcholine was measured as described under Methodology and is expressed as in the legend to Fig. 2.

shown). The next two panels (*c, d*) of Fig. 7 show that the activity of C3 convertase, assessed either by measurement of residual C3 in a C3-dependent haemolysis assay (Kabat & Mayer, 1948) (panel *c*), or by measuring the cleavage of ¹²⁵I-labelled C3 (Kerr, 1980) (panel *d*), is unaffected by Ca²⁺ up to 10 mM. The bottom panel (*e*) shows that there is no consistent effect of Ca²⁺ on the binding of residual ¹²⁵I-labelled C3 to the lymphocytes. Although the result of panel *b* in Fig. 7 suggests that Ca²⁺ does not act by removing C9, experiments to examine this possibility more specifically are under way; preliminary results with C9-depleted serum show that C9-dependent leakage is indeed prevented by Ca²⁺.

Of course, one indication that Ca²⁺ acts at as late a stage in complement-induced leakage as in virally-induced leakage would be the demonstration that Ca²⁺ is able, at 4 °C, to inhibit leakage from cells in which pore-formation has been induced at 37 °C, as is the case with Sendai virus- or influenza virus-induced leakage (Patel & Pasternak, 1985). Such an experiment is shown in Fig. 8. Moreover, just as virus, in the presence of Ca²⁺, is able to induce potential pores at 37 °C which can then be 'opened' by EGTA at 4 °C (Patel & Pasternak, 1985), so the same is true of complement (Fig. 9). Analogous data for inhibition by Zn²⁺ and for reversal by EGTA, at 4 °C are presented in Bashford *et al.* (1985*b*).

These results, taken together with the demonstration (a) that Ca²⁺ does not affect the binding of melittin to Lettré cells (Pasternak *et al.*, 1985), and (b) that Ca²⁺ is able to block the salt-mediated stimulation of leakage from pores partially induced in low ionic media by melittin, α-toxin, or polylysine (Pasternak *et al.*, 1985; Fig. 3 and unpublished observations), make it likely that Ca²⁺ acts

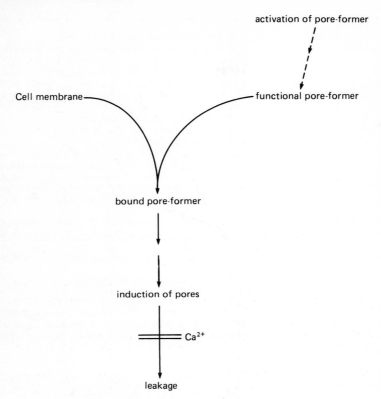

Scheme 1. *Sequence of events in pore-formation*

at a stage subsequent to the binding of functional pore-complexes to membranes, as indicated in Scheme 1.

Conclusion

The data here presented may be summarized as follows. First, they extend the similarity noted by Bhakdi & Tranum-Jensen (1983, 1984) between pore-formation by *S. aureus* α-toxin, *Streptococcus* O lysin and complement, to pore-formation by melittin and viruses (paramyxoviruses at pH 7, other enveloped viruses at pH 5.5), and in so doing emphasize the toxin-like nature of these viruses (Pasternak, 1983, 1984*a*, *b*). Second, they reveal a common sensitivity of pore-formation by such agents to extracellular Ca^{2+} and Zn^{2+}, which may have physiological implications in disease. Third, the demonstration that a cationic protein as simple as polylysine is able to induce the same effects in Lettré cells as viruses, toxins or complement, opens up the possibility of investigating the molecular basis of these effects and of their inhibition by divalent cations in relatively simple systems.

We are grateful to many colleagues for advice and for the gift of materials, to Mrs. Vivienne Marvell and Mrs. Barbara Bashford for the preparation of this paper, and to the MRC, SERC and Cell Surface Research Fund for financial support.

References

Bashford, C. L. & Pasternak, C. A. (1984) *J. Membr. Biol.* **79**, 275–284

Bashford, C. L., Micklem, K. J. & Pasternak, C. A. (1983) *J. Physiol. (London)* **343**, 100P–101P

Bashford, C. L., Alder, G. M., Patel, K. & Pasternak, C. A. (1984) *Biosci. Rep.* **4**, 797–805

Bashford, C. L., Micklem, K. J. & Pasternak, C. A. (1985*a*) *Biochim. Biophys. Acta* **814**, 247–255

Bashford, C. L., Alder, C. A. & Pasternak, C. A. (1985*b*) *Biochem. Soc. Trans.* in the press

Bashford, C. L., Alder, G. M., Gray, M. A., Micklem, K. J., Taylor, C. C., Turek, P. J. & Pasternak, C. A. (1985*c*) *J. Cell Physiol.* **123**, 326–336

Bhakdi, S. & Tranum-Jensen, J. (1983) *Biochim. Biophys. Acta* **737**, 343–372

Bhakdi, S. & Tranum-Jensen, J. (1984) *Phil. Trans. Roy. Soc. B* **306**, 311–324

Borsos, T. & Rapp, H. J. (1967) *J. Immunol.* **99**, 263–268

Boyle, M. D. P., Langone, J. J. & Borsos, T. (1979) *J. Immunol.* **122**, 1209–1213

Campbell, A. K., Daw, R. A., Hallett, M. B. & Luzio, J. P. (1981) *Biochem J.* **194**, 551–560

Chiu, F. J. & Müller-Eberhard, H. J. (1984) *Fed. Proc. Fed. Amer. Soc. Exp. Biol.* **43**, 1449

Dawson, C. R., Drake, A. F., Helliwell, J. & Hider, R. C. (1978) *Biochim. Biophys. Acta* **510**, 75–86

Diem, K. & Lentner, C. (eds.) (1970) *Documenta Geigy Scientific Tables*, 7th edn, pp. 565–567, Geigy Pharmaceuticals, Macclesfield

Dufton, M. J., Cherry, R. J., Coleman, J. W. & Stanworth, D. R. (1984) *Biochem. J.* **223**, 67–71

Füssle, R., Bhakdi, S., Sziegoleit, A., Tranum-Jensen, J., Kranz, T. & Wellensiek, H. J. (1981) *J. Cell Biol.* **91**, 83–94

Getz, D., Gibson, J. F., Sheppard, R. N., Micklem, K. J. & Pasternak, C. A. (1979) *J. Membr. Biol.* **50**, 311–329

Harris, E. J. & Pressman, B. C. (1967) *Nature (London)* **216**, 918–920

Huang, R. T. C., Rott, R. & Klenk, H.-D. (1981) *Virology* **110**, 243–247

Humphrey, J. H. & Dourmashkin, R. R. (1969) *Adv. Immunol.* **11**, 75–115

Impraim, C. C., Micklem, K. J. & Pasternak, C. A. (1979) *Biochem. Pharmacol.* **28**, 1963–1969

Impraim, C. C., Foster, K. A., Micklem, K. L. & Pasternak, C. A. (1980) *Biochem. J.* **186**, 847–860

Kabat, E. A. & Mayer, M. M. (1948) *Experimental Immunochemistry*, Charles C. Thomas, Springfield, Illinois

Kerr, M. A. (1980) *Biochem. J.* **189**, 173–181

Knutton, S. (1978) *Micron* **9**, 133–154

Knutton, S., Jackson, D., Graham, J. M., Micklem, K. J. & Pasternak, C. A. (1976) *Nature (London)* **262**, 52–54

Lenard, J. & Miller, D. K. (1981) *Virology* **110**, 479–482

Lettré, R., Paweletz, N., Werner, D. & Granzow, C. (1972) *Naturwissenschaften* **59**, 59–63

McNiven, A. C., Owen, P. & Arbuthnott, J. P. (1972) *J. Med. Microbiol.* **5**, 113–122

Meech, R. W. (1976) *Calcium in Biological Systems* (Duncan C. J., ed.), pp. 161–191, Cambridge University Press, Cambridge

Mehta, S., Bashford, C. L., Knox, P. & Pasternak, C. A. (1985) *Biochem. J.* **227**, 99–104

Micklem, K. J. & Pasternak, C. A. (1977) *Biochem. J.* **162**, 405–410

Micklem, K. J., Nyaruwe, A., Alder, G. M. & Pasternak, C. A. (1984*a*) *Cell Calcium* **5**, 537–550

Micklem, K. J., Alder, G. M. & Pasternak, C. A. (1984*b*) *Cell Biochem. Function* **2**, 249–253

Micklem, K. J., Nyaruwe, A. & Pasternak, C. A. (1985) *Mol. Cell. Biochem.* **66**, 163–173

Müller-Eberhard, H. J. (1975) *Annu. Rev. Biochem.* **44**, 697–724

Pasternak, C. A. (1983) *Phil. Trans. R. Soc. B.* **303**, 176

Pasternak, C. A. (1984*a*) *Membrane Processes: Molecular Biological Aspects and Medical Applications* (Benga, G., Baum, H. & Kummerow, F., eds.), pp. 140–166, Springer-Verlag, New York

Pasternak, C. A. (1984*b*) *J. Biosci.* **6**, 569–583

Pasternak, C. A. & Micklem, K. J. (1973) *J. Membr. Biol.* **14**, 293–303

Pasternak, C. A., Bashford, C. L. & Micklem, K. J. (1985) *J. Biosci.* in the press

Patel, K. & Pasternak, C. A. (1983) *Biosci. Rep.* **3**, 749–755

Patel, K. & Pasternak, C. A. (1985) *J. Gen. Virol.* **66**, 767–775

Poste, G. & Pasternak, C. A. (1978) *Cell Surface Revs.* **5**, 306–349

Ramm, L. E. & Mayer, M. M. (1980) *J. Immunol.* **124**, 2281–2285

Sato, S. B., Kawasaki, K. & Ohnishi, S.-I. (1983) *Proc. Natl. Acad. Sci. U.S.A.* **80**, 3153–3157

Skehel, J. J., Daniels, R., Hay, A., Wang, M. & Wiley, D. (1985) *Vaccine* in the press

Tosteson, M. T. & Tosteson, D. C. (1981) *Biophys. J.* **36**, 109–116

Tschopp, J. (1984) *Fed. Proc. Amer. Soc. Cell Biol.* **43**, 1450

Tschopp, J. Müller-Eberhard, H. J. & Podack, E. R. (1982) *Nature (London)* **298**, 534–538

Varley, H. (1969) *Practical Clinical Biochemistry*, 4th edn., pp. 278–279, William Heinemann–Medical Books Ltd., London

Wyke, A. M., Impraim, C. C., Knutton, S. & Pasternak, C. A. (1980) *Biochem. J.* **190**, 625–638

Subject Index

First and last page numbers of papers to which entries refer are given.